Ecological Studies, Vol. 111

Analysis and Synthesis

Edited by

O.L. Lange, Würzburg, FRG
H. A. Mooney, Stanford, USA
H. Remmert †, Marburg, FRG

Ecological Studies

Volumes published since 1989 are listed at the end of this book.

E. Paavilainen   J. Päivänen

Peatland Forestry

Ecology and Principles

With 87 Figures and 28 Tables

Springer

Professor Dr. EERO PAAVILAINEN

The Finnish Forest Research Institute
Department of Forest Ecology
Box 18
FIN-01301 Vantaa
Finland

Professor Dr. JUHANI PÄIVÄNEN

University of Helsinki
Faculty of Agriculture and Forestry
Department of Forest Ecology
Box 24
FIN-00014 Helsinki
Finland

ISBN 3-540-58252-5 Springer-Verlag Berlin Heidelberg New York

Library of Congress Cataloging-in-Publication Data. Paavilainen, E. Peatland forestry: ecology and principles / E. Paavilainen, J. Päivänen. p. cm. -- (Ecological studies; v. 111) Includes bibliographical references and index. ISBN 3-540-58252-5 (Berlin). -- ISBN 0-387-58252-5 (New York) 1. Peatland forestry. I. Päivänen, J. (Juhani), 1941-. II. Title. III. Series. SD401.7.P33 1995 634.9'5--dc20 952115

Printed in Germany

Typesetting: U. Kunkel, Reichartshausen
SPIN 10087965 31/3137-5 4 3 2 1 0 - Printed on acid-free paper

Preface

Peatlands (or mires) cover at least 550 million ha globally, of which the greater part is situated between 50° and 70° N. Although the majority of peatlands are not used for human needs, there are large areas where agriculture, peat or wood production has been practiced. The suitability of peatlands for forestry differs from country to country depending on climatic conditions, raw wood demand, silvicultural management practice and tradition, as well as the infrastructure in the remote areas considered.

Peatland utilization for forestry may be divided into three efficiency categories: (1) **exploitation** (harvesting of trees with inadequate attention to regeneration) leading to reduction in the renewable resource; (2) **silvicultural management** (harvesting of trees with natural or artificial regeneration) aimed at maintaining the renewable resource in a sustainable way; and (3) **progressive forest management** (drainage, fertilization, afforestation, thinning, ditch maintenance, final harvest and regeneration) aimed at increasing the renewable resource. In North America, forestry utilization is changing from exploitation towards a kind of silvicultural management whereas progressive forest amelioration activities on forested peatlands and waterlogged mineral soils are part of the normal forestry practices in Fennoscandinavia, the Baltic countries and Russia. In the British Isles afforestation of open peatlands is associated with drainage.

This monograph is a review and synthesis of peatland forestry on northern peatland (mire) ecosystems. It covers peat soil properties, mire hydrology, carbon and nutrient cycling, and classification of mire sites. The emphasis, however, is on peatland forests as a renewable natural resource. The approach originated in northern Europe. This is because in Fennoscandia, especially Finland, operational scale forest drainage has a long tradition based on research aimed at maintaining and increasing the wood production of peatlands. Whenever relevant, a closer look is also given to other countries in northern Europe, Canada, the USA, and to the British Isles in the temperate zone.

Forest amelioration has been shown to be profitable when only directed towards appropriate sites. Possible environmental consequences have to be taken into consideration – in both the planning and realization of measures to minimize harmful effects on the site, locally downstream and even globally on the atmosphere. Despite steadily increasing scientific progress, there are still gaps in our knowledge. Both basic and applied research, based on international cooperation, are needed to achieve further levels of silvicultural and progressive forest management on peatlands. We hope that this review will promote research activities in the fields concerned. By introducing progressive forest management to relatively small areas, it is possible to preserve untouched mires and old forests for future scientific research, teaching and recreation.

Our work would not have been possible without a close cooperation with colleagues not only with our home research and teaching units, the Finnish Forest Research Institute and University of Helsinki, but also with scientists and practical managers in both Finland and abroad. Also, the knowledge obtained during our visits to all countries where peatland forestry has been practiced, at least on an experimental scale, was useful in the preparation of this review. The original drafts of Chapters 1, 2, 3.1, 3.2, 3.4, 3.5, 4 and 6 were prepared by Juhani Päivänen and Chapters 3.3, 5, 7 and 8 were prepared by Eero Paavilainen. However, both authors take the responsibility for the final version.

We would like to thank the following Finnish colleagues who read drafts of this work and provided suggestions, comments and information that resulted in substantial improvement: Erkki Ahti, Leena Finér, Seppo Kaunisto, Raija Laiho, Antti Reinikainen, Michael Starr and Harri Vasander. We are also very grateful to Björn Hånell (Sweden), Finn H. Braekke (Norway), Uno Valk (Estonia), Peteris Zalitis (Latvia), Juozas Ruseckas (Lithuania), Stanislav Vompersky (Russia), Hedwig Roderfeld (Germany), D.G. Pyatt and Russell Anderson (UK), E.P. Farrell (Ireland), Elon S. Verry (USA.), and Fred Haavisto and John Jeglum (Canada) for the valuable information received. Special thanks go to Hannu Nousiainen for the cover layout, to Mrs. Sari Rainansalo for text processing and to Mrs. Raija Linnainmaa for making the drawings. We also appreciate Michael Starr's contribution in revising the English.

The work was supported by a grant from the Finnish Cultural Foundation.

January 1995 EERO PAAVILAINEN JUHANI PÄIVÄNEN

Contents

1 Peat and Peatlands

1.1 Definitions

The most recent definition of wetlands is given by Lugo (1990, p. 2): "**Wetlands** are areas that are inundated or saturated by surface water or groundwater, at such a frequency and duration that under natural conditions they support organisms adapted to poorly aerated and/or saturated soil." This definition uses the following criteria: saturated or poorly aerated soil, and biotic response to the soil condition. According to the Canadian definition (Zoltai and Pollett 1983), wetlands are areas where wet soils are prevalent, having a water table near or above the soil surface for most of the thawed season, supporting hydrophilic vegetation, and they include pools of open water (less than 2 m deep). This means that shallow open water is also included in wetlands (Zoltai and Pollett 1983).

Forested wetland is defined to be any wetland with a significant component of woody vegetation, regardless of the size of the plants (Lugo 1990). Traditionally, standing water was required in order to be classified as forested wetland. However, saturation of soil, or even a rising water table, may affect trees as much as flooding.

Peat is defined as organic matter derived from vegetation having 25% or less inorganic matter on a dry mass basis (e.g. Andrejko et al. 1983). Peat typically consists of more or less fragmented plant residues sequently deposited. The presence of plant fibers distinguishes peat from organic aquatic sediments (Okruszko 1979).

Organic soil is material which has 17% or more carbon by mass (e.g. Canada Soil Survey Committee, Subcommittee on Soil Classification 1978). These definitions suggest that the terms peatland and organic terrain are not the same, the latter being a broader concept encompassing both peat and carbonaceous (having 25 to 75% organic) deposits (Jeglum 1985). Organic soils have also been classified as **histosols** (Soil taxonomy 1975).

Muck is dark, well-decomposed organic material with a high mineral matter content which has accumulated under conditions of imperfect drainage. In the USA, peats or sediments with 20 to 50% organic matter content are referred to as mucks (Staneck and Worley 1983). The term muck has also been used in connection with the transformation of fibrous peat into amorpho-granulated material after drainage (Okruszko 1972, p. 312; Okruszko 1979, p. 44).

A variety of **aquatic sediments** are formed beneath standing water (see Pakarinen 1984, p. 45). They have been classified as follows:

- **Dy** is a deposit of muddy acid sediments consisting mainly of amorphous precipitated humus colloids. It is biologically relatively inert and forms in waters originating from peatlands and paludified forests. It is characterized by a high C/N ratio.
- **Gyttja** is an aquatic sediment that is predominantly composed of dead plankton, detritus and bottom fauna. It is grey-brown to blackish in colour and formed under aerated conditions sufficiently rich in nutrients and microorganisms. Deposits of gyttja may form the bottom layer in peatlands.
- **Marl** is a loose deposit with a high calcium and magnesium carbonate content. It can be formed in freshwater ponds or flark fens fed by carbonate-rich waters.
- **Sapropel** is composed of plankton and benthic organisms and partly of remains of higher plants. It is formed under anaerobic, nutrient-rich conditions. Sapropels (defined sensu lato) up to 40 m thick have been found under certain peat deposits in the former USSR (Neishtadt 1968). Sapropels are related to gyttjas and "muds", and sometimes these terms overlap.

Peatlands or **mires** are wetland ecosystems that are characterized by the accumulation of organic matter, which is produced and deposited at a greater rate than it is decomposed, leading to the formation of peat (Gore 1983). Peatlands and mires are usually supported by a humid climate and high water table levels, leading to low-level microbial activity in the soil (Reinikainen 1976). Peatland ecosystems are also characterized by a high degree of interaction between living organisms and the environment. Thus, the vegetation plays a decisive role in modifying edaphic conditions.

The term **mire** can probably be considered as a slightly wider concept than **peatland**, because it encompasses all peat-forming habitats and this feature distinguishes mires from all other ecosystems (Moore 1984). Gore (1983) used the term mire to cover all different kinds of ecotypes fitting the definitions for both peatland and mire. In this book, the terms peatland and mire are synonymous. In an ecological context we have used the term mire, whereas in connection with drainage or forest management the more traditional term peatland is used.

It must be stressed, however, that the terms peatland and mire are not consistent. In Finland, peatland or mire is defined by the presence of peat-forming plants. A virgin or undisturbed peatland or mire has greater than 75% coverage of *Sphagnum* moss. There is no requirement for minimum thickness of peat. This is a very liberal definition of peatland or mire compared with some other countries where a minimum value for peat thickness is a criterion: e.g. 20–30 cm in Germany (Schneider 1976; Benzler et al. 1982), 30 cm in Sweden (Hånell 1991a), 40 cm in Canada (Zoltai and Pollett 1983), and 45 cm in Ireland (Hammond 1979).

The term **muskeg** is used in Canada to designate peat-supporting organic terrain, especially in connection with engineering and geotechnics (Radforth 1969).

The word is Indian in origin and has been used widely in North America as a synonym for peatland. The terms peatland and **organic terrain**, on the other hand, are not the same. The latter is a broader concept encompassing both peat and carbonaceous (25 to 75% organic) deposits (Jeglum 1985).

Bog, fen, swamp and **marsh** are more specific terms than peatland, mire and muskeg. They are strongly linked to the concepts of trophy and nutrient status of the site (see later in this chapter).

It is evident that water quantity and quality are of prime importance to the development of mire vegetation. Depending on the quantity of water or the characteristics of the water table, mire vegetation has been divided into four groups (e.g. Masing 1975): (1) **Limnetic** associations living under conditions of permanent inundation, (2) **telmitic** associations of periodically inundated sites, (3) **semiterrestrial** associations in sites where the excess of water occurs mainly as a result of a high subsoil water level and only for short periods of inundation, and (4) **terrestrial** associations which develop on uninundated surfaces where the degree of soil moisture depends on the subsoil water level.

The other fundamental criterion determining the development of mire vegetation is the quality of the water feeding a mire. The nutrient, humic acid and free oxygen contents and the rate of the water flow are important properties describing the water quality.

Trophy describes the nutrient status of mire vegetation and the substrate. The term **ombrotrophic** describes a site receiving nutrients only in the form of precipitation or wind-borne dust (e.g. Sjörs 1948; Eurola et al. 1984; Jeglum 1991b). Ombrotrophic conditions represent the extreme situation at the lower end of nutrient availability, and ombrotrophic sites usually have a thick peat layer.

An ombrotrophic situation develops in the centre of a mire complex when it grows above its margins. Ombrotrophic mires are characterized by low pH levels as well as by low electrolyte and calcium contents (e.g. Tolonen and Hosiaisluoma 1978; Eurola et al. 1984). However, since ombrotrophic species can also grow on minerotrophic sites, attention has to be directed to the use of sensitive plant indicators of minerotrophy. It has also been suggested that ombrotrophy should be defined in terms of hydrologic criteria (Damman 1990).

A site is **minerotrophic** when the nutrient concentration in the surface peat and mire water is significantly higher than that of precipitation. This may happen where the depth to underlying mineral soil is shallow or if nutrients are being supplied by mineral-rich surface water or groundwater.

Ombrotrophy has usually been strictly separated from the three classes of minerotrophy: **oligotrophic** (minero-oligotrophic, poor minerotrophic), **mesotrophic** (intermediate minerotrophic), and **eutrophic** (rich minerotrophic) (Laine et al. 1986; Jeglum 1991b). However, sometimes it is said that ombrotrophic mires are highly oligotrophic (Eurola et al. 1984).

The terms **ombrogenous** and **geogenous** reflect the genesis of peat. While some scientists use the terms ombrotrophic and ombrogenous as synonyms (e.g. Jeglum 1991b), others (e.g. Damman 1986) consider them as referring to clearly

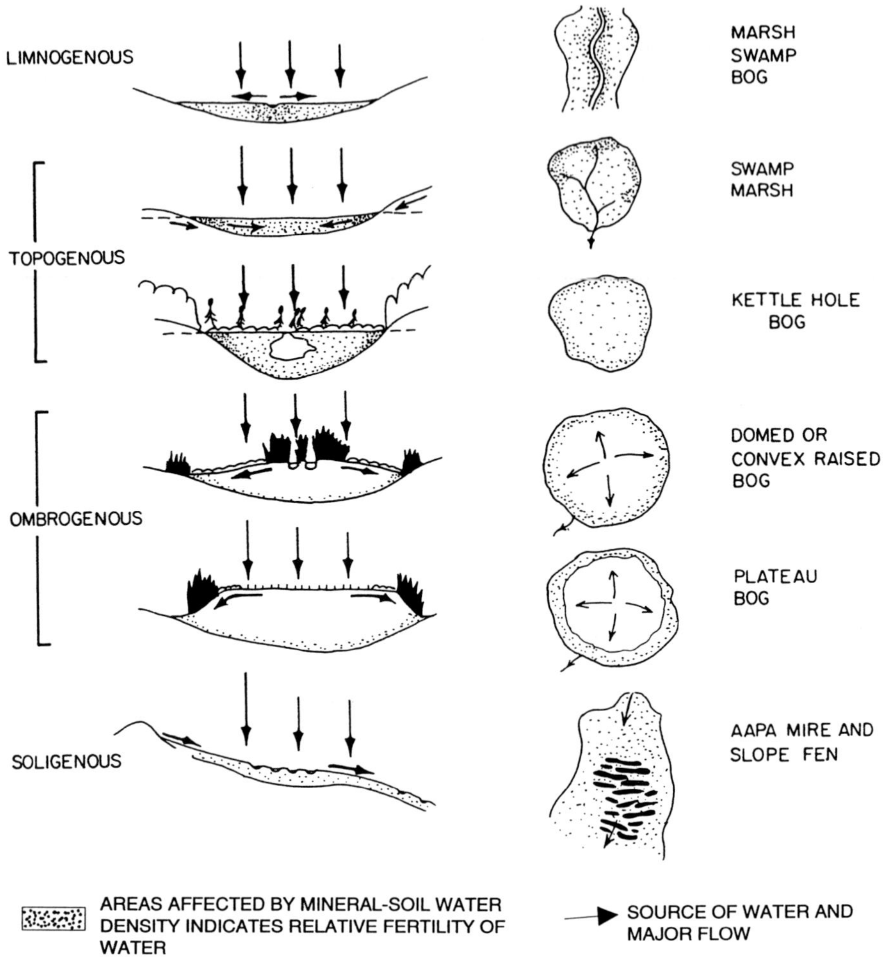

Fig. 1.1. Types of peatland development in relation to source of water. *Stippling* in the cross sections and aerial views indicates areas affected by mineral soil water, the density of stippling showing the relative fertility of the water. *Arrows* indicate source of water and major flow. (Redrawn from Damman 1986)

different conditions. An ombrogenous mire is a landscape unit which includes the ombrotrophic (only rain-fed) centre as well as the minerotrophic border separating the mire from the surrounding upland (Fig. 1.1). The term ombrotrophy is restricted to that part of an ombrogenous mire not influenced by inflowing mineral-rich, i.e. geogenous, waters (Damman 1986, p. 384).

The term **geogenous** includes waters of limnogenous, topogenous and soligenous origin. **Limnogenous** indicates that the water is derived from lakes or rivers, while **topogenous** and **soligenous** refer to water derived from mineral soil, respectively static and flowing (Gore 1983; Damman 1986).

Eurola et al. (1984, p. 18–25) suggested separating mire sites into those dependent only on an inherent nutrient influence (i.e. nutrients already present in the peat or supplied by precipitation) and those receiving supplementary nutrient effects. In the latter case, the site may be affected either by groundwater, surface water, meltwater or a mineral soil water.

There are several words in common use for mire, each of them implying some special characteristic connected to the site. The following definitions are used in this book:

- **Bogs** are ombrotrophic sites influenced solely by water that falls directly as rain or snow. The peat is predominantly *Sphagnum* moss peat.
- **Fens** are at least slightly minerotrophic sites influenced by water derived predominantly from outside their own immediate limits. The peat is formed mainly from *Carex, Phragmites* etc. residues.
- **Swamps** are well-wooded, minerotrophic wetlands or mires, where the peat layer thickness varies or can even be totally absent.
- **Marshes** are grassy wet areas, periodically inundated with standing or slowly moving water. The substratum usually consists of mineral or organic soils with a high mineral content, but there is little peat accumulation.

In a recent book about forested wetlands (Lugo et al. 1990), the terms **bog, swamp** and **alluvial formations** were defined in the included map as follows (Matthews 1990): "Bogs are peat-rich ecosystems found primarily in the high latitudes on waterlogged soils; swamps are peat-poor systems generally occupying seasonally inundated soils in tropical and subtropical regions; alluvial formations are peat-poor forests along rivers that flood adjacent lands, either intermittently or for several months, during the rainy season." These definitions bind the terms to latitude and to peat-rich or peat-poor ecosystems, and the earlier well-established meanings are lost.

1.2 Mire Development

1.2.1 Development of Mire Vegetation

Elements of mire ecosystems and the main environmental formation factors are schematically presented in Fig. 1.2. The main environmental factor influencing the initiation of a mire vegetation is water – its quantity and quality. Rybnicek (1985) emphasized the importance of water and considered mire vegetation to depend only indirectly on the solid mineral surroundings.

The conditions for wetland establishment are well described by Verry (1988):

1. A physiography or climate that is conducive to slow water movement. Suitable physiography includes flat areas such as glacial lake beds, sandy outwash plains, and river floodplains. Depressions with no or limited outflow such as

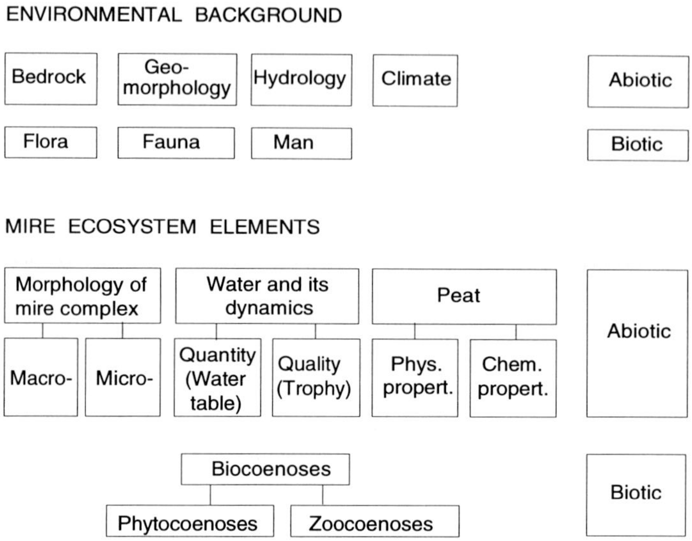

Fig. 1.2. Elements of mire ecosystems and their main environmental formation factors. (Redrawn from Rybnicek 1985)

ice block depressions are also favourable sites. However, peatlands can also develop on slopes where precipitation is much greater than evapotranspiration.

2. Water available at the surface of the earth. Sufficient water is available at the earth's surface where annual precipitation exceeds evapotranspiration. Water is also available at the surface where the basin area to wetland area ratio is large (3–20), so that precipitation and runoff are greater than evapotranspiration. Wherever groundwater rises near the surface, or discharges at the surface, there will be sufficient water for wetlands to form if flow rates are slow.

Kangas (1990) outlined two types of wetland (or mire) formation. In the **classical model**, the direction of the succession is from aquatic towards terrestrial systems. It starts with open water and, through drying energies (sediment buildup, evapotranspiration, drainage), progresses towards the climax state of mesic forest (Kangas 1990). This type of mire development has also been called **terrestrialization** (Malmer 1985, see also Fig. 1.3).

Constant environmental conditions are implied by the classical model of wetland and mire succession. However, this assumption is seldom justified because many environmental conditions vary with time. A disturbance may cause the succession to stagnate or even reverse its direction. If the disturbance has a periodic nature, the ecosystem may develop towards cyclic climaxes. Kangas (1990) mentions possible disturbances such as fire, extremely dry or wet spells, windthrow, disruptions by animal populations, and the actions of man. In the northern hemisphere the extreme cold periods may also disturb the succession. In the long

Fig. 1.3. The boreal landscape consists of three main elements: forest, mire and water. Terrestrialization of open water by sedges, herbs, and *Sphagnum* mosses. (Photo: J. Päivänen)

term, mire formation may be considered as the anti-cycles of glaciers (Verry 1988). Usually, the glacier strips away the earlier organic layers, but in Finnish Lapland a 1.5-m-thick peat layer has been found beneath four superimposed till beds, indicating several glaciers since the formation of the peat deposit in question (Hirvas 1991).

Paludification refers to the conversion of a mineral soil site to a mire due to a rise in the groundwater table. The direction of the succession is from a dry site towards a wet one, which is opposite to the classical model (Kangas 1990). It is primarily a northern phenomenon, with a spreading of *Sphagnum* mosses and an accumulation of peat. The *Sphagnum* mosses and peat act as water-holding reservoirs (Fig. 1.4). The water stagnates and the oxygen content in the soil decreases, resulting in death of the trees and accumulation of peat. Peat accumulation (see Sect. 3.1) occurs when the rate of addition of dry matter exceeds that of decay (Clymo 1983). The term paludification has come to include the formation of peat directly, or via terrestrial humus, on podzolized soils and even bare rock surfaces (Gore 1983).

Primary peat formation is also an important way of initiating wetland and mire formation in the boreal zone. This is a process whereby the surface is occupied by mire vegetation immediately after the retreat of water or glacial ice. This type of mire formation has been typical along emerging coastlines (Huikari 1956).

Fig. 1.4. *Sphagnum* carpet dominates the ground layer of a mire ecosystem. *Sphagnum riparium* in a wet, meso-oligotrophic site. (Photo: J. Päivänen)

1.2.2 Mire Complex Types

In the northern hemisphere in particular, climatic and topographical factors affect the rate of primary production and decay, and thus peat accumulation. Temperature sums, annual minimums and maximums, precipitation and evapotranspiration interactions, freezing, freeze-thaw fluctuations and other factors determine the **wetland form** (Verry 1988). However, where climate is optimal for plant growth, as in the temperate zone and the moist tropics, factors other than climate may be of prime importance for the development of mires (Frenzel 1983).

Plant communities dominated by mire plants (site types) form **mire complexes** which are affected by the climatic and topographical factors mentioned above. Thus, a mire complex is a particularly coherent mire region with all its mire types (Cajander 1913). Mire complexes occurring in the same geographical region, showing similar combinations of vegetation (and site types), and having approximately the same surface relief and peat stratigraphy, form a **mire complex type** (Ruuhijärvi 1982a).

The main mire complex types are: (1) **limnogenic mires** in the southern and central parts of the temperate zone; (2) **ombrotrophic raised bogs** in the northern temperate zone and southern boreal zone (Figs. 1.5 and 1.6); (3) **minerotrophic aapamires** in the central and northern parts of the boreal and subarctic zones; (4) **palsa mires** characterized by peat mounds with permafrost

Fig. 1.5. Ombrotrophic raised bog in the southern boreal zone, southern Finland. (Photo: J. Päivänen)

Fig. 1.6. Concentric raised bog, Newfoundland, Canada. (Photo: J. Päivänen)

lens; and (5) **arctic polygon mires** in the islands of the Arctic Ocean and tundra region of northernmost Eurasia (Moore and Bellamy 1974; Ruuhijärvi 1982a, 1983; Eurola et al. 1984). In extremely wet maritime climates, **blanket mires**, a subvariant of ombrotrophic raised bogs, are formed (Moore and Bellamy 1974). The maritime-continental gradient is well shown in the change in raised bog development both in North America (Damman 1979) and Fennoscandia (Finland, Sweden and Norway) (Ruuhijärvi 1982a). The mire complexes of remote areas in northern Eurasia are not as well known as those in Fennoscandia or the eastern part of North America (see Botch and Masing 1983).

For a detailed presentation of mire complex types and their distribution, see Moore and Bellamy (1974), Botch and Masing (1983), Frenzel (1983), Gore (1983), Ruuhijärvi (1983), Sjörs (1983), and Moore et al. (1984).

1.3 Areal Distribution of Peatlands

Because the definitions of wetland and peatland (or mire) vary from country to country, the estimates of the global coverage of these ecosystems differ considerably. According to figures collected by Lugo et al. (1990), the total area of **wetlands** (non-forested and forested) in the world is between 200 and 530 million ha, or about 3% of the land surface. Maltby (1988a,b) estimates that wetlands occupy 6% of the world's land surface and that **peatlands** cover at least 500 million ha (3.8%), from tundra to tropical environments. Matthews and Fung (1987) give a total global wetland area of 526.3 million ha, of which about half lies between 50 and 70°N. There is a secondary concentration (about 35%) in tropical/subtropical latitudes (20°N – 30°S).

The literature review compiled by Kivinen and Pakarinen (1981) shows how the estimates for the total area of peatlands in the world have changed during this century as more inventory results have become available. The earliest estimates were around 100 million ha, but with time this figure has increased; first to 150 million (Tibbetts 1968; Schneider 1976), then to 230 million (Moore and Bellamy 1974) and further to 420 million (Kivinen and Pakarinen 1980) and 450 million ha (Kivinen and Pakarinen 1981). Quite recently, it has been suggested that the final figure may be around 500 million ha (Heathwaite 1993; Pfadenhauer et al. 1993).

As mentioned previously, there is no fixed minimum thickness of peat for defining peatlands. Kivinen and Pakarinen (1981) stated that they had – as far as possible – used a minimum peat thickness of 30 cm to calculate the total peatland area of 450 million ha. However, for Finland (10.4 million ha) they seem to have used the Finnish biological and ecological definition for peatlands which does not have a minimum thickness requirement.

From the point of view of peatland forestry, this latter approach makes more sense and we have tried to use it to prepare the figures in Table 1.1. For Sweden, for example, the peatland area includes wet mineral soils with a *Sphagnum* moss cover (Hånell 1989). However, for the Baltic countries the figures represent the mire area having a minimum peat thickness of 30 cm (Table 1.1), because it was impossible to determine the share of the total wetland areas (Estonia 1 million ha; Latvia 1.1 million ha; Lithuania 3 million ha) occupied by wet mineral forest soils (Valk 1988; Kaunisto et al. 1991; Ruseckas 1991). This means that the figures for the peatland areas of the Baltic countries are underestimated compared to those of Finland and Sweden.

Table 1.1. Estimates of mire areas of the world

Region and country		Area 10^6 ha	Source
Nordic countries			
Finland		10.400	Kivinen and Pakarinen (1981)
Sweden			
– Peatlands	6.368		
– Wet mineral			Hånell (1989)
soils	2.024	8.392	
Norway		3.000	Braekke (1990b)
Denmark		0.060	Gore (1983)
Iceland		1.030	Gore (1983)
Baltic countries			
Estonia		0.907	Valk (1988), Zobel (1992)
Latvia		1.140	Kaunisto et al. (1991)
Lithuania		0.606	Ruseckas (1991)
Central and South Europe			
Poland		1.300	Okruszko (1976)
Germany			
– The former FRG	1.110		
– The former GDR	0.550	1.660	Kivinen and Pakarinen (1981)
Czech Republic		0.031	Kivinen and Pakarinen (1981
Switzerland		0.055	Kivinen and Pakarinen (1981
Austria		0.022	Kivinen and Pakarinen (1981
Netherlands		0.015	de Bakker (1982)
Belgium		0.018	Kivinen and Pakarinen (1981)
France		0.040	Francez et al. (1992)
Hungary		0.100	Kivinen and Pakarinen (1981)
Romania		0.007	Kivinen and Pakarinen (1981)
Yugoslavia (former)		0.100	Kivinen and Pakarinen (1981)
Bulgaria		0.001	Kivinen and Pakarinen (1981)
Greece		0.005	Kivinen and Pakarinen (1981)
Spain		0.006	Kivinen and Pakarinen (1981)
Israel		0.005	Kivinen and Pakarinen (1981)
Italy		0.120	Kivinen and Pakarinen (1981)
United Kingdom			
– England	0.362		
– Scotland	0.821		
– Wales	0.159		
– Northern Ireland	0.167	1.509	Taylor (1983)
Republic of Ireland		1.176	Taylor (1983)

Table 1.1. Cont.

Region and country		Area 10^6 ha	Source
Eastern Eurasia			
CIS (the former USSR) "Excessively moist forests and peatlands"		245.000	Vompersky (1991)
China		3.500	Zhao and Päivänen (1986)
Japan		0.200	Miyazawa (1992)
North America			
Canada			
– Peatlands	111.327		Haavisto and Jeglum (1991)
– Other wetlands	58.673	170.000	Zoltai and Pollett (1983)
USA			
– Peatlands and other wetlands S of 49°N	17.387		
– "Muskeg" and "grassy marshes" in Alaska	44.500	61.887	Hofstetter (1983)
Tropical peatlands			
Africa		3.312	Shier (1985)
Asia		34.002	Shier (1985)
Central America		2.546	Shier (1985)
South America		3.977	Shier (1985)

The figure of 245 million ha for Russia and the other states of the former USSR includes "excessively moist forests and peatlands" (Vompersky 1991), and the figure of 170 million ha for Canada includes peatlands and other wetlands (Haavisto and Jeglum 1991). This figure for Canada may be too low, however, because the statistics often do not include merchantable wetland forests (Haavisto and Jeglum 1991).

The estimate for the peatland area of the coterminous United States (17.4 million ha) is taken from Hofstetter (1983, p. 212). We have left out the figures for "seasonally flooded basins" from the total wetland area and added the figure of 44.5 million ha for Alaska (Hofstetter 1983, p. 216).

For most of the European countries, the peatland area figures have been taken from Kivinen and Pakarinen (1981). The peatland area for the Netherlands has been lowered to 0.015 million ha after de Bakker (1982), and that for France to 0.040 million ha after Francez et al. (1992).

The percentage cover of mires for countries in the northern hemisphere is as follows:

	%
Finland	31
Estonia	21
Sweden	20
Canada	19
Latvia	17
Republic of Ireland	17
Russia	11
Norway	9
Lithuania	9
UK	6

The figures for tropical peatlands in Table 1.1 are taken from Shier (1985). In the tropics and subtropics, the largest peatland areas are confined to coastal, river valley and mountain plateau areas. Lowland mire formation occurs only under the influence of limnic water. Tropical mires and peat resources are, in general, poorly investigated and documented. New resources are continually being reported.

The criteria used to define peatlands in the tropics are unspecific, e.g. "sufficient" depth of accumulated organic material and "sufficient" organic content in the deposit. Shier (1985, p. 31) mentions that the total area of African tropical swamps has been estimated at 34 million ha, but he himself accepts only less than 10%. Similarly, Thompson and Hamilton (1983, p. 350–352) report that there are over 6 million ha of permanent swamps and almost 40 million ha of seasonally inundated swamps in Africa, but the true peatland area is very small indeed. It is restricted to the areas with the highest rainfall but, more specificially, to the regions with the least probability for dry seasons.

The areas of tropical peatlands are underestimated compared to those outside the tropics. It is unfortunate that there is no comparable data available for the world's peatland area.

1.4 Peat Resources

Kivinen and Pakarinen (1981) have given some estimates for peat resources in tons of peat with 35–40% moisture content for different countries. Furthermore, they calculated the world's peat resources at about 400×10^9 t (40% moisture content), having an energy content of 5×10^9 GJ.

Mires in a natural state act as sinks of atmospheric carbon, through fixation by plants via photosynthesis. Thus, they are accumulators for organic carbon (Sect. 3.1).

There are several sources of uncertainty when estimating the peat resources of the world. First of all, as seen in the previous section, estimates of peatland area vary. In addition, values for the average peat depth and bulk density have to be agreed upon. Nowadays peat resources are more often expressed as stores of carbon. Therefore, accurate carbon content data need to be available.

Both bulk density and carbon content depend on the type and degree of decomposition of the peat deposit. According to peat resource inventories of Finnish peats, the average bulk density values are between 0.090 and 0.094 g cm^{-3} (Lappalainen 1982; Lappalainen and Uhlgren 1991). For the surface peat layers of both virgin peatlands and those drained for forestry, the following bulk densities have been determined (Päivänen 1973b):

Peat type	Average	Min. – Max.
	g cm^{-3}	
Sphagnum peats	0.085	(0.030–0.190)
Carex peats	0.121	(0.050–0.205)
Woody peats	0.140	(0.090–0.210)

For virgin peats in New Brunswick, Klemetti and Keys (1983) have presented bulk densities similar to those from Finland, the mean value being 0.080 g cm^{-3} and ranging between 0.015 and 0.144 g cm^{-3}. However, in different climatic regions or near sea level or delta regions, the bulk density may be much higher. For example, a mean bulk density value as high as 0.320 g cm^{-3} has been reported for the San Joaquin Delta region in central California (Marachi et al. 1983).

The carbon content of peat is often assumed to be about 50% of the dry matter content. However, the decomposition of peat obviously affects the carbon content (Naucke 1990, p. 246).

The following figures illustrate the wide range of estimates for the total carbon in the world's peat:

Total carbon (Pg)[a)]	Source
41.5	Buringh (1984)
165	Moore and Bellamy (1974)
180	Olson et al. (1983)
210	Oechel (1989)
250	Lappalainen (1992)
300	Sjörs (1981)
455	Gorham (1991)
489	Schlesinger (1977)

[a)] 1 Pg = 10^{15} g

The moderate estimate of 300 Pg given by Sjörs (1981) is about 13% of the terrestrial carbon in the biosphere. Gorham (1991) reached the figure of 455 Pg by using a total area of virgin boreal and subarctic peatlands of 342 million ha, a mean depth of peat of 2.3 m, a bulk density of 0.122 g cm^{-3} and a carbon content of 51.7%.

2 Utilization of Peatlands

2.1 Early Forms of Utilization

The mires of Central Europe were left undisturbed for a long time because of their high water content, poor accessibility, and man's inability to farm them using primitive means. Man generally found mires unpleasant. Criminals took refuge in peatland areas and the condemned were executed there (Lüttig 1989). Extremely well preserved bodies have been found in peat deposits (Glob 1969; Hayen 1990) – the most famous of them being the 2000-year-old "Tollund man" in Denmark. These people appear to have been sacrificed in some ritual, perhaps to ensure luck and fertility for the peasant community in the coming year (Glob 1969).

The use of peatlands for **hunting** is probably the oldest form of mire usage. The **picking of wild berries** and **harvesting of wild rice** from peatlands are traditions which still have local economical value. Herb-rich fens form an important **natural summer pasture** for animals such as reindeer (*Rangifer tarandus*) and its wild ancestors caribou (*R. caribu*) and wild forest reindeer (*R. tarandus fennicus*) in northern latitudes.

Sedge fens were traditionally used for **collecting wild forage** for cattle. The remnants of hay barns on mineral soil isles in minerotrophic tall-sedge fens in Finland can still be found. In some parts of the country, artificial flooding of these sites was still practiced in the late 19th century. Water and ice cover during winter were used to level the surface of the mire, to kill undesired elements in the vegetation (trees, bushes, mosses) and to increase the production of forage plants, e.g. *Carex aquatilis, C. lasiocarpa, C. rostrata, Equisetum fluviatile* (Vasari 1988).

2.2 Agriculture

The Pontine mires (*Paludi Pontine*) south-east of Rome were one of the first **peatlands** to be **drained for agriculture**. This happened more than 2500 years ago. However, the drainage canals deteriorated during the Roman Empire and the area repaludified. It was not until after the World War I that it was reclaimed again for agriculture.

The Dutch were probably the first to systematically reclaim peatlands for agriculture on a large scale. Their expertise in land and water management was later used to develop unsettled areas in a number of neighbouring countries, e.g. the Spree and Havel regions in present Germany (Lüttig 1989). In France, large-scale mire draining programmes were also begun as early as at the turn of the 16th century (Francez et al. 1992).

The first subsurface drains, at the beginning of the Christian era, were made of stones and wood. However, subsurface drainage had to be reinvented in the 17th century. The use of brick drain pipes was developed in England in 1840. The use of subsurface drainage in connection with agricultural utilization of peatlands expanded rapidly all over Europe afterwards (Eggelsmann 1973).

The organized drainage of lands in Russia began during the reign of Peter the Great. The earliest drainage works were associated with both the building of cities and the drainage of agricultural land (Skoropanov 1968). It has been estimated that there are about 1 million ha of mires in agricultural use in the former USSR (Kivinen and Pakarinen 1981).

In most cases, the reclamation of peatland for farming in Germany started with peatland burning followed by different kinds of cultivation methods (see Overbeck 1975; Göttlich and Kuntze 1990). Especially the mixed sand cultivation method has considerably extended the agricultural area in Germany (e.g. Kuntze et al. 1988). In fact, agriculture has become the main use of peatlands in the former Federal Republic of Germany. Today, 75% of the raised bog and 95% of the fen area can be considered cultivated (Lüttig 1989). In the former German Democratic Republic, 87% of the peatland area, mostly fen peat, is used for agriculture (Illner 1983; Wojahn and Schmidt 1987; Lorenz et al. 1992).

The situation in Poland is very much the same as in Germany. The share of the peatland area reclaimed and cultivated is 82%, most of which is used as grasslands (Okruszko 1976). For some other central European countries, it is now difficult to distinguish which part of the arable land was originally peat (e.g. Bather and Miller 1991; Francez et al. 1992).

Peatlands have also been used for agriculture in the Baltic countries. In Estonia (Zobel 1992) about 84 000 ha (9.3% of the total peatland area) and in Lithuania (Ruseckas 1991) about 194 000 ha (32.0%) of the peatlands have been drained and reclaimed for agricultural purposes. There are no corresponding figures available for Latvia.

In Sweden at the turn of the century about 600 000 ha of cultivated land was on drained peatland (Hånell 1990). Peatland exploitation for farming was an important element in the colonization of northern Sweden and fundamental for maintaining a population in the rural parts of the country. Nowadays, in a period of agricultural over-production, peatland farming attracts little interest (Pettersson 1990) and the cultivated peatland area is now only about 300 000 ha (Hånell 1990). In Norway, the area of peatland under agricultural utilization is estimated to be about 200 000 ha (Braekke 1990b).

In Finland, the reclamation of peatlands for agriculture had already started in the 1600s. At first, cultivation was based on burning the surface peat layer in spring. This resulted in relatively high yields of crops, but was exploitative because of the loss of the organic layers. Later, the agricultural utilization of peatlands was based on the plowing technique. The practice of adding mineral soil (clay, sand) and mixing it into the surface peat layer has been typical in Finnish peatland cultivation (Valmari 1982). The addition of mineral soil to the cultivated peat causes changes in soil albedo, in the thermal conductivity of the soil, its specific heat, its total porosity and its moisture retention (Pessi 1956). The addition of mineral soil also seems to reduce the hydraulic conductivity of peat (Myllys 1994).

There was considerable reclaiming of peatlands for agricultural purposes after World War II, when a tenth of the Finnish population had to be resettled. In fact, peatlands account for the greatest part of the arable land (300 000 ha) which has been reclaimed in Finland since 1945. This is particularly the case in the northern part of the country where, in many cases, more than half of a farm's arable land was originally peatland (Pessi 1972).

During the 1960s, it was estimated that about 700 000 ha or 30% of the arable land in Finland was on drained mires (Kurki 1963). However, like in Sweden, the area of cultivated peatlands in Finland has drastically decreased during the last decades. Today, the area of the arable land on peat is somewhere between 420 000 and 260 000 ha (Kanninen 1992). The decrease partly reflects the surplus of agricultural products and the general decrease in the total arable land area. Another reason is that effective drainage, cultivation and fertilization have led to increased microbial activity in the surface peat layers and thus to a loss of organic matter. Therefore, the share of arable land classified as peat has also decreased.

In North America over 30 million ha of land has been drained, but only a small portion of this consists of organic soils. It is estimated that only 1.7 million ha of organic soils have been drained in all of North America, and mainly for agriculture (Armentano and Menges 1986). About 0.7 million ha of that are situated in the boreal zone (Gorham 1991).

Estimates of the area of peatland under agricultural utilization in the world vary between 6 million (Kivinen and Pakarinen 1981) and 14 million ha (Armentano and Menges 1986). Because of their physical and chemical properties, fen peats are better suited to in situ agricultural utilization than *Sphagnum* peats (Hammond 1975). Recently, there has been concern about the release of carbon dioxide associated with the agricultural utilization of mires (Armentano and Menges 1986).

2.3 Peat Harvesting

As early as the first century **peat was cut and dried for fuel** along the coast of the North Sea. The impetus for this may have been the shortage of forests in this area (van de Griendt 1992). During the 12th and 16th centuries peat was burnt in order to extract sea salt (Overbeck 1975; Lüttig 1989).

The use of peat for fuel in Germany increased heavily during the seventeenth and eighteenth centuries. Peat cutting was mostly carried out more or less unsystematically by local farmers for private consumption. However, vast areas were used for fuel peat harvesting, particularly near Bremen, so that burning caused notable air pollution (Lüttig 1989). Nowadays in the former Federal Republic of Germany peat is harvested on about 35 000 ha, most of which (90%) is situated in Lower Saxony. The bulk of the harvested peat (80%) is primarily used for horticulture. The rest is mostly used as sod fuel peat to be refined into peat coke and activated carbon. Small amounts are used for balneological and therapeutic purposes (Günther and Schmilewski 1986; Eichelsdörfer 1990).

In Ireland, peat has been used as fuel since prehistoric times and still has a high economic value. After the woodlands were cleared in the 16th and 17th centuries, a large part of the Irish population was dependent on peat as a source of energy (Cooke 1976). With an annual production of about 5 million t of peat, Ireland ranks second, after the former USSR, in the harvesting of peat. The total area used for peat harvesting was more than 80 000 ha by the early 1980s, and it was planned to increase the area up to 100 000 ha (van Eck et al. 1984). To these figures, an area of 172 000 ha of older cutover raised bogs has to be added (Hammond 1979).

The former USSR possesses the world's largest peat reserves, estimated at 200 000–250 000 million t. It is estimated that peat deposits suitable for future utilization cover 60.7 million ha (Markov and Khoroshev 1986). Kivinen (1981) mentions that the area involved in peat harvesting is more than 4 million ha. The peat harvest totalled 170 million t in 1984, most (88%) being used in agriculture. There are already more than 600 000 ha of cutover areas in the European part of the country. These have mainly been changed into agricultural land (Sokolov and Yampolski 1986).

In Finland, peat harvesting for fuel was initiated in 1876 when peat was mined for an iron works in the eastern part of the country. However, the quantities remained insignificant since wood was the primary fuel and sufficiently available. The use of peat as fuel increased in the 1920s when the Finnish railway began burning peat in its engines, and during the World War II when peat was relied upon for domestic fuel (Kalmari 1982; Suoninen 1982). However, it was not until the 1970s in connection with the energy crises that the government set a target of increasing the annual production of fuel peat to 20 million m^3. In 1992, the peat harvest reached 21.5 million m^3, 93% being fuel peat and the rest horticultural peat. The area under harvest was 47 000 ha (Turvetuotanto ... 1992). It has been estimated that the total peatland area suitable for the harvesting of fuel peat is 500 000 ha (Lappalainen and Uhlgren 1991).

Worldwide, peat harvesting is carried out on about 1 million ha (Armentano and Menges 1986). According to Cantrell(1991), the world harvest of peat in 1989 was 217.5 million t.

Between 1985 and 1995, the global demand for peat was projected to grow at an average of 1% per year (Cantrell 1991). Globally, supply and demand are easily balanced, but peat reserves are not evenly distributed. In some countries, the peat reserves have been exhausted but the demand is high. For example, the Netherlands is the biggest importer of horticultural peat with an annual demand of 2.5 million m^3 (van de Griendt 1992). It has been argued that, except for a few countries (e.g. Finland, Canada, Ireland, the former USSR), peat cannot assume a significant role as an energy supplier. Instead, the possibilities for peat utilization are primarily in the sector of special technologies (Lüttig 1983).

2.4 Forestry

2.4.1 Early Beginnings

Some mires support tree growth naturally, but forest drainage enables economic wood production from a much more extensive area of peatlands. In the case of virgin mires, excessive water in the substrate checks root growth and microbial activity, and may even lead to unfavourable biochemical phenomena (see Sect. 3.5.1). The most important objective of drainage is therefore to adjust the water content of the soil to a level which ensures sufficient aeration. In this chapter, a short history of forest drainage is given.

The first mire drainage operations from which the response of trees to drainage have later been detected were not originally planned to increase tree growth. In **Russia**, for example, large-scale drainage was done in the 17th century to benefit water mills and timber floating (Lukkala 1931). However, the idea of draining peatlands to promote tree growth occurred surprisingly early: "After fen and bog soil has been drained through ditching... it can profitably be used for forest growth..." (Gadd 1773, p. 268, citation translated from Swedish).

Tanttu (1915) has published a comprehensive literature review concerning the oldest forest drainage undertakings in Russia, the Baltic countries and **Germany**. It appears that draining of mires to increase tree growth was well known in Russia in the mid-19th century. The growth response had been found to depend on the fertility of the site, peat depth, drainage intensity, tree species and tree age. The very first attempts to afforest a mire in **Great Britain** were done as early as in 1730, and in **Belgium** in 1836 (Zehetmayr 1954).

The Baltic countries were also among the first to start forest drainage. In present-day **Latvia,** forest drainage has been practiced since the early 19th century (Bush 1964). In **Estonia,** systematic forest drainage began in the 1840s, although some drainage ditches had been dug even earlier (Valk 1988).

In **Sweden,** recommendations to use mires for wood production were made in the middle of the 19th century (Obbarius 1845). Forest drainage gained many advocates – some stated that this was one of the most important means of promoting silviculture in Sweden (Bohnstedt 1852). Forest drainage on a practical scale started in Sweden in the middle of 19th century (Malmström 1958). Statistics concerning forest drainage activity in Sweden go back to 1873. However, the period of most intensive drainage activity occurred during the economic depression in the 1920s and 1930s, when private landowners were given state subsidies to carry out forest drainage in order to reduce unemployment. A peak was reached in 1933, during which almost 10 000 km of ditches (corresponding to a drainage area of 50 000 ha) were dug (Hånell 1991c).

In **Norway,** the first documented drainage for forest production took place in 1852. Drainage activity was relatively low until the turn of the century. The highest activity took place during the decade before World War II when 15 500 ha were drained annually (Braekke 1990b).

In **Finland,** agriculture was encouraged in the late 1700s by means of cleaning rapids in brooks and lowering the surface of lakes. The idea was to gain new arable land (Tuokko 1992). Mires were also drained during the famine years (1866–1868). The target was both to provide work for the unemployed and to occupy new arable land. The bulk of these drained mires was never reclaimed for agriculture, but came to form valuable sites for later tree growth studies on drained peat in Finland (see Seppälä 1969).

The other purposes of early peatland drainage operations were either to prevent paludification of surrounding mineral soils or to diminish the incidence of spring frost (Tirkkonen 1952). Thus peatland drainage was considered not only as a soil melioration of the site, but also as a necessary protective measure for the surroundings. This has also been stated in Sweden (Malmström 1928).

More systematic drainage aimed at increasing the growth of tree stands on peat soil or wet mineral soil started on state-owned lands in 1908. At about the same time, many timber companies also began to drain their forested wetlands. Private owners did not start forest drainage activity until 1928, when the first Forest Improvement Law was introduced. This law made it possible to direct public funds to forest improvement operations on privately owned land (see Tirkkonen 1952; Heikurainen 1982c).

Up to World War II (prior to the territorial cession) the following drainage work had been carried out in the mire and wetland area of Finland (Tirkkonen 1952):

Ownership category	Length of ditches (km)	Drained area (ha)
State-owned land	33 330	221 485
Privately owned land	44 833	286 135
Timber companies	31 350	239 272
Church	1 406	9 000
Total	110 919	755 892

Before World War II all the ditches were dug manually, as in the other countries of Fennoscandia.

2.4.2 Extent and Development since World War II

Forest drainage activity in Europe was completely interrupted during World War II. After the war drainage work was severely limited by a shortage of capital and it was not until the 1950s that these activities gained new importance. New ditching methods, including blasting, were tested before plowing and mechanized excavating techniques were developed.

Forest amelioration in waterlogged areas is now part of the normal forestry practice in Fennoscandia, the Baltic countries, Russia, the British Isles and in some parts of the USA (Päivänen and Paavilainen 1990). In Canada, there are large-scale experiments, but forest drainage is still not widely practiced. Currently, about 15 million ha of peatlands and wetlands have been drained for forestry in the boreal and temperate zones (Table 2.1); more than 90% is situated in Fennoscandia (Finland, Sweden and Norway) and the former USSR. In the following, a review of the the extent and development of forest drainage in countries where this kind of forest amelioration is considered important is given.

Table 2.1. Estimates of the area drained for forestry in different countries

Country	Area 10^3 ha	Source
Finland	5 900	Yearbook...(1992)
Sweden	1 410	Hånell (1990)
Norway	420	Braekke (1990b)
Estonia	460	Kaunisto et al. (1991)
Latvia	500	Zalitis (1990)
Lithuania	590	Ruseckas (1991)
Russia	3 800	Vompersky (1993, pers. inf.)
Belorussia	280	Vompersky (1993, pers. inf.)
Poland	120	Okruszko (1976)
Germany	110	Kivinen and Pakarinen (1981)
UK	600	Taylor (1983); Cannell et al. (1993)
Republic of Ireland	210	Farrell (1990)
Republic of China	70	Lappalainen (1991)
Canada	25	Haavisto and Jeglum (1991)
USA	400	Duncan and Terry (1983)
Total	14 895	

Finland

Drainage work picked up again from about 1950. However, it was not until the 1960s that forest drainage activity developed into a nationwide campaign to increase forest growth. The area annually drained increased steadily up to 1969, when 295 000 ha were drained. The total forest drainage area (about 6 million ha by 1991), according to national forestry statistics (Aarne 1993), is greater than that found in the national field inventories (Paavilainen and Tiihonen 1988). This is partly explained by the fact that the forestry statistics include old drainage areas which have had complementary ditching and thus been counted twice. In addition, some of the drained sites originally had only a shallow peat layer, and have later been classified as mineral soil.

In 1986 a national programme, "Metsä 2000" (Forest 2000 Programme . . . 1986), gave targets for national wood production for the period 1986–2005. These targets were revised in 1992 (Metsä 2000 -ohjelman . . . 1992). According to these revisions, the annual forest drainage area will decrease during the first 10-year period to 15 000 ha. The annual need for **maintenance of drainage areas (cleaning of old ditches** and **complementary ditching)** is estimated to be 150 000 ha. So far, maintenance of drainage areas has been far less than needed – reaching only 35 000 ha per year in the 1980s. However, in 1991 an annual figure of 64 000 ha was reached (Aarne 1993).

The extend and practice of forest drainage in Finland are thoroughly discussed in Chapter 4.

Sweden

The first peak in forest drainage in Sweden occurred in 1933. A second peak occurred around 1980. By the end of the 1980s, about 15% (1 410 000 ha) of Swedish peat-covered wetlands had been drained for forestry. About two thirds were peatlands and one third was wet mineral soils. At present, about 17% of the current annual increment in the Swedish forests comes from peatland forests. It is estimated that 50% of Sweden's peat-covered wetlands could be used for forestry (Hånell 1990, 1991c). However, since 1986 **traditional forest drainage** can only be carried out with a permit from the County Government Board and is no longer supported by subsidies.

During the last decade, a new type of forest drainage – **remedial ditching** – has come into practice in Sweden. Remedial ditches are dug in connection with forest regeneration to prevent the groundwater from rising after clear-felling (Hånell 1991c). In Finland, this type of amelioration measure would more likely be called site preparation for regeneration (Kaila and Päivänen 1981).

Norway

The total area drained for forestry in Norway is about 420 000 ha, which is about 19% of the total wetland area (2.2 million ha) below the timberline. After the first intensive period of drainage just before World War II, forest drainage activity increased again in the 1950s and 1960s. Nowadays drainage is directed towards paludified forests and tree-covered mires, but complementary ditching and ditch cleaning in old drainage areas are given priority. There are still about 1.2 million ha of paludified mineral soils and potentially productive mires which could be improved with drainage. It has been estimated that if this area was drained the annual increment could be increased by 6.4 million m^3 (Braekke 1990b).

Estonia

There has been a long tradition of forest drainage in the Baltic countries. In Estonia, the total area drained for forestry is about 463 000 ha. Until recently, the annually drained area has been 15 000 ha. It has been estimated that 13% of the drained area requires renovation, mainly ditch cleaning (Valk 1988; Kaunisto et al. 1991; Zobel 1992).

Latvia

The total area drained for forestry amounts to 500 000 ha, including both mires and forests on peaty soils. The potential yield capacity achieved with forest amelioration is, however, inadequately used. The tree stands exhibit optimum yield capacity on only 16% of the drained area, which means that species composition and growth characteristics are unlikely to meet timber demands (Zalitis 1990b).

Lithuania

The peatland area drained for forestry is 190 000 ha. Furthermore, some 400 000 ha of forested wetlands have also been drained for forestry (Ruseckas 1991).

The Former **USSR**

The total area drained for forestry in the former USSR was about 5.5 million ha. This accounted for only 2.2% of the total area of wetlands (Sabo 1988; Vompersky 1991). Most of the drained area has been drained only since 1954 (Konstantinov and Stoliarov 1988). The rate of forest drainage increased sharply between 1960 and 1980, reaching an annual maximum of more than 260 000 ha around the end of this period. Since then, the annually drained area decreased to 149 000 ha in 1988. Very often the efficiency of forest drainage has been found to be low mainly due to insufficient intensity of drainage, insufficient nutrient status of the drained sites and lack of maintenance of the drainage ditches. Research and development are directed to monitoring of drained areas by means of remote sensing and to solving the problems connected with ditch cleaning (Vompersky 1991).

When taking into consideration the new independent countries, the forest drainage area in Russia can be estimated at about 3.8 million ha (Table 2.1). However, accurate figures are not available for every country.

United Kingdom

The afforestation achievements since 1919 have been considerable in Great Britain. The forested area is now over 2.1 million ha or 10% of the total land area (Malcolm 1991).

The afforestation activity has mainly concentrated on bare heath land, peatland and rough pasture of low value for agriculture, particularly in the uplands of Scotland, the north of England and Wales (Holmes 1979). Many of the soils are initially too wet and have inadequate aeration for the growth of commercial tree species. Toleman (1973), for instance, estimated that 74% of the areas to be afforested by the Forestry Commission in the 1970s were either mineral soils suffering from excess water or peatlands. The two main purposes of draining are to improve tree growth, particularly in the first 10 years, and to increase the root depth and thus reduce the risk of wind-throw (Pyatt 1990a).

It has been reported that some 800 000 ha have been drained for forestry in Great Britain, and 250 000 ha in northern Ireland (Heikurainen 1982a). These figures may, however, be over-estimates. Forestry Commission data show that the drainage activity on true peat soil may have been much smaller (Bather and Miller 1991). The true figure for the UK may be nearer 600 000 ha, including 505 000 ha in Great Britain (Cannell et al. 1993) and about 80 000 ha in Northern Ireland (Taylor 1983).

The long-term prospects for forests on peatland are reported to be good. Yields from the second rotation seem to be higher than the first generation, even when using less fertilizer (Pyatt 1990b).

Ireland

At the beginning of this century the total forested area in the Republic of Ireland was less than 50 000 ha. It has now increased to 410 000 ha (Farrell 1990).

Because of the open nature of peatlands and the almost total loss of old forests, peatland forestry in Ireland is totally dependent on drainage and afforestation. In 1982, it was estimated that there were about 170 000 ha man-made forests on state-owned peatlands (van Eck et al. 1984). This figure may now be around 210 000 ha, which corresponds to less than 18% of the total mire area of the republic. The tree stand establishment technique is based on double mouldboard plowing, which forms a 2-m inter-turf distance and 4-m furrow spacing. Exotic tree species such as *Picea sitchensis* and *Pinus contorta* have mainly been used. Industrial cutaway peatlands also offer great prospects for forestry. It is assumed that a significant proportion of the more than 80 000 ha of peatlands currently harvested for fuel peat will later become available for afforestation (Farrell 1990).

United States

There are many mires in the region from the Lake States to the Atlantic Ocean. These mires are mainly ombrotrophic, raised bogs which display features of a continental-marine gradient (Damman 1979). A very small area of the mires has been drained using **traditional forest drainage** techniques, although an estimated area of at least 1.6 million ha could be profitably drained for forestry (Stoeckeler 1961).

In the Lake States the so-called **prescription drainage system** (Terry and Hughes 1978) has been practiced on poorly drained mineral soil sites. This means a low-intensity ditch system with which the natural drainage ways are opened and the locally high water table controlled (Trettin 1986). This prescription drainage has been an important forest management practice for sustaining a commercially viable forest on a 12 000 ha tract in northern Michigan (Trettin et al. 1991).

In the southeastern states two groups of mires and wetlands can be distinguished: **bottomlands and swamps** are overflow wetlands in which clear-cutting and natural regeneration of decidious species are practiced; **flatwoods and pocosins** are broad, flat areas that have poor surface drainage and include both mineral and peat soils. The latter wetland sites have been converted into the most productive forest land, growing *Pinus taeda* and *P. elliottii* with a site-specific **water management** system which includes drainage, bedding, and fertilization treatments (Duncan and Terry 1983; Allen and Campbell 1988). The area under this kind of intensive forestry may be several 100 000 ha (Schlaudt 1955; Maki 1959; Duncan and Terry 1983).

Canada

Drainage and amelioration of peatlands for forestry purposes have rarely been practiced in Canada. Until now, extensive and untapped forest resources have readily been available, the harvesting of which has been cheaper than the creation of additional forest areas (Stanek 1968a). However, this situation is gradually changing. Haavisto and Jeglum (1991) have summarized the areal extent of forest drainage activities in Canada on a provincial and territorial basis. If not otherwise indicated, the following review is based on this information.

In the **Yukon Territory** and the **Northwest Territories** no forest amelioration activities have taken place on peatlands. In **British Columbia**, very little commercial forest land is found on peat although the potential for improving forest production on some of the wet sites is considerable. However, it is assumed that there is very little reason to manage peatland sites to increase tree growth.

In **Alberta**, only half of the allowable cut of coniferous tree species is annually harvested. However, the authorities are interested in increasing the productivity of the forests in the province. The tree stands on the 12.7 million ha of peatlands (20% of the land area) are considered to be an important renewable natural resource, the productivity of which could be increased through drainage (Päivänen 1985; Hillman 1987). It has been estimated that an area of 4 million ha could be drained to improve tree growth or to convert the area into productive forest land (Hillman 1987). Since 1975, eight systematic peatland drainage trials, covering an area of 1144 ha, have been established to ascertain the most suitable ditch depths and spacings on a variety of peatland sites (Hillman 1991).

Although the area of peatland in both **Saskatchewan** (about 9 million ha) and **Manitoba** (over 20 million ha) is considerable, forestry activity has been insignificant on peatlands. No effort has been made to drain peatland forests in these provinces.

In **Ontario**, there are 22.6 million ha of peatlands (25% of the total land area). Black spruce (*Picea mariana*) is the most important tree species in the province, half of the growing stock of which is on peatlands (Ketcheson and Jeglum 1972). Some drainage trials have been established for research purposes. The earliest of these was established in 1929 (see Payandeh 1973). In the same region of northern Ontario, another drainage experiment was established in 1961 using dynamite to excavate parallel lateral ditches spaced 45–60 m apart (Stanek 1968a). Some local, readily available machinery equipped with a backhoe has also been tested for digging drainage ditches on forested wetlands (Haavisto and Atkinson 1977). However, it was not until in 1984 that the first properly designed, operational forest drainage experiment was planned and carried out (Koivisto 1985; Jeglum 1991c). The drainage project at Wally Creek, in northeastern Ontario on an area of 410 ha, was both planned and executed using Finnish techniques. The main goal of the project is to evaluate the costs of drainage and growth response of black spruce to drainage on different site types. Based on the results from Wally Creek, Rosen (1989) has compiled a forest drainage manual for Ontario. Altogether about 1000 ha have been drained in Ontario for forestry purposes.

The first forest drainage experiment in **Quebec** was established in 1968 (Stanek 1970). The forest drainage activity on a larger experimental scale as well as an operational scale started in the 1980s. By the end of 1988, a total of 13 500 ha had been drained (Trottier 1991a,b). The province of Quebec has provided a monetary incentive for private forest land owners to undertake drainage on peatlands. Also, the companies can carry out drainage projects in public forests to pay their stumpage dues. The increase in the productivity of an area under a timber supply and forest management agreement means an increase in annual allowable cut. This has to be considered as the start of an official drainage policy in the province. A forest drainage manual summarizing operational drainage methods in Quebec has also been published (Guide sur le Drainage . . . 1989).

In **New Brunswick**, some experimental drainage work has been done to obtain basic information on how tree growth responds to drainage. On waterlogged mineral soils some drainage has been done, more for the purpose of providing access to the wet sites for harvesting and silvicultural work than for increasing the productivity of the trees. In **Nova Scotia**, no large-scale operational drainage has been undertaken, but some field experiments (a total area of about 60 ha) have been established to determine the optimal ditch spacing at different kinds of sites, to determine the drainage effect on plantation survival and growth, and to assess the capability of excavators and the Finnish Lännen S-10 tractor digger under Nova Scotia conditions (Neily 1991). No forest drainage projects have been reported from **Prince Edward Island**.

The whole province of **Newfoundland** has about 6.4 million ha of peatland from which about 2.5 million ha are on the island part of the province. The forestry utilization of these peatlands has been small, although some basic features of the mires and their suitability for tree growth have been evaluated (Heikurainen 1968; Päivänen and Wells 1977). Over the last 25 years, only 50 ha of open peatlands have been drained and afforested on an experimental scale. Many of the tree species proved unsuccessful because of the poor quality of the planting stock, poor choice of species and, in certain cases, poor site selection and inadequate drainage design (Wells 1991). The windy and exposed areas have been drained mainly according to Scottish standards, using water furrows without a proper design. However, through more careful selection of sites and more suitable drainage techniques and machinery, peatland forestry could yield better results. Considerable areas of wet mineral soils would also be suitable for forest drainage (Päivänen 1985).

Because the area and distribution of the types of mires economically feasible for forestry purposes in Canada are not readily available, it has been urged that a complete inventory should be undertaken (Haavisto and Jeglum 1991). Until now, less than 25 000 ha (Jeglum 1990; Haavisto and Jeglum 1991) have been drained in Canada for forestry purposes. The earlier estimate – 100 000 ha – given by Kivinen and Pakarinen (1981), was evidently far too high.

It has been considered that forest drainage offers important opportunities in Canada (Haavisto and Wearn 1987). However, knowledge has to be increased at least in three areas: (1) growth responses of local tree species must be better documented according to site types; (2) economic assessments must be made for investment decisions; and (3) environmental impact assessments must be carried out (Jeglum 1991a).

2.5 Peatland Conservation

As discussed in Section 1.1, mires or peatlands are a special type of wetland. Based on the knowledge of the total mire area and the area under economic utilization, it is obvious that the bulk of the world's mires is, in fact still intact and the share of peatlands drained by man is only a small percentage:

World peatland area			
Total	Drained (10^6 ha)	Drained (%)	Source
450	15.5	3.4	Kivinen and Pakarinen (1981)
349	23.1	6.2	Armentano and Menges (1986)
346	11.5	3.3	Gorham (1991)

These estimates cover different zones of the world, and the definitions of peatland, mire, or organic soil-wetland may differ from each other. Nevertheless, they give a picture of the magnitude of peatland utilization in the world.

The total world area of mires is so large that these ecosystems, taken as a whole, are not in danger (Goodall 1983). However, the picture changes when examined country by country. The proportion of peatland area that is virgin in different countries varies from over 99% (Canada) to almost 0% (the Netherlands) (Kivinen and Pakarinen 1981; Johnson 1985). Thus, it is understandable that the concern for mire protection is a local problem. The local interest in an area may raise the conservation value of a mire complex or site far beyond that judged by its position in the range of ecological variation on a regional or global scale (Lindsay 1992).

For instance, it has been stated that 54% of the wetland area of the coterminous United States has been lost since colonial times and that this phenomenon is repeated throughout the developed world, although there has been no comprehensive survey of the details (Maltby 1988a). The difficulties and inaccuracies in estimating the loss of wetland area have been discussed by Kuzila et al. (1991).

It has been claimed that, over the past two centuries, 90% of western Europe's lowland mires has been destroyed by drainage and conversion to farmland or forestry (Sadler 1992). This may be true if the estimate covers only the continental part of central-western Europe. Lüttig (1991), for instance, admits that in Germany only a minute portion of the original raised bogs, certainly under 5%, is truly in a natural condition.

Lindsay (1992) describes the problems connected with the peatland conservation as "a Cinderella syndrome" which has held back mire protection compared to the protection of other kinds of sites. This is partly due to the poor public image of peatlands. National inventories of valuable natural resources have often considered them as wasteland (Maltby 1986). However, this attitude is rapidly changing. A number of countries already have a well-developed mire conservation policy.

Conservation of mires in Finland started in 1965 with the formation of a committee for mire protection. As a result, the Finnish Government in 1979 and 1981 approved in principle the basic protection plans for an area of 0.5 million ha. In addition, national parks and strict nature reserves contain 0.2 million ha of mires. The aim is to preserve about 7% of the original mire area. During the 1980s, some 400 000 ha have been preserved on state-owned land. About 85 000 ha of privately owned mires, which were included in the protection plan, still remain unprotected (Ruuhijärvi 1982b, 1989). The conservation of eutrophic fens in the southern part of the country is perhaps the most necessary (Heikkilä 1992). It should be noted that in addition to the protected mire area, about 3 million ha of virgin peatlands in Finland will remain outside of the interests of agriculture, forestry or peat harvesting.

In the British Isles, mire conservation is also advancing. The Irish Government has recently defined goals for mire protection. The aim is to preserve 10 000 ha of raised bogs and 40 000 ha of blanket bogs. These figures represent 4% of the original total area of peatlands in Ireland (Cross 1992; Foss and O'Connel 1992). So far, about one third of this goal has been achieved (Salminen 1991). According to Bather and Miller (1991), current estimates indicate that there are over 20 000 ha of raised mire in Great Britain having a conservation value, of which more than 13 000 ha already have "Sites of Specific Scientific Interest" status. However, the need for mire conservation may be even greater (Kaakinen et al. 1990). It has also been suggested that any new afforestation in the lowlands should avoid semi-natural habitats. Therefore, the nature conservation value of potential afforestation sites first has to be assessed (Watkins 1991).

In the former FRG, about 12% of the approximately 250 000 ha of existing raised bogs in Lower Saxony are to be placed directly under nature conservation.

An additional amount of 32 000 ha is to be rewetted after peat harvesting, with the aim of restoration (Niedersächssisches . . . 1981; Roderfeld 1993). Restoration of former agricultural fields and pasture on drained peat is also being considered in the former GDR (Schmidt and Scholz 1993).

In Canada, there are several inventory and evaluation programmes to support mire conservation policy. Wetlands were even identified as a priority national conservation issue by Canada at the International World Conservation Strategy Conference in Ottawa in 1986 (Rubec et al. 1988).

Mires can be categorized according to their conservation value in different ways. A four-level classification has been used in Finland: (1) internationally, (2) nationally, (3) regionally and (4) locally important mires. This has been done within each mire complex region (Ruuhijärvi 1982b). In Great Britain, selection of mires for conservation requires an assessment of their scientific importance. Proposed sites are judged against other examples of the same type using attributes such as size, naturalness and species rarity and diversity (Fojt 1992). In northern Ireland, mires should be intact, and then both physical and floristic criteria are used to select the objects meriting conservation measures (Faulkner et al. 1992).

The area to be protected should contain the whole mire complex, i.e. a hydrologically independent area. Mire site types are best conserved within large mire complexes, otherwise their original nature may not be retained (Ruuhijärvi 1982b). A new threat to mire conservation – even to raised bogs – is air pollution. The deposition of nutrients and air pollutants is changing the mire ecosystem even though the local hydrological conditions may remain the same.

In some European countries, e.g. in the Netherlands (Joosten 1992) and Germany (Günther and Schmilewski 1986; Terkamp 1992), regeneration (rehabilitation or restoration) of cutaway mires is considered as mire protection. In Great Britain, the water table in virgin mires near drained ones has been kept high by damming the water tracks (Ogilvie 1990). In Germany the use of mole drains to rewet former agricultural fields on fen peat has been studied (Schmidt and Scholz 1993). Even in Finland, which still has many virgin peatlands, rehabilitation of drained peatlands by damming ditches has been examined. This has been done in quite recently established national parks, where the former drained mires will be converted back to virgin ecosystems (Vasander et al. 1992).

3 Bases of Peatland Forestry

3.1 Carbon Cycling in Mire Ecosystems

3.1.1 Virgin Mires

The primary conditions necessary for mire formation and peat accumulation were discussed in Section 1.2.1. Peat accumulation results when the rate of addition of dry matter exceeds the rate of decay. Conceptually, the function of a mire ecosystem can be presented in the form of energy or carbon flow (Pakarinen 1976; Fig. 3.1). **Accumulation of carbon** in peat is typical for mire ecosystems, and the determination of the rate of accumulation is a key element in any research on the **carbon balance** in mires. Mosses, mainly *Sphagnum* mosses in bogs and fens or hypnoid mosses in rich fens, add new organic matter to the surface (Clymo 1983, 1984; Tolonen and Vasander 1992), but vascular plants produce rhizomes and roots, and these add matter to deeper layers in the peat also (Saarinen et al. 1992). However, most of the organic matter is – at least in ombrotrophic bogs – added to the top 10 cm (Clymo 1983).

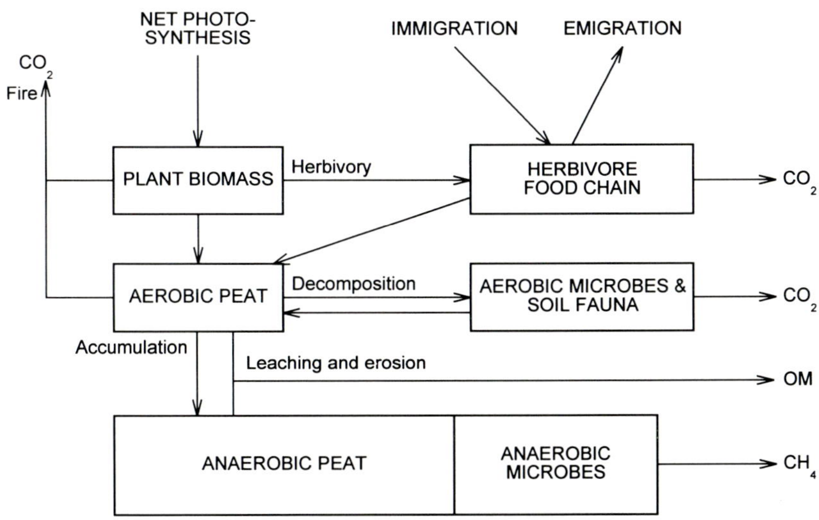

Fig. 3.1. Conceptual model of a mire ecosystem; *OM* organic matter. (Redrawn from Pakarinen 1976)

Depending on the time scales chosen, it has been suggested to differentiate between **long-term** (over millenia) and **short-term** (over past decades or a few centuries) **rates of carbon accumulation** (Tolonen and Vasander 1992). The **true rate of carbon accumulation** (Clymo 1992), also called the **actual rate of net accumulation** (Tolonen and Vasander 1992), is, however, smaller than both the long- and short-term accumulation rates since carbon loss is taking place in the whole peat column, including its anaerobic part. Thus, vertical growth in a mire (especially in an ombrotrophic bog) is dependent on the relationship between net production in the surface layer and decomposition in the peat horizons below it. Lateral extension is regulated by the hydrological properties of the mire and its physical surroundings (Clymo 1984; Korhola 1992).

The models developed by Clymo (1992) assume that the surface layer of a mire is in a steady state: fixing carbon, losing some by decay, and passing some on to the underlying peat proper. Most of the decay takes place in the well-aerated surface layer (the **acrotelm**; see Sect. 3.5.1). Below the stagnant water level (the **catotelm**) the decay process – mainly by bacteria – is anaerobic. Anaerobic decay is much slower than aerobic, and it is this slowness of decay that is the direct cause of peat accumulation (Clymo 1992). The pictorial process model developed by Clymo (1992) shows schematically that if the rate of input is constant but decay is proportional to the whole mass accumulated so far, the true rate of accumulation decreases with time and eventually approaches zero (Fig. 3.2). This suggests an asymptotic approach to limiting maximum peat depth (Clymo 1983, 1992; Warner et al. 1993).

Most peat accumulation studies have been conducted on ombrotrophic bogs, and very limited data are available for fens. Because of the greater decomposition activity related to minerotrophy, the accumulation of organic matter is lower

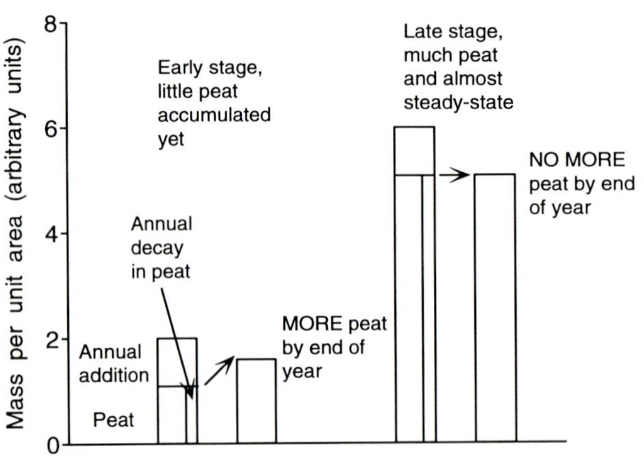

Fig. 3.2. Schematic diagram indicating how the true rate of peat accumulation in an ombrotrophic bog decreases with time. At an early stage of development, the annual rate of addition exceeds the combined loss by decay from all depths. At a late stage the annual addition at the surface is exactly the same but losses occur throughout a much greater mass of peat, resulting in no net accumulation. (Redrawn from Clymo 1992)

Table 3.1. Net primary production (P) and peat accumulation (A) according to different studies

Site and location	P $g\ m^{-2}a^{-1}$	A	A/P (%)	Source
Continental raised bog, Manitoba	710	36	5	Reader and Stewart (1972)
Tundra mire, Devon Island, Canada	185–280	5–10	2–5	Pakarinen (1973), Bliss (1975)
Blanket bog, UK	447	29	7	Heal et al. (1975)
Blanket bog, Ireland	523	32	6	Moore et al. (1975)
Raised bogs, South Finland	300–420	25–48	6–16	Tolonen (1977)
Mean value, Karelia	300	50	16	Kozlovskaya et al. (1978)
Raised bog, South Finland	500	50	10	Reinikainen (1981)
Raised bog, South Estonia	330	13–40	4–12	Ilomets (1982)
Raised bog,	760	71	9	
Forested mire, Karelia	750	57	8	Elina et al. (1984)
Raised bog, Maine, USA	344–700	35–55	5–16	Tolonen et al. (1988)
Raised bog, Central France	922	106	12	Francez (1991)
Sedge fen, Central Finland	1088	23	2	Laine et al. (1992b)

in minerotrophic than in ombrotrophic sites. The relationship between accumulation and net primary production in virgin mire ecosystems varies between 2 to 16% (Table 3.1, see also Päivänen and Vasander 1994). A very preliminary mean of true carbon accumulation value in Finland is 14.7 g C $m^{-2}a^{-1}$ for *Sphagnum* bogs and 9.5 g C $m^{-2}a^{-1}$ for sedge mires (Tolonen and Vasander 1992). However, knowledge about carbon accumulation rates of mires, particularly in minerotrophic mires, is still limited.

The role of carbon cycling in mire ecosystems in relation to global climatic change has gained in importance lately (e.g. Sjörs 1981; Armentano and Menges 1986; Gorham 1991; Laine and Päivänen 1992; Vasander and Starr 1992). Virgin mires are unusual ecosystems in the sense that while they are net sinks for **carbon dioxide** (CO_2), they also emit large quantities of both CO_2 and **methane** (CH_4). On some fertile sites, a third very effective "**greenhouse gas**", **dinitrogen oxide** (N_2O), may also be emitted.

The greenhouse effect of a gas depends not only on its ability to absorb infrared radiation but also on its lifetime and chemical reactions in the atmosphere. According to Houghton et al. (1990), the global warming potential of various greenhouse gases for a period of 100 years (i.e., the warming effect of an emission of 1 kg of each gas relative to that of carbon dioxide) is:

CO_2	1
CH_4	11
N_2O	270

If the climate becomes warmer or drier, then the rate of aerobic decay and the depth of the acrotelm (see Sect. 3.5.1.2) might increase even in virgin mires. Existing mires might cease to grow and new peat deposits might begin to form in regions with hitherto unsuitable climates (Gorham 1991; Warner et al. 1993).

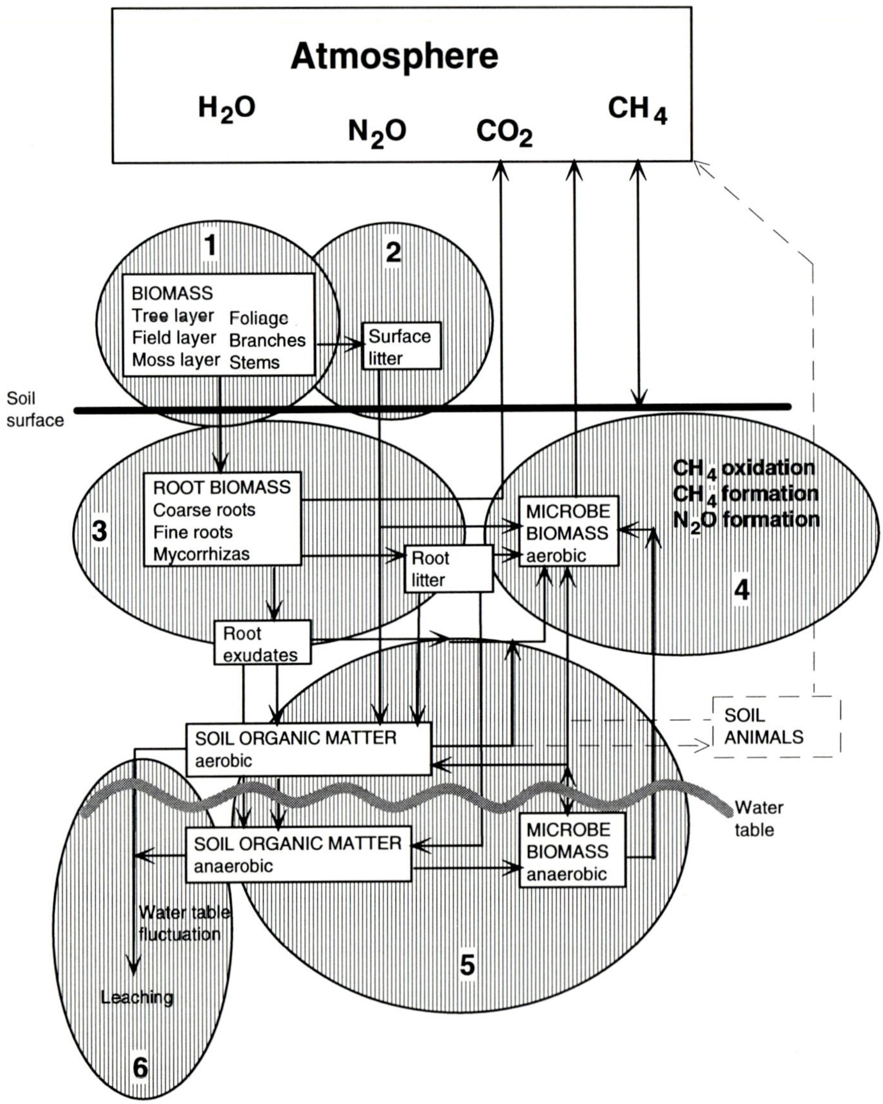

Fig. 3.3. Schematic presentation of the carbon cycling ecosystem model in mires. *Boxes* denote carbon stores; *arrows* the material flow between the stores; the *shadowed areas* beneath refer to different submodel modules. (Courtesy of Jukka Alm)

When studying the carbon balance of peatlands, empirical data are first collected for the development of submodels that will subsequently be linked as modules to produce a complete peatland carbon cycling model (Fig. 3.3). This framework is used in the Finnish study project, Carbon Balance of Peatlands and Climatic Change (see the progress report edited by Kanninen and Anttila 1992, p. 189–242 as well as Sect. 3.1.2).

3.1.2 Drained Mires

3.1.2.1 General Effects of Drainage

Drainage is a prerequisite for most forms of peat and mire utilization. The lowering of the water table through drainage makes aerobic microbial activity possible in the deeper surface peat layers and, consequently, fluxes of carbon dioxide (CO_2) to the atmosphere are increased. At the same time, the increased oxidation of methane (CH_4) to carbon dioxide means that methane emissions are decreased and may even be terminated. On some sites, drainage has even led to the situation that the surface peat layer acts as a sink for methane, when bacteria which oxidize methane consume it from the atmosphere (Martikainen et al. 1992).

3.1.2.2 Mires Drained for Agriculture

A mire drained, reclaimed and used for agriculture, acts as a strong source that releases carbon dioxide and dinitrogen oxide into the atmosphere (Armentano and Menges 1986; Martikainen et al. 1992). Cultivation and repeated application of fertilizers accelerate the decomposition of peat. Conversely, by maintaining high water levels, the peat can be preserved longer. The subsidence and decomposition of peat depend mainly on the intensity of drainage, the initial peat depth and the time elapsed since drainage (Nesterenko 1990). The share of oxidation of the peat subsidence has been found to be higher in minerotrophic than ombrotrophic peats in northwestern Germany (Eggelsmann 1976).

3.1.2.3 Peat Harvesting Sites and Peat Use

Mires drained and used for peat mining will be transferred into net carbon sources for the period of time they are under this utilization. In peat burning, most of the carbon contained in the fuel is released in the form of carbon dioxide.

In order to estimate the greenhouse effects of peat harvesting and use, the initial state of the mire with the instantaneous and cumulative greenhouse effects associated with the harvesting and burning of peat as well as the subsequent use of the cutover areas have to be taken into consideration (Hillebrand 1993). Theoretical calculations have suggested that the smallest cumulative greenhouse effect is caused when a mire used earlier for agriculture is put into peat production, and the area is paludified when the production has come to an end (Hillebrand 1993).

3.1.2.4 Mires Drained for Forestry

The objective of forest drainage is to increase the plant biomass production in the tree layer. Many studies have shown the response of tree stands to drainage (see Chap. 7). The field and ground layers also respond to drainage; they are replaced by a vegetation resembling that typical for mineral soil sites (see Sect. 3.4.2). If the site type is fertile enough, the above-ground biomass production on a drained mire site exceeds that of a virgin site (Reinikainen 1981; Table 3.2). The share of tree layer increases and that of the field and ground layer decreases in the annual above-ground biomass production (see also Sakovets and Germanova 1992). On the poorest sites, i.e. ombrotrophic bogs, the total biomass production may, however, even decrease after drainage (Vasander 1982).

One has to be careful with the nomenclature dealing with biomass studies. The term **whole tree** includes all the components of the tree biomass above the stump cross section. The entire tree including the stump and roots is referred to as the **complete tree** (Hakkila 1989, p. 11–12).

Little is known about the biomass production of the below-ground parts of tree stands growing on drained peatlands. Estimates of fine root production have usually either been based on: (1) a growth percentage of the above-ground parts (e.g. Mälkönen 1974) or (2) an assumption of a certain turnover rate for the roots (Finér 1989). For a Scots pine (*Pinus sylvestris*) stand on drained dwarf shrub pine bog in eastern Finland, these kinds of estimations indicated that below-ground parts accounted for between 16% (method 1) and 46% (method 2) of the biomass production of the total stand (Finér 1990). In mineral soil stands, fine root production may account for 5 to 75% of the net primary production of the stands (see references in Finér et al. 1992).

Forest drainage has been claimed to increase carbon dioxide emissions to the atmosphere and thus contribute to the greenhouse effect (e.g. Armentano and Menges 1986; Silvola 1986, 1988). However, whether a mire drained for forestry changes from being a net accumulator of carbon to a net source of carbon depends on the ratio of the biomass accumulation (mainly in the form of mer-

Table 3.2. Above-ground biomass and its production on a virgin and two drained dwarf shrub pine bogs in southern Finland. (Values from Reinikainen 1981)

	Virgin	50 Years after drainage	74 Years after drainage
Biomass, t ha^{-1}	19	77	118
Tree stand, %	81	96	99
Field layer, %	15	3	0.1
Ground layer, %	4	1.2	0.4
Annual production, t $ha^{-1}a^{-1}$	2.3	4.4	6.6
Tree stand, %	49	85	97
Field layer, %	38	10	0.5
Ground layer, %	13	5	2.5

chantable timber, litter and root biomass) to the increased mineralization of carbon in the peat.

In earlier soil respiration measurements from forest drainage areas, the contribution of roots to the release of CO_2 from peat was not separated, and therefore the results may overestimate the CO_2 emission due to accelerated peat decomposition (Silvola 1986, 1988). However, the major part (perhaps 60–90%) of the CO_2 released from peat drained for forestry seems to originate from the microbial decomposition of the organic matter accumulated earlier (Silvola et al. 1992).

Forest drainage evidently increases the decomposition of peat, measured as CO_2 production (Silvola and Alm 1992; Martikainen et al. 1994), while CH_4 emissions generally decrease after a lowering of the water table (Martikainen et al. 1992; Martikainen et al. 1993). If the water table is lowered enough, the site can even be a net sink for CH_4. Only sites rich in nutrients usually show an increase in N_2O emission after drainage (Fig. 3.4; see also Table 3.3). The effect of gases on climate warming depends on their capacity to absorb infrared radiation and on their lifetime in the atmosphere. Thus the measured amounts have to be corrected as mentioned in Section 3.1.1.

Table 3.3. Rough estimates for the net accumulation of carbon dioxide, methane and dinitrogen oxide emissions in virgin and for forestry drained peat soils in Finland

	CO_2 net accumulation C, g $m^{-2}a^{-1}$	CH_4 emission C, g $m^{-2}a^{-1}$	N_20 emission N, g $m^{-2}a^{-1}$
Virgin mire			
Ombrotr.	5–120[a]	2– 5[b]	< 0.004[b]
Minerotr.	4– 60[a]	18–42[b]	< 0.004[b]
Drained			
Ombrotr.	−100[c]	1– 2[b]	< 0.004[b]
Minerotr.	+ 60[c]	0[b]	0.05– 0.13[b]

[a] (Tolonen and Vasander 1992).
[b] (Martikainen et al. 1994).
[c] (Laine et al. 1992b).

To derive reliable estimates for carbon cycling in mire ecosystems the gas emissions should be verified and compared with the changes in the carbon store in peat and vegetation, using various time scales. This has been attempted by determining bulk density and carbon content profiles along transects running from the undrained part to the drained part of a mire (Laine et al. 1992a,b). The undrained and drained areas should have had the same kind of vegetation (site type) and peat profile at the time of drainage; any differences later will be due to drainage. The carbon stores of the peat in the undrained and drained areas are calculated for the peat layer in which drainage is considered to have caused changes (Anderson et al. 1992; Laine et al. 1992a,b).

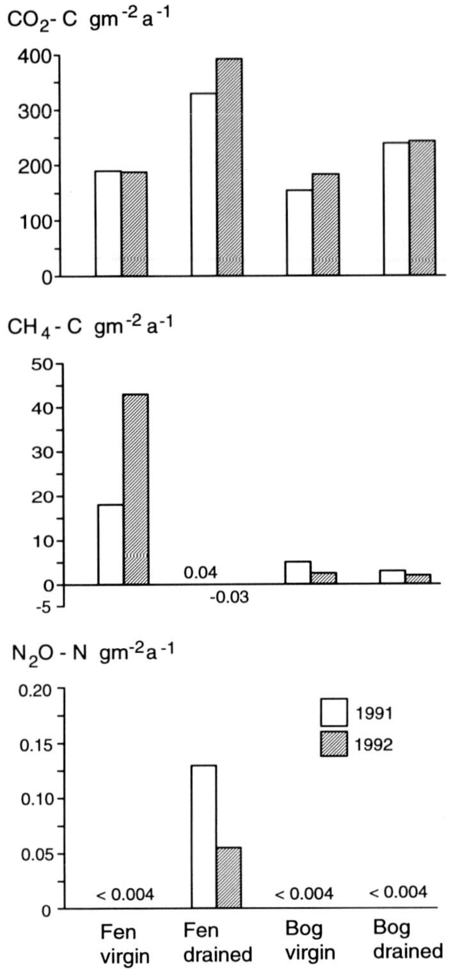

Fig. 3.4. The annual emissions of CO_2, CH_4 and N_2O (g m^{-2} a^{-1}) at some virgin and drained sites in southern Finland. (Redrawn from Martikainen et al. 1994)

Preliminary results dealing with only the changes in **carbon stores in peat** indicate that sites where the response of the tree stand to drainage is weak (ombrotrophic sites) may become sources of carbon to the atmosphere. However, in minerotrophic sites where the tree growth response is stronger, the change is negligible or even more carbon may accumulate than is emitted (Laine et al. 1992b), at least for some decades after drainage when the tree layer is in the most productive phase. The increased litter production of the tree layer (above and, especially, below ground) after forest drainage largely determines the carbon balance of a site (Finér 1990; Finér et al. 1992; Laine et al. 1992b).

For maritime conditions, contradictory results about the carbon balance of drained peatlands have been presented. Braekke (1987) reports for Norway (59°16'N, 10°19'E) annual carbon losses as high as 290 g m^{-2} due to forest drainage. In contrast, Anderson et al. (1992) in Scotland (58°26'N, 3°25'W) could not detect any siginificant changes in the mass budgets of blanket peat after 22 years of drainage.

The changes in the **carbon store of the whole mire ecosystem** (i.e., soil and tree stand) due to drainage can also be estimated. The average net increase in carbon store in biomass allocated to the tree layer due to drainage is mainly formed during the first rotation (Cannell et al. 1992; Cannell et al. 1993). After that carbon input and output throughout the tree stand is more or less in balance. Of the carbon in stem wood removed in harvestings, only the part that is stored in the products of the wood processing industry remains. The amounts of carbon stored and the residence times of carbon in the products are still insufficiently understood.

Laine and Laiho (1992) have calculated that the annual post-drainage aboveground tree stand biomass production (stem wood) in Finland would compensate for the loss of carbon resulting from the oxidation of the peat, as estimated by Armentano and Menges (1986), for about 200 years.

Similar results from different parts of Russia have also been presented. Vompersky and Smagina (1984) concluded that 10-year-old forest drainage areas on peat at Zapadnaya Dvina, some 350 km west of Moscow (56°N, 32°E), still act as sinks for carbon, although at a reduced rate of accumulation, and that it is the increased litter production from the tree layer that compensates for the increased mineralization of peat, even though below-ground litter production was not included in their study. From the same area, Vompersky et al. (1992) described the post-drainage succession of two mire ecosystems (mesotrophic and oligotrophic site) towards a semi-climax forest on drained peat: the organic matter balance remained positive, although the merchantable stem wood was assumed to be harvested using a 100-year rotation.

Glebov and Melentyeva (1992) studied the organic matter balance of a 30-year-old, mesotrophic drainage area growing a birch-dominated stand in the Tomsk region in Russia (57°N, 85°E). Peat decomposition increased after drainage, but the carbon mass in the ecosystem had not changed; the carbon release due to peat oxidation was compensated for by the dead phytomass (litter) production.

Information given about the effect of forest drainage on the carbon balance of mires in Russian Karelia is inconsistent. For a mesotrophic site drained 20 years earlier and growing a mixed stand of Scots pine (*Pinus sylvestris*) and birch (*Betula pubescens*) in Karelia (62°N, 32°E), Sakovets and Germanova (1992) reported that the carbon store of the peat was reduced by 0.32 t $ha^{-1}a^{-1}$ due to increased decomposition. However, if the increased phytomass production had also been taken into consideration, the total carbon accumulation in the mire ecosystem would have been calculated to increase by 1.23 t ha^{-1} a^{-1}. On the other hand, Makarevskii (1992) has calculated that on average the additional increment in tree stands on drained areas does not compensate for carbon losses from peat in Karelia.

In the scenarios of the carbon balance for drained mires the time scale used is one of the dominating factors. This might not have been fully understood or taken into consideration in the above-mentioned studies. In the short term, that is over the next 50–200 years, consideration of global warming is of greatest relevance. Cannell et al. (1993) have done theoretical calculations for drained and afforested peat soils under the British conditions, assuming an average yield class growing *Picea sitchensis*: If the rates of peat oxidation are slow, drainage and afforestation will increase carbon accumulation. If the rate of oxidation is about 150 g C $m^{-2}a^{-1}$, afforestation will give a net benifit for only 50–100 years, but if the rate is > 300 g C $m^{-2}a^{-1}$, there may be no net benefit of afforestation (Cannell et al. 1993).

It seems to be too early to generalize the effects of forest drainage to the carbon balance of mire ecosystems. However, although drainage may accelerate the mineralization of peat, resulting in a loss of carbon to the atmosphere, this does not necessarily mean a net loss from the ecosystem, or even from its below-ground component.

3.2 Physical Properties of Peat Soil

3.2.1 Peat Structure

3.2.1.1 Peat Type

Peat is formed from more or less fragmented plant residues sequentially deposited. The presence of plant fibres is a distinguishing feature (Okruszko 1979).

Classifications of **peat type** are usually based on the macroscopic determination of the relative occurrence of various **peat constituents** (groups of plant residues) in peat samples (Heikurainen 1964). If necessary, a microscope may be used to determine the peat type for samples in an advanced stage of decomposition (Heikurainen and Huikari 1952). The peats are usually divided into three groups based on the dominant peat constituent as follows: *Sphagnum* peats (S peats), *Carex* peats (C peats) and woody peats (L peats).

However, it has to be kept in mind that the difficulties associated with the classification of peats are related to the heterogeneity and diversity of the organic matter in terms of physical, chemical, and micromorphological features. Therefore, no single measurement or observation can adequately differentiate between all peats (Lévesque et al. 1980).

3.2.1.2 Decomposition

Peat, as a dead plant material, undergoes deformation processes collectively called decomposition, decay, breakdown or humification. The processes encompassed are: (1) loss of organic matter, as gas or in solution; (2) loss of physical structure; and (3) change in chemical state (Clymo 1983).

In terms of the physical properties of peat, we are mostly interested in how to describe the stage of decomposition. The classification most frequently employed in the field is based on an arbitrary scale presented by von Post (1922). The method determines the stage of decomposition by means of an ocular study of peat samples. Peat is squeezed in the hand and the fluid which escapes and the nature and amount of peat substance remaining in the hand are examined. The sample is then allocated to one of ten **von Post humification classes**. H1 refers to fully undecomposed plant material and H10 to fully decomposed peat.

For more detailed assessments, it is often desirable to define the stage of decomposition of peat in a more accurate and objective way than is possible using von Post's method. In such cases, laboratory measurements are usually necessary. In the former Soviet Union, **Pyavchenko's decomposition percentage** (e.g. Sarasto 1960) was previously in general use, but later the **centrifugation method** was considered as a standard (State Standard of...1976). In Central Europe, the **Vertorfungsgrad** method (Segeberg 1956), originally developed by Keppeler, has been widely applied. In the USA, the decomposition of peat has been generally described by means of **fibre content** (Farnham and Finney 1965).

The different methods of describing the stage of decomposition have been compared in several studies (e.g. Segeberg 1956; Sarasto 1960; Karesniemi 1972; Raitio and Huttunen 1976; Schneekloth and Schneider 1976; Tolonen and Saarenmaa 1979; Roderfeld 1992). Due to the rapid determination of von Post humification classes in the field, this method has been found to be superior to several other time-consuming laboratory methods if both accuracy and duration are taken into consideration (Malterer et al. 1992; Bohlin 1993).

3.2.1.3 Bulk Density

As peat decomposes, it also becomes denser. The **bulk density** (D_b) of peat is determined by dividing the oven-dry (105 °C) mass of peat (M_s) by the volume of the undisturbed (saturated) peat sample (V_t)(see Fig. 3.5):

$$D_b = \frac{M_s}{V_t}.$$

Because the bulk density of peat is an important property, great efforts have been made to develop accurate methods to obtain volumetric peat samples both from the surface layers (e.g. Baden and Segeberg 1949; Boelter 1969; Päivänen 1973b; Stanek and Silc 1976; Peat testing...1979; Westman 1981; Jeglum et al. 1991; Laiho and Laine 1992b) and deeper layers (Korpijaakko 1981; Tolonen and Ijäs 1982). Accurate bulk density determinations are needed, for example, in water regime studies, to convert the water content of peat measured in mass units into volume percentages (Boelter and Blake 1964; Päivänen 1973b), or for fertility studies to convert basic concentration data into absolute amounts (Erviö 1970; Westman 1981; Kaunisto and Paavilainen 1988). However, volumetric

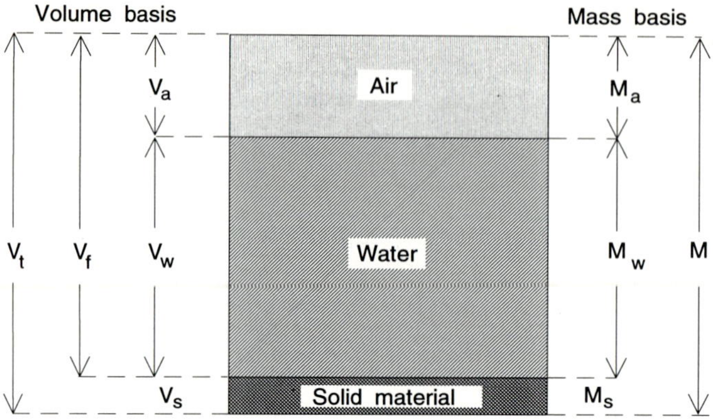

Fig. 3.5. Air, water and solid material, and their relationships in peat, volume and mass basis. Symbols: V_t total volume of undisturbed (fresh) peat sample; V_f total porosity; V_a volume of air; V_w volume of water; V_s volume of solids; M_t total fresh mass of peat; M_a mass of air; M_w mass of water; M_s oven-dry (105°C) mass of peat

peat sampling started in Finland in connection with tree fine root studies on drained peatlands in the 1950s (Heikurainen 1955).

In *Sphagnum* peats, there is a positive and approximately linear regression between bulk density and the von Post humification scale value (e.g. Päivänen 1969; Korpijaakko and Radforth 1972; Tolonen and Saarenmaa 1979). In *Carex* peats, the correlation may be weaker because of the different type of plant material accumulation (Tolonen and Saarenmaa 1979).

For fully saturated peats with average ash contents, bulk density values can be estimated from their water content either on a volume or mass basis (Korpijaakko et al. 1981; Laine and Päivänen 1982). A significant negative correlation has even been established between the energy content of in situ peats and their water content (Tolonen et al. 1982).

Bulk density values for Finnish peats from undisturbed and drained areas vary from 0.04 to 0.20 g cm^{-3} (Päivänen 1973b). Values for Minnesota peatlands range from 0.02 to 0.26 g cm^{-3} (Boelter 1964b). Values as high as 0.2 to 0.4 g cm^{-3} have been reported for Central European peats under "Niedermoor" conditions (Kuntze 1984). For comparison, the bulk density of mineral soils varies between 0.8 and 2.0 g cm^{-3} (e.g. Holstener-Jørgensen 1958).

3.2.1.4 Density of Solids

The **density of solids** (D_s) refers to the mass of dry solids divided by the volume of the solids. The density of solids in a peat soil is typically about 1.5 g cm^{-3} (Puustjärvi 1970; Päivänen 1973b). For comparison, the density of solids of mineral soils is usually 2.6–2.8 g cm^{-3} (Elonen 1971).

3.2.1.5 Total Porosity

The bulk density (D_b) and density of solids (D_s) determine the **total porosity** (V_f):

$$V_f\,(\%) = \frac{D_s - D_b}{D_s} 100.$$

The total porosity of peat is high, about 97% for undecomposed and 81–85% for highly decomposed peats (Boelter 1969; Päivänen 1973b). In mineral soils the total porosity is much lower, varying between 30 and 65% (e.g. Holstener-Jørgensen 1958; Todd 1959).

3.2.1.6 Volume Changes

Surface fluctuation in the height of the surface of virgin and drained mires has also been called "mire breathing" ("Atmen der Mooroberfläche") (Uhden 1956; Baden and Eggelsmann 1964). In undisturbed Finnish mires, the daily fluctuation is low, generally 0.5–1 mm, but the amplitude for the whole growing season may range from 18 to 45 mm (Kurimo 1983). The long-term surface fluctuation seems to follow the corresponding fluctuation of the groundwater table, but delayed and on a reduced scale (Kurimo 1983).

During drying, in situ peats show both surface subsidence and "fibre" shrinkage (Ahti 1987). Pyatt and Craven (1979) have even reported surface peat cracking in drained Scottish peatlands. However, cracking has not been reported elsewhere.

Forest drainage causes a significant **subsidence** of the peat surface. For drained mires in southern Finland, Lukkala (1949) measured the following magnitudes of subsidence over 36 years: spruce mires 21–32 cm, dwarf-shrub pine bogs 24–27 cm and tall-sedge fens 31–41 cm. More recent measurements indicate that subsidence in different sites varies between 15 and 40 cm, 50 to 60 years after drainage (Laine and Laiho 1992). In the beginning, most of the subsidence is caused by the physical collapse of the peat matrix when the water is removed. Later, the subsidence is increasingly related to the oxidation and decomposition of the peat. Oxidation percentages of 35% (Braekke 1987) and 14% (Armentano and Menges 1986) are, however, uncertain.

Because of the **shrinkage-swelling properties** of organic soils, bulk density varies with the water content of the soil. The tendency to shrink upon drying is particularly connected to humus particles. Plant fibres exhibit a different tendency. The changes occurring in the course of drying, manifested in the form of peat shrinkage, depend on the mutual relations and interaction forces of the two peat components (Okruszko 1960).

The shrinkage of peat samples upon drying is greater when the peat is more decomposed (e.g. Päivänen 1982b). Thus, the shrinkage of peat samples has also been found to increase with increasing sampling depth (Päivänen 1982b). These findings are probably due to the colloidal properties of the highly decomposed peat material. In contrast, the shrinkage of undecomposed Sphagnum peat during drying has frequently been noted to be low. This property is the main reason why peat moss is used as bedding material for domestic animals and as a growth substrate in greenhouses (Puustjärvi 1968). Information concerning the shrinkage of peat upon drying and factors influencing this process are particularly important for the harvesting and technical application of peat.

3.2.1.7 Geotechnical Properties

Bearing capacity or soil strength is defined as the resistance of the soil surface to pressure. The bearing capacity of peat increases with decreasing water content. Indeed, it is the moisture tension or the negative soil water pressure that determines soil strength (Schothorst 1982). The relative bearing capacity can be measured using methods based on penetration resistance (Saarilahti 1982; Schothorst 1982), vane shear strength (Saarilahti 1982) or radiowave probes (Saarilahti 1982; Rummukainen 1984). For example, the ground pressure of excavators and forwarders to be used in summer conditions on peatlands should not exceed 40 kPa (Eeronheimo 1991).

3.2.2 Water Retention

When considering living plants, the soil water condition is best described by means of soil water energy relationships. The tension with which water is held in the soil is termed the **matric potential** (**matric suction** or **soil water suction**). It is a composite of the capillary and adhesive forces produced by the soil matrix (Hillel 1971). The **water retention of soil** refers to the relationship between water content and matric suction. It is also referred to by a number of terms, including **moisture characteristics, water sorption curve,** and **pF curve** (Peerlkamp and Boekel 1960; Richard 1963; Puustjärvi 1969). The form of the curve describing this relationship is determined by the structure and the pore size distribution of the soil.

Matric suction and the corresponding water content of peat samples can be determined in the laboratory by applying various suction or pressure methods (Boelter 1964b; Päivänen 1973b). Only undisturbed fresh samples in which the natural structure is maintained should be used (Boelter 1964a). The direction of the change in the soil water content also influences the interrelationship between matric suction and the water content. This phenomenon is called **hysteresis** (e.g. Poulovassilis and Childs 1971). It is also possible to measure the actual matric suctions prevailing in the peat in situ using **tensiometers** (e.g. Ahti 1971, 1978; Laine and Mannerkoski 1975). However, tensiometers cannot measure soil water tensions higher than about 800 kPa (Hendrickx 1990).

The **water retention capacity of peat soils** is generally greater than that of mineral soils at low (< pF 1.5) matric suction values (Fig. 3.6). In the case of larger matric suction values, too, coarse mineral soils usually retain less water than slightly decomposed *Sphagnum* peat, the water retention capacity of clay

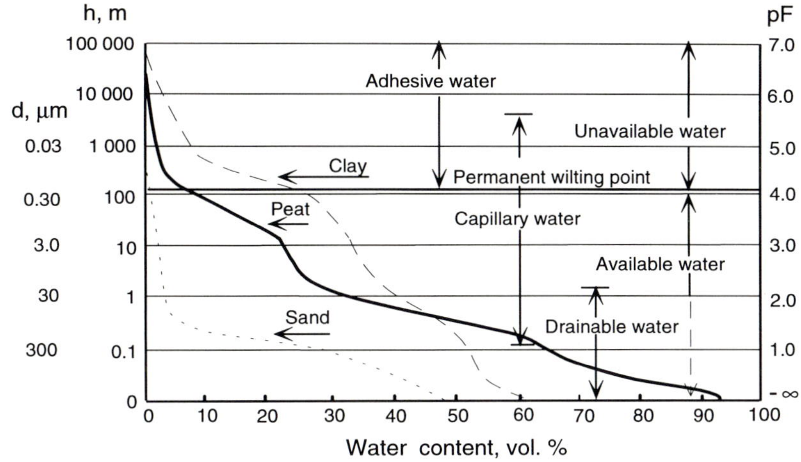

Fig. 3.6. Water desorption characteristics for some different kind of soil materials. Symbols in y-axis: *d* diameter of water-filled pore; *h* water suction as the height of a water column; *pF* the logarithm of the height of the water column in centimetres. (Redrawn from Andersson and Wiklert 1972)

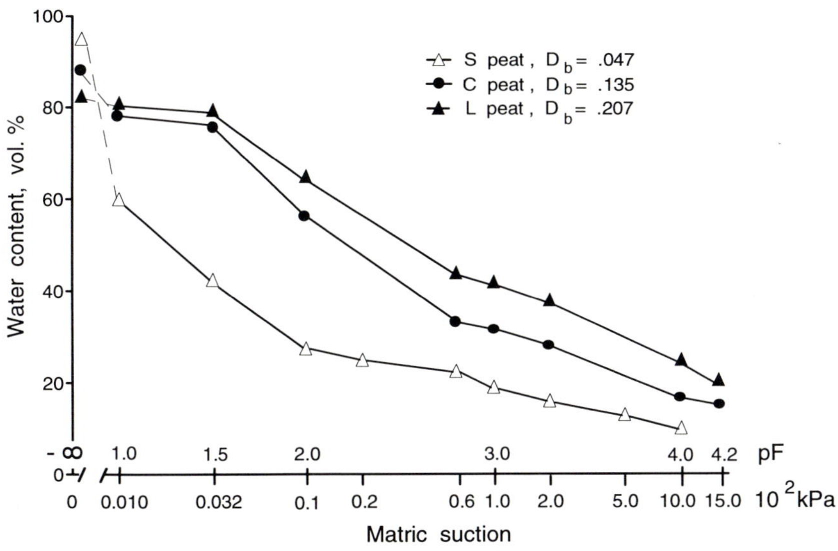

Fig. 3.7. Water desorption characteristics for different peat types; D_b bulk density. (Redrawn from Päivänen 1973b)

being of similar magnitude to that of moderately decomposed peat (e.g. Andersson and Wiklert 1967). Slightly decomposed peats contain a large quantity of water at saturation, but the water is released relatively easily with increasing matric suction (Fig. 3.7). In peats which have reached a more advanced stage of decomposition, the water contents at saturation are lower, but the loss of water with increasing matric suction is also smaller (Boelter 1964b; Päivänen 1973b; Myllys 1992).

The concept of **field capacity**, as explained for mineral soils (e.g. Veihmeyer and Hendrickson 1949, p. 75), is not very useful for peat soils where the groundwater table is relatively near the ground surface and the matric suction values corresponding to equilibrium are considerably lower than those usually presented for mineral soils (Päivänen 1973b).

It is possible to estimate from the water retention curve the quantity of water that can be removed from a sample of peat soil by draining. The water retained in a peat soil sample which is in a state between water saturation and matric suction pF 2 can be assumed to be that quantity of water which is removable by drainage alone. This amount of water has been called the **water yield coefficient** (Boelter 1969; see also the discussion by Ingram 1983). The water yield coefficient decreases with increasing peat bulk density (Päivänen 1973b):

Bulk density ($g\ cm^{-3}$)	Water yield coefficient ($cm^{-3}cm^{-3}$)
0.05	0.60
0.10	0.36
0.15	0.22
0.20	0.18

The **pore size distribution** of peat can also be determined from the water retention curve (e.g. Hartge 1965). It has been found that the quantity of large (> 30 µm) pores rapidly decreases and that of medium-sized (30–0.2 µm) and small (< 0.2 µm) pores increases with increasing bulk density (Päivänen 1973b).

3.2.3 Hydraulic Conductivity

Permeability is a property of a porous medium and refers to the ease with which gases and liquids can pass through it. When dealing with the permeability of a peat soil to water, we are mainly interested in conditions prevailing in saturated soil. According to Darcy's law, the **hydraulic conductivity** (coefficient K) indicates the flux of water per unit gradient of hydraulic potential, and it is usually expressed as $cm\ s^{-1}$ (Glossary of Soil . . . 1979).

The hydraulic conductivity of saturated peat soil layers located below the groundwater table has been quantified using the **piezometer method** or seepage tube. This measures the flux of water in the horizontal direction (e.g. Boelter 1965; Päivänen 1973b; Rycroft et al. 1975a). The piezometer method may, however, overestimate the hydraulic conductivity of peat soils (see Rycroft et al. 1975b). The other method frequently applied to measure hydraulic conductivity in saturated peat soils is the **auger hole method** (van Beers 1958), which gives the average hydraulic conductivity for the whole distance between the groundwater table and the bottom of the groundwater hole.

Table 3.4. Hydraulic conductivity (K) values for undisturbed peat soils measured with piezometer techniques

Material and location	K (cm s^{-1})	Source
Undecomposed *Sphagnum* peat, Minnesota	4×10^{-2}	Boelter (1965)
Moderately decomposed herbaceous peat, Minn.	7.5×10^{-6}	Boelter (1965)
Blanket bog, UK	2×10^{-7}	Ingram (1967)
Blanket bog, UK	9×10^{-8}	Ingram (1967)
Reclaimed peat, Ontario	1.8×10^{-3}	Irwin (1968)
Well-decomp.,herbaceous peat, Wyoming	2.5×10^{-7}	Sturges (1968)
Ombrotrophic bog	6×10^{-3}	Dai and Sparling (1973)
Undecomposed *Sphagnum* peat, Finland	2.2×10^{-3}	Päivänen (1973b)
Well-decomposed *Sphagnum* peat, Finland	1×10^{-5}	Päivänen (1973b)
Raised bog, 3 m below the peat surface, Minn.	5×10^{-3}	Chason and Siegel (1986)
Black spruce bog, Alberta	5×10^{-4}	Toth and Gillard (1988)

The range in hydraulic conductivity of undisturbed peat soils is known to be large: from 4×10^{-2} to 9×10^{-8} cm s^{-1} (Table 3.4). Usually, a clear tendency of decreasing values with increasing decomposition has been established, irrespective of whether decomposition is determined as the degree of humification (von Post 1922) or in terms of bulk density values (Baden and Eggelsmann 1963; Päivänen 1973b; Burt et al.1990; Plamondon and Belleau 1991). The hydraulic conductivity is generally greater for woody and *Carex* peats than for *Sphagnum* peats at the same stage of decomposition (Päivänen 1973b). The decrease in hydraulic conductivity observed with increasing peat depth (Huikari 1959a; Päivänen 1973b; Bélair 1991) is again largely a function of the degree of decomposition, the frequent occurrence of macropores in the topmost peat layer (commonly caused by a number of phenomena, viz. movement induced by tree roots and channels resulting from decaying rhizosphere roots or the activities of soil invertebrates) and the formation of irreversible colloids. It has recently been suggested that the anomalously low hydraulic conductivity values in moderately decomposed catotelm peat (see Sect. 3.5.1) is caused by methane gas generated in situ by anaerobic respiration (Bélair 1991; Mathur et al. 1991; Reynolds et al. 1992). However, high hydraulic conductivity values have been reported for partially decomposed peat even 3 m below the peat surface, especially if wood and rootlet residues are present (Chason and Siegel 1986).

3.2.4 Thermal Properties

3.2.4.1 Specific Heat and Thermal Conductivity

Specific heat is the amount of heat which is needed to raise the temperature of 1 g of a material by 1°C. The specific heat of water is very high, being about four or five times that of most dry soil material; at a temperature of 15 °C it is 4.12 J $g^{-1}°C^{-1}$ (Lutz and Chandler 1965). The specific heat of peat and sand in different moisture conditions is as follows (Kivinen 1948):

	$J\ g^{-1}°C^{-1}$
dry peat	0.61
dry sand	1.19
saturated sand	2.84
saturated peat	3.91

Thus, wet peat needs more heat for the same increase in temperature than does wet mineral soil (see also Eggelsmann et al. 1993).

Thermal conductivity describes the amount of heat that is conducted in 1 s through 1 cm of material when the temperature difference on both sides of the material in question is 1°C. The thermal conductivity of water is 0.00696 J $cm^{-1}s^{-1}$ $°C^{-1}$. The thermal conductivity of peat soil is heavily dependent on the water content; increasing water content improves the thermal conductivity (Eggelsmann et al. 1993).

3.2.4.2 Temperatures in Surface Peat

Besides the net radiation, temperatures in surface peat also depend on water content. The water content, in turn, affects evaporation as well as the thermal conductivity and specific heat of the peat. Drainage affects surface peat temperature conditions chiefly through changes in the water content of peat, and later also through changes in peat structure and vegetation (tree stand included).

Heikurainen and Seppälä (1963) found that the daily average surface peat temperature decreased during the growing season when the drainage intensity was increased from a ditch spacing of 30 to 10 m, but not when the spacing was changed from 90 to 30 m. The temperature difference was the greatest at 20–cm depth, and then decreased towards the surface. However, the temperature differences resulting from different drainage intensities seldom exceeded 2°C. Similar trends were also found by Mannerkoski (1988). The mean daily temperatures in the surface peat layers of an ombrotrophic low-sedge bog drained for forestry were the lower, the drier the peat. There was also a clear negative correlation between the mean growing season temperature (at 5- and 15-cm depth) and the distance measured from the peat surface to the groundwater table. However, the effect of drainage was less than 2°C in every case. Smaller temperature changes in surface peat layers, caused by forest drainage, have also been reported (Latja and Kurimo 1988).

Differences in the surface peat temperature conditions due to drainage are most pronounced immediately after ditching. Later, the differences decrease because of subsidence of the peat and changes in vegetation. In an old forest drainage area (dwarf-shrub pine bog), the difference in mean daily temperatures (at 10-cm depth) between poorly and well-drained experimental plots was 0.6°C in southern Finland and 1.5°C in northern Finland. Expressed in terms of temperature sum (threshold value +5°C), the corresponding difference during a growing season would be 74 and 147 dd°C, respectively (Hytönen and Silfverberg 1991). **Temperature sum** is a parameter that simulates the joint effect of time and temperature (on biological processes). A linear temperature sum is a summation of the fractions of daily mean temperatures that exceed the applied base temperature (+ 5°C). The unit is degree day (d.d.) (see Sarvas 1967).

The surface peat temperature during the growing season is lower in drained than in virgin sites. The further decrease in temperature with increasing intensity of drainage is not as distinct. The anaerobic conditions prevailing in a virgin mire is, however, the minimum factor for tree growth. Drainage intensity can therefore be based on criteria other than the temperature regime.

3.2.4.3 Soil Frost

Soil frost refers to frozen soil water. Groundwater near the soil surface in virgin mires acts as a significant sink for heat. This is especially true for fens receiving a large amount of groundwater input (Verry 1991). The formation of soil frost in peat is dependent on a number of factors. The weather conditions prevailing in early winter are of great importance, particularly the date of the first heavy frosts and forming of snow cover, and the thickness of the snow cover during winter.

A comprehensive survey carried out over 30 years on both mineral soil and virgin black spruce bog in Minnesota showed that the late-December snowpack thickness had a marked effect on the maximum annual penetration depth of soil frost (Verry 1991):

Snowpack (cm)	Max. frost depth (cm)
< 10	30–56
10–25	15–30
25–40	5–15
> 40	< 5

Lowering the water table by means of drainage has a complex effect on the energy balance of peatlands. Drying of the surface peat decreases its specific heat, and temperature ranges become more extreme. However, at the same time, the decrease in thermal conductivity slows down the energy flux into and from the deeper peat layers (Laiho and Laine 1992a). The closing tree stand canopy after drainage reduces the depth of the snow cover (Päivänen 1973a).

Soil frost may start to form earlier in drained peat than in virgin peat soil. The results on the effect of drainage on the penetration depth of soil frost in peat soils are still partly ambiguous. There are observations showing that soil frost penetrates deeper in drained than in virgin peat soils (Eurola 1968, 1975; Hannelius 1975), but contradictory results have also been presented (Swanson and Rothwell 1989; Laiho and Laine 1992a). The lower soil frost penetration in the drained peats is most probably due to the formation of a relatively dry, insulating peat layer on the surface of the mire (Laiho and Laine 1992a). Spring thaw has also been observed to occur significantly later in drained areas than in virgin mires (Latja and Kurimo 1988; Swanson and Rothwell 1989).

3.3 Nutrient Regimes of Peatlands

3.3.1 Input of Nutrients to Peatland Ecosystems

The most important aspect of the nutrient status of virgin peatlands is whether the input of nutrients comes from the atmosphere alone or also from the surrounding mineral soils and possibly from the soil under the peat layer (Fig. 3.8). The fertility of a site is especially increased by the addition of nutrients to water coming from mineral soils. Therefore, the **minerotrophic** peatlands affected by

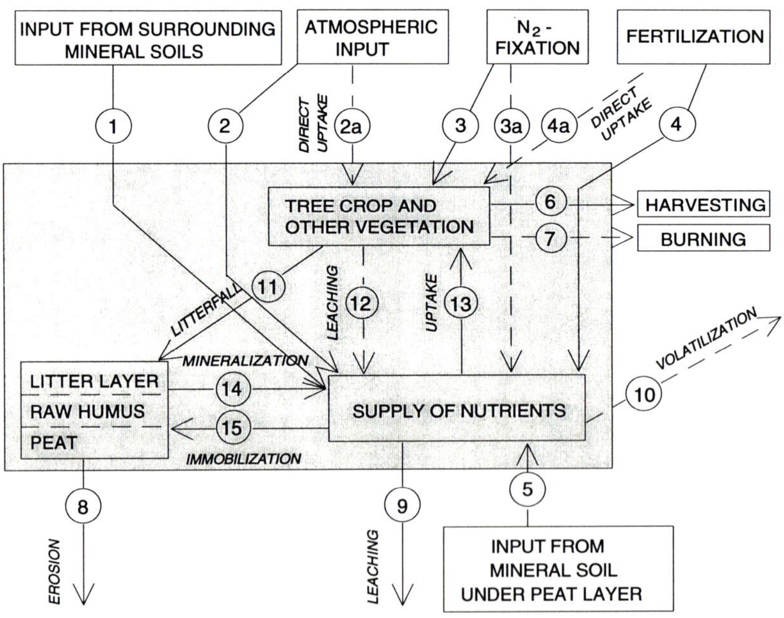

Fig. 3.8. Major components of the nutrient budget in peatland forest ecosystem. *1–5* Inputs to the ecosystem; *6–10* outputs from the ecosystem; *11–15* nutrient cycling in the ecosystem; *solid line* major effect; *dotted line* minor effect

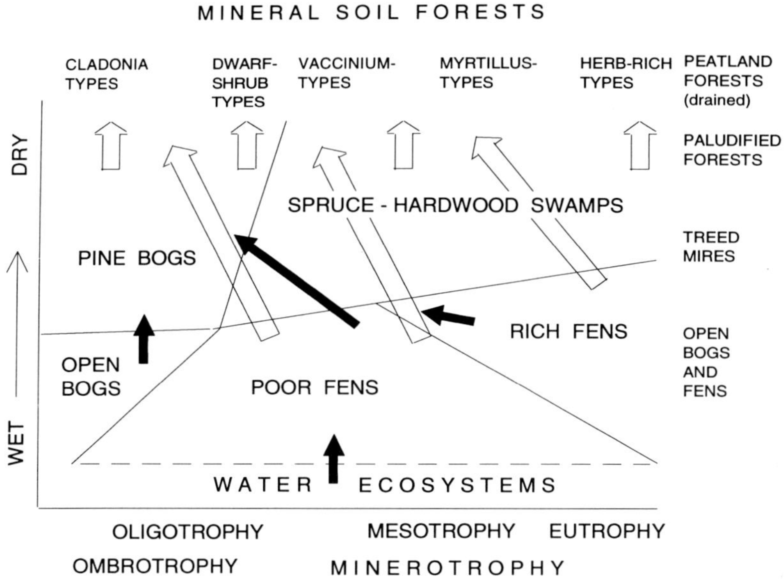

Fig. 3.9. General succession trends of boreal mire ecosystems presented in a schematic ordination of mire and forest site types. *Black arrows* natural succession; *open arrows* amelioration succession. (Redrawn from Reinikainen 1990)

mire margins are more fertile and suitable for forest production than **ombrotrophic** peatlands where the only input of nutrients to the ecosystem comes as **atmospheric deposition**. Forest drainage, by partly cutting the nutrient flow from surrounding mineral soils, has, in this respect, a negative influence on the nutrient status of mire ecosystems.

In boreal mire ecosystems, the site fertility of originally treeless minerotrophic fens decreases gradually as a result of peat accumulation. On drained deep-peated soils, the succession of vegetation also tends towards **oligotrophy** (Fig. 3.9).

The atmospheric deposition of nutrients, especially nitrogen and sulphur, has increased over the last few decades due to air pollution. As a result, preconditions for forest production, both on ombrotrophic and minerotrophic peatlands, are affected. For example, in the long term high nitrogen deposition may not only improve the growth of trees but also cause an imbalance between nutrients.

Precipitation chemistry varies considerably with distance from the sources of emission. For example, nitrogen deposition in Central Europe is some tens of kilograms annually, whereas in northern Europe it is less than 10 kg ha^{-1} (Bujsman and Erisman 1988; Tuovinen et al. 1990).

There are also differences in deposition between maritime and continental areas. The rainfall chemistry of a mire ecosystem in Finland and a more maritime site in Sweden is compared in Table 3.5. The maritime effect is strongest for chlorine and sodium. Boron deficiency is common for continental sites (Huikari

1977a; Braekke 1979, 1983a; Kolari 1983; Veijalainen et al. 1984) and boron often needs to be supplemented on drained peatlands by adding fertilizer.

The amount of many nutrients reaching the ground is greater under a tree canopy than in the open (e.g. Päivänen 1974b; Helmisaari and Mälkönen 1989; Hyvärinen 1990). The leaching of nutrients from the canopy is illustrated by the data for a Scots pine (*Pinus sylvestris*) stand on a drained mire (Table 3.6). Potassium was leached from the canopy more easily than the other studied nutrients.

Nutrient input from nutrient-rich mineral soil underneath a shallow peat layer can greatly increase site fertility. Studies carried out on peat cutover areas show that trees are able to exploit the nutrient stores of mineral soil when the depth of the peat layer is less than 30–40 cm (e.g. Kaunisto 1985b, 1987b; Saarinen 1988; Kaunisto and Viinamäki 1991).

The **biological fixation** of nitrogen increases the nitrogen stores of mire ecosystems. The annual dinitrogen fixation by surface peat and *Sphagnum* in an ombrotrophic bog has been estimated by Chapman and Hemond (1982) at 1 N g m^{-2}. Other sources of nitrogen input for certain peatland sites may also be the fixing of nitrogen by alders (e.g. Wheeler et al. 1986; Smolander 1990) and the **non-symbiotic heterotrophic N_2-fixation** common in boreal forest litter (Weber and Sundman 1986).

Table 3.5. Mean chemical properties of rainfall from November 1985 to October 1988 in study areas in Finland (62°14´N, 29°50´E, 80 m above sea level) and Sweden (57°07´N, 13°13´E, 120 m above sea level)

Property		Finnish site	Swedish site
pH		4.67	4.72
N-total	kg $ha^{-1}a^{-1}$	8.06	13.39
K	kg $ha^{-1}a^{-1}$	1.94	2.75
Ca	kg $ha^{-1}a^{-1}$	3.01	5.75
SO_4-S	kg $ha^{-1}a^{-1}$	7.00	6.17
Cl	kg $ha^{-1}a^{-1}$	6.16	22.60
Na	kg $ha^{-1}a^{-1}$	2.75	7.91

Source: Paavilainen (1990b).

Table 3.6. The amount of nutrient depositions under canopy and in the open in a Scots pine (*Pinus sylvestris*) stand on a drained mire during the period June 4 – Oct. 4, 1971

	g ha^{-1}		
	Inorganic N	Phosphorus	Potassium
Under canopy	1950	125	1741
Open	1653	108	1191
Difference	+297	+17	+550

Source: Päivänen (1974b).

3.3.2 Nutrients in Soil

3.3.2.1 Chemical Properties of Peat and Raw Humus

Surface peat and overlying **raw humus**, sometimes together with litter, form a store of nutrients for trees and ground vegetation. Nutrients are released from this store through **mineralization** and added to it through **litterfall** and bound by **immobilization** processes (Fig. 3.8).

The chemical properties of peat are primarily determined by the original chemical composition of peat-forming plant communities (e.g. Kivinen 1933) and the degree of decomposition.

There are several classifications of peat which describe peat chemistry. The classification proposed by the International Peat Society in 1979 (Proposal ... 1979, p. 48) is based on: (1) **botanical composition**, (2) **degree of decomposition** and (3) **trophic status** of peat. Various types of peat are included in the different classification categories as outlined below:

1. The following three peat types based on the botanical composition were agreed upon:
- *Sphagnum* peat (S) – composed predominantly of the remains of *Sphagnum* and other mosses;
- *Carex* peat (C) – composed predominantly of the remains of sedges (*Carex*), grasses (*Molinia, Phragmites*), herbs (*Dryopteris, Equisetum*) and related species;
- Woody peat (L) – composed of at least one third tree and woody shrub remains.

All other peat types are to be regarded as mixtures of these three types.

2. Peats are divided into three categories according to their degree of decomposition:
- weakly decomposed – humification degree H = 1–3 (see von Post 1922);
- medium decomposed (H = 4–6);
- strongly decomposed (H = 7–10).

3. Peats are divided into three groups according to their trophy:
- **oligotrophic**;
- **mesotrophic**;
- **eutrophic**.

There are clear differences in the chemical properties *of Sphagnum, Carex* and woody peats (Table 3.7).

The *Sphagnum*-dominated peats are characterized by low pH values and low nitrogen and extractable calcium contents but a high C/N ratio and a slightly higher content of soluble phosphorus compared to *Carex*-dominated peats. The potassium contents are almost the same in both types. The calcium, potassium

Table 3.7. Some average chemical properties of the 0-20 cm layer of *Sphagnum* (S), woody (LS, CL, LC) and *Carex* (C) peats (oven-dry milled peat)

Peat type group	pH	C/N	N (%)	Ca[a)]	K[a)] (mg l^{-1})	P[a)]
S	3.99	60	0.81	207	31	4.0
LS	4.05	41	1.13	295	38	4.3
CL, LC	4.49	23	1.93	750	44	3.2
C	4.79	23	1.97	568	34	2.6

Source: Urvas et al. (1979).
[a)] Ammonium acetate (pH 4.65) extractable mg l^{-1} refers to one litre (= dm^3) of ground dry peat.

and phosphorus contents are somewhat higher in woody peats than in the corresponding *Sphagnum* or *Carex* peats.

Bohlin et al. (1989) used **principal component analysis** to classify and characterize some peats from Sweden on the basis of botanical composition, particle-size distribution, elementary constituents, degree of decomposition, calorific value, and concentrations of monosaccharide, amino sugar, amino acid constituents, and bitumen. The first principal component (accounting for 48% of the variance) separated the peat samples mainly on the basis of their botanical composition and the second principal component (16% of the variance) was correlated to the degree of decomposition and the nanolignid (*Ericaceous* dwarf shrub) content. *Carex* peats have higher pH, ash content, and nitrogen, sulphur, amino acid and amino sugar concentrations than *Sphagnum* peats.

As could be expected, the chemical properties of surface peat are correlated to the peatland site type. For example, herb-rich sedge mires have considerably higher peat pH values and nitrogen and phosphorus contents than do less fertile tall-sedge or low-sedge mires. The potassium content of peat is, however, not correlated with the trophy of the site (Table 3.8).

In sedge mires, the potassium content is higher and nitrogen content lower in the 0–10 cm surface layer than in the 10–20 cm layer. This difference is typical of undisturbed mires (e.g. Holmen 1964; Pakarinen and Tolonen 1977; Kaunisto and Paavilainen 1988).

On drained peatlands, usually of low or medium fertility, a loose raw humus layer of litter and the remains of *Pleurozium schreberi* can develop. The thickness of this layer varies and its **bulk density** may be less than 0.01 g cm^{-3}. The total nitrogen concentration of this layer is usually somewhat lower, but the concentrations of phosphorus, potassium, calcium and micronutrients are higher than those of the underlying peat layer (Kaunisto and Paavilainen 1988). Because of low bulk density, the amounts of nutrients in raw humus are, however, very small compared with peat.

Table 3.8. Total content of nutrients in peat from virgin sedge mires

Peat layer (cm)	Herb-rich sedge mire	Tall-sedge mire	Low-sedge mire
	Nitrogen, mg g^{-1}		
0–10	16.8	11.8	10.0
10–20	21.7	15.4	12.5
	Phosphorus, mg g^{-1}		
0–10	0.96	0.67	0.45
10–20	1.00	0.74	0.46
	Potassium, mg g^{-1}		
0–10	1.05	1.69	1.22
10–20	0.48	0.69	0.42
	Calcium, mg g^{-1}		
0–10	5.28	2.86	2.09
10–20	5.05	3.06	2.18
	Magnesium, mg g^{-1}		
0–10	1.18	0.75	0.64
10–20	0.87	0.68	0.51

Source: Westman (1981).

3.3.2.2 Amounts of Nutrients in Surface Peat and in Litterfall

In virgin and drained peatlands, most of the roots responsible for nutrient uptake are in the topmost (0–20 cm) peat layer (Heikurainen 1955; Paavilainen 1966a, 1967; Finér 1989, 1991b; Håland and Braekke 1989). The total amount of nitrogen in this layer is of the order of thousands of kg ha^{-1}, whereas the amounts of other main nutrients, especially potassium, are much smaller. The micronutrient amounts range from some hundreds of grams (B) to kilograms (Cu, Zn) or more (Mn, Fe) per ha (e.g. Holmen 1964; Paavilainen 1980; Vasander 1981; Westman 1981; Kaunisto and Paavilainen 1988; Finér 1989, 1991a; Ipatiev 1990).

In the following Sections, some Finnish examples of nutrient stores and cycles are presented. The sites are located on both drained ombrotrophic and minerotrophic mires (Table 3.9, Fig. 3.10).

The amounts of nutrient elements in the surface peat of the sites studied, except for potassium and zinc, are of the same orders of magnitude as presented by Kaunisto and Paavilainen (1988), and the amounts of nutrient elements are considerably lower for ombrotrophic than for minerotrophic mires (Table 3.10).

The amounts of nitrogen and phosphorus are clearly higher, but the amount of potassium is only half of that reported by Westman (1981) for undisturbed mires of corresponding site types (Fig. 3.11). Laiho (1991) found that the amounts of nitrogen and phosphorus in surface peat increased with increasing age of drainage, especially for infertile peatland site types. The amount of potassium remained almost the same in the better site types and increased somewhat in the less fertile types with increasing drainage age. With increasing drainage age, **subsidence** of the peat brings peat nutrients from deeper layers to the surface layer (Laiho and Laine 1990, 1994).

Table 3.9. Some characteristics of drained, unfertilized mire sites for which nutrient stores and cycles have been calculated

Site No.	Original peatland site type[a)]	Tree species	Stand: Volume (m^3ha^{-1})	Stand: Annual increment ($m^3ha^{-1}a^{-1}$)	Stand: Age (a)	Reference
		Ombrotrophic sites				
1	IR	*Pinus sylvestris*	80	5.9	85	Braekke and Finér (1991): Nutrients in peat, Finér (1991a): other peat characteristics
2	IR	*Pinus sylvestris*	116	4.5	95	Paavilainen (1980)
		Minerotrophic sites				
3	VNR	*Pinus sylvestris*	48	2.7	40–50	Finér (1989)
4	RhNR	*Pinus sylvestris* &	72	2.0	40–60	Finér (1989)
		Betula pubescens mixed stand	39	2.0	40–60	
5	MK	*Picea abies*	162	4.0	> 100	Finér (1989)
6	RhNR	*Pinus sylvestris*	100		60	Paavilainen (1990b)
7	RhK	*Betula pubescens*	70		60	Paavilainen (1990b)

a) IR = dwarf-shrub pine bog, VNR = tall-sedge pine fen, RhNR = herb-rich sedge pine fen, MK = *Vaccinium myrtillus* spruce swamp, RhK = herb-rich-hardwood spruce swamp (English names according to Table 3.17).

Table 3.10. Amounts of nutrient elements (kg ha^{-1}) in the 0-20 cm peat layer of the drained, unfertilized study sites. Material from old drainage areas are given for comparison

Element	Ombrotrophic sites[a)] 1	2	Minerotrophic sites[a)] 3	4	5	Old drainage areas[b)]
N	1912	1629	2570	5880	4970	3000 –7000
P	105	222	125	380	310	90 – 250
K	64	89	90	220	235	30 – 65
Ca	392	460	830	1650	1800	500 –1800
Mg	69	–	140	200	250	50 – 200
Mn	10.9	–	9	19	55	3.5– 25
B	0.36	–	–	–	–	0.2– 0.6
Cu	1.1	–	1.0	2.5	1.9	0.7– 3.2
Zn	4.2	–	17	16	23	1 – 3

a) See Table 3.9.

b) (Kaunisto and Paavilainen 1988).

Fig. 3.10 a–d. Tree stands in selected study sites (see Table 3.9): **a** Site 1, dwarf-shrub pine bog (Photo: L. Finér); **b** site 4, herb-rich sedge pine fen (Photo: L. Finér); **c** site 5, *Vaccinium myrtillus* spruce swamp (Photo: L. Finér); **d** site 7, herb-rich spruce-hardwood swamp (Photo: J. Issakainen)

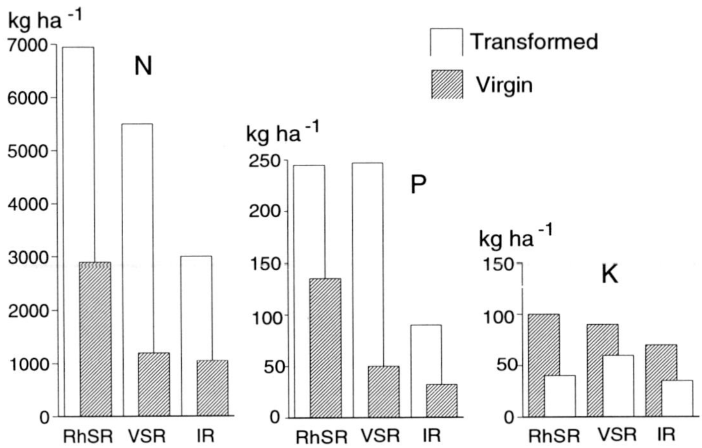

Fig. 3.11. The amounts of nitrogen, phosphorus and potassium in surface peat (0-20 cm) on undrained pine mires (Westman 1981) and on drained peatlands (Kaunisto and Paavilainen 1988) developed from corresponding peatland site types. The transformed peatlands had been drained about 75 years before peat samples were taken

Table 3.11. Average amounts of nutrient elements returning to the soil annually as tree litterfall in the drained, unfertilized study sites

Element	Ombrotrophic sites[a]		Minerotrophic sites[a]				
	1	2	6	7	4		
	Pinus sylv.	*Pinus sylv.*	*Pinus sylv.*	*Betula pub.*	*Pinus sylv.*	*Betula pub.*	
					Mixed stand		Total
	Litterfall (kg ha^{-1}a^{-1} dry mass)						
	1739	2360	1600	1900	1065	815	1880
	Amount of nutrients						
N, kg ha^{-1}	10.61	13.39	9.87	25.34	7.58	10.09	17.67
P, kg ha^{-1}	0.70	0.91	0.86	2.99	0.38	0.56	0.94
K, kg ha^{-1}	1.37	1.47	1.08	2.88	0.83	1.39	2.22
Ca, kg ha^{-1}	8.64	8.22	6.93	17.77	3.70	6.56	10.26
Mg, kg ha^{-1}	1.12	0.95	1.44	4.82	0.52	1.88	2.40
Mn, kg ha^{-1}	1.23	0.74	0.91	2.96	0.49	1.00	1.49
B, g ha^{-1}	31.0	20.1	13.6	31.2	–	–	–
Cu, g ha^{-1}	68.0	11.4	4.5	11.9	–	–	–
Zn, g ha^{-1}	95.0	–	–	–	–	–	–

a) See Table 3.9.

Nutrients taken up by vegetation are returned to the soil partly in the form of above-ground litterfall. The nutrient content of broad-leaved tree litter is higher than that of conifer litter (e.g. Mälkönen 1974, 1977; Berg and Staaf 1986; Paavilainen 1984a, 1987; Finér 1989). Table 3.11 shows the differences in litterfall chemistry between some sites and tree species.

Ground vegetation also produces litter. For example, field layer litterfall for a drained dwarf shrub pine bog (site 2, Table 3.9) during one growing season accounted for 6.4–10.3 kg ha^{-1} nitrogen, 0.5–0.8 kg ha^{-1} phosphorus, and 0.8–2.1 kg ha^{-1} potassium (Paavilainen 1980).

The large amount of nutrients returned to the soil in the form of litter is one of the biological reasons for the favourable influence of hardwoods in mixed stands. For example, the growth of pine on nutrient-poor peat is increased when alder (*Alnus incana*) is also present (Kaunisto and Viinamäki 1991).

In addition to litterfall from the above-ground parts of vegetation, nutrients are also returned to the soil in the form of dead roots. Finér (1989) estimated the annual nutrient amounts of fine-root litter per ha based on calculations for different model stands (sites 3–5, Table 3.9) as follows: 12.3–18.0 kg N, 0.7–1.5 kg P, 0.8–2.2 kg K, 3.3–8.3 kg Ca, 0.7–1.0 kg Mg and 0.1–0.4 kg Mn.

3.3.2.3 Mineralization and Immobilization of Nutrients

Only part of the total nutrient content of peat soils is readily available to the trees (Holmen 1964; Starr and Westman 1978; Paavilainen 1980; Braekke 1987; Wiklander and Nömmik 1987; Kaunisto and Paavilainen 1988). Usually less than 1% of the total nitrogen is in an available form. The proportion of **soluble** phosphorus is less than 10% and **exchangeable** calcium about two thirds of the total amounts of these nutrients. Potassium is the most easily available main nutrient to the trees. In surface peat almost all potassium is in the exchangeable form.

Drainage and application of fertilizer or wood ash also improve the living conditions of peat-decomposing microbes, thus accelerating the mineralization of nutrients from the organic material (e.g. Paarlahti 1964; Lähde 1966, 1969; Karsisto 1979a,b; Paavilainen 1980; Tuominen 1981; Braekke and Finér 1990). For example, in study area 3 (Table 3.9), the annual amount of nitrogen mineralized after drainage was estimated to be about 100 kg ha^{-1} and that of phosphorus about 5 kg ha^{-1} (Silvola 1988). Under very favourable conditions on a drained and forested fen the annual mineralization rate corresponded to 0.6% of the total nitrogen content in the 0–30 cm peat layer (Wiklander and Nömmik 1987).

The nutrients released from peat are partly „biologically" bound into the microbial biomass and partly chemically bound into different compounds; the rest are available to trees and other vegetation. The immobilization rate is dependent on the chemical properties of peat. For example, the higher the peat-C/N ratio, the more the nitrogen released during decomposition is bound by the microbes. Biological fixation is especially efficient in undecomposed *Sphagnum* peat, which consists primarily of cellulose and hemicellulose. In partly decomposed *Sphagnum* peat, the share of cellulose and hemicellulose drops to less than half (Isotalo 1951).

Drainage also increases the amount and biomass of invertebrates important for the decomposition of peat. For example, in a study site in the Tomsk region of Russia, invertebrate numbers in a grass-rich mire were 8000 per m^2, compared to 18 000 per m^2 following drainage (Kozlovskaja 1974).

The release of nutrients from tree litter is dependent on species and site factors. The release of nutrients during the first few years is more rapid for hardwood than coniferous litter (e.g. Berg and Staaf 1981, 1986; Bogatyrev et al. 1983). In the litter layer of pine and spruce stands on afforested peatlands, large proportions of K, Ca, Mg and Mn are extractable and available to trees. The extractable part of total P is also relatively high, 15–21%, whereas Fe and Cu, for example, are strongly bound in the litter layer (Braekke 1988b).

The release of nutrients from Scots pine needle litter during one growing season on a drained and unfertilized dwarf shrub pine bog (site 2, Table 3.9), expressed as percent of the initial contents, was (Paavilainen 1980): nitrogen 27–49%, phosphorus 62–63%, and potassium 52–89%. Karsisto (1979a,b) showed that the microorganisms most positively affected by drainage and fertilization are aerobic, ammonification bacteria.

3.3.3 Nutrients Bound by Vegetation

3.3.3.1 Tree Crop

The total amounts of nutrient elements bound in the tree biomass vary from some hundred grams (B, Cu) to more than 300 kg (N, Ca) per ha (Table 3.12).

A considerable part of the nutrients retained by trees, especially nitrogen, is accumulated in the crowns, particularly in the foliage. The root systems also account for a large proportion of stored nutrients. Small and fine roots, in particular, contain relatively large amounts of nutrients (Paavilainen 1968, 1969a, 1980; Finér 1989, 1991b, 1992; Braekke 1990a). There is, however, no clear

Table 3.12. The amount of nutrients (kg ha^{-1}) bound in the tree biomass on drained, unfertilized experimental sites

Element	Ombrotrophic site[a)]	Minerotrophic sites[a)]				
	2	3	4			5
	Pinus sylvestris	*Pinus sylvestris*	*Pinus sylvestris*	*Betula pubescens*		*Picea abies*
			Mixed stand		Total	
N	203.3	130.5	83.7	100.7	184.4	315.1
P	21.4	13.6	5.4	5.6	11.0	23.2
K	75.8	32.6	18.8	21.3	40.1	96.0
Ca	123.2	61.2	49.7	55.7	105.4	337.4
Mg	36.0	13.6	11.3	10.8	22.1	32.1
Mn	8.9	3.4	4.1	4.6	8.7	28.1
B	0.32	0.17	–	0.15	–	0.67
Cu	0.19	0.10	0.10	0.09	0.19	0.33
Zn	2.67	0.67	0.31	0.99	1.30	2.67

[a)] See Table 3.9.

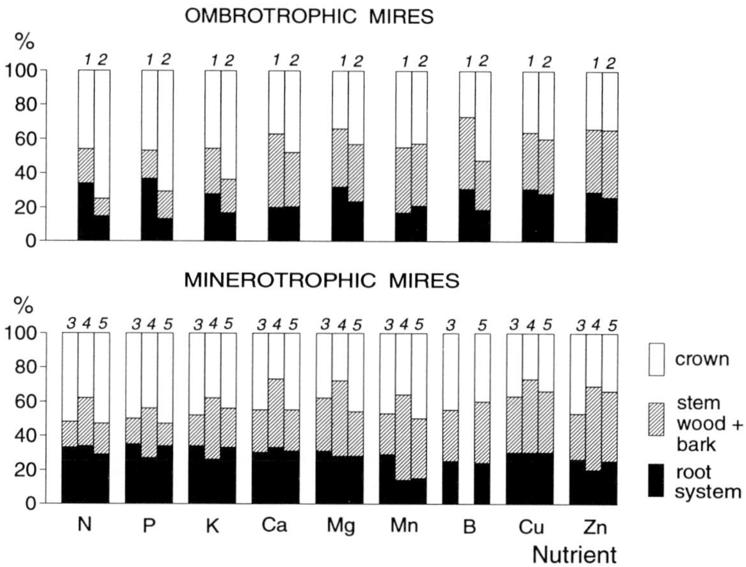

Fig. 3.12. The distribution of nutrients into different parts of trees. *1–5* Sites studied (see Table 3.9)

Table 3.13. The annual nutrient uptake by trees in drained and unfertilized experiment sites

Element	Ombrotrophic site[a)]	Minerotrophic sites[a)]				
	2	3	4			5
	Pinus sylvestris	*Pinus sylvestris*	*Pinus sylvestris*	*Betula pubescens*		*Picea abies*
			Mixed stand		Total	
	Annual increment, $m^3ha^{-1}a^{-1}$					
	4.5	2.7	2.0	2.0	4.0	4.0
	Amount of nutrients, $kg\ ha^{-1}a^{-1}$					
N	40.78	26.25	19.15	23.59	42.74	30.83
P	3.65	2.54	1.05	1.27	2.32	2.41
K	10.50	4.56	2.00	3.31	5.31	5.29
Ca	18.10	11.93	8.02	13.01	21.03	22.43
Mg	4.20	2.24	1.32	2.97	4.29	2.28
Mn	1.59	0.72	0.71	1.40	2.11	1.74
B	0.050	–	–	–	–	–
Cu	0.025	–	–	–	–	–

[a)] See Table 3.9.

difference between ombrotrophic and minerotrophic sites in the allocation of nutrients to various parts of trees (Fig. 3.12). The distribution of nutrients in different parts of a tree is more dependent on tree species and on the structure and developmental stage of the stand.

The annual nutrient uptake by trees varies according to the site type, tree species and stand age and volume. In the experimental sites, these relationships were, however, not very clear (Table 3.13).

In mixed stands, birch takes up more nutrients per mass of biomass than pine. Spruce consumes more calcium, manganese and boron than pine, but not as much as birch (Finér 1989).

Some of the nutrients bound in the foliage, fine roots and other tree tissues are **retranslocated** and used to meet the requirements for new biomass production. Nitrogen, phosphorus and potassium are clearly mobile nutrients (e.g. Helmisaari 1990). In the experimental sites (sites 3–5, Table 3.9) studied by Finér (1989), the share of nitrogen retranslocation was 15–29% of the total annual consumption of nitrogen for conifers and 46% for birch. The corresponding values for phosphorus were 21–32% and 46%, and for potassium 27–38% and 68%.

3.3.3.2 Ground Vegetation

On many peatland sites, large amounts of nutrients are bound in the ground vegetation. On a drained ombrotrophic pine bog (site 2, Table 3.9), the amount of some nutrients (e.g. N and Ca) bound in the field layer vegetation was more than one fifth of the amounts bound in the total phytomass of the tree stand and the field layer (Fig. 3.13; see Paavilainen 1980).

The amounts of nutrients in different compartments of vegetation on a very nutrient-poor pine bog in the virgin state are presented in Fig. 3.14. Drainage and fertilization decrease the share of ground vegetation biomass and, therefore, also the comsumption of nutrients.

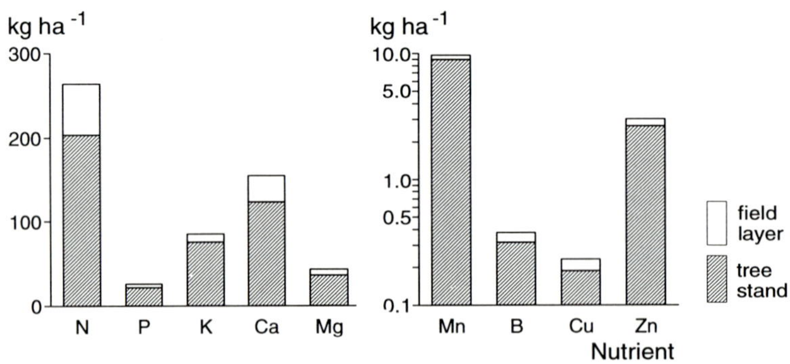

Fig. 3.13. The amounts of nutrients bound in the tree stand and field layer. Site 2, see Table 3.9

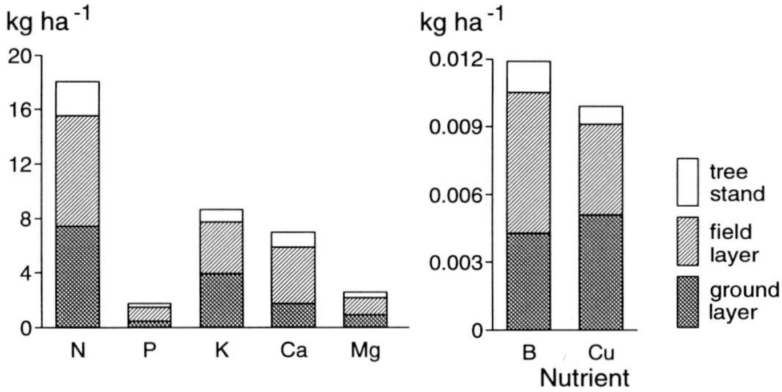

Fig. 3.14. The amounts of nutrients bound in the tree stand, field layer and ground layer of a nutrient-poor bog. (Vasander 1981)

3.3.4 Nutrient Balance

Balanced mineral nutrition is essential for sustainable peatland forestry. Problems arise most often either on very poor sites or on nutrient-rich peatlands where the ratio between the stores and the uptake of nutrients is imbalanced.

The lack of available nutrients is, in spite of the influence of atmospheric deposition, often a growth-limiting factor on ombrotrophic bogs. Drainage alone does not alter this situation much because the net mineralization rate of the surface peat is still too slow to release enough nutrients for a sufficient increase in wood production. Drainage may even weaken the productivity of ground vegetation on these sites so much that the increase in tree growth cannot compensate for this loss, and the total biomass production of the drained bog is thus smaller than that of a virgin bog (e.g. Reinikainen 1981; Reinikainen et al. 1984; Vasander 1982, 1990).

An improvement in tree growth on nitrogen-poor pine bogs after drainage is only possible when additional nutrients are given. Fertilization of these sites is, however, usually not economically profitable (e.g. Heikurainen and Seppälä 1973, see also Chap. 5) because of the short duration of its effect and the small total increase in volume growth.

On the best ombrotrophic and on the most minerotrophic mires, the amount of available nutrients in the peat is usually high enough to support a large enough increase in tree growth to economically justify draining (see Chap. 5). Fertilization is often worthwhile on these sites, to ensure an additional increase in wood production or to correct a possible nutrient imbalance after drainage and harvesting of the tree stand.

Leaching, in some cases also erosion, and harvesting are the most important forms of nutrient output from the peatland ecosystem. Forest fires, which occur seldom on peatlands, and volatilization of nutrients are of minor importance to the nutrient balance.

On undrained bogs, the atmospheric input on nutrients compensates to a great extent for the losses caused by leaching with on without erosion. From drained bogs, especially Mg and Ca are leached in amounts which are far greater than the amounts provided by deposition (Sallantaus 1992; see Table 3.14).

The atmospheric input of nutrients may be significant for biomass production. For example, studies in Norway show that the input of macronutrients, except phosphorus, was rather large compared to net uptake (Table 3.15).

Table 3.14. Annual input (deposition) and output (leaching) from virgin and drained bog catchments. (Data from Sallantaus 1992)

	$kg\ ha^{-1}a^{-1}$				
	N	P	K	Ca	Mg
Input	7.20	0.12	0.80	1.20	0.80
Output					
Drained	2.10	0.14	1.20	6.70	1.80
Natural	1.40	0.04	0.79	1.70	0.76

Table 3.15. Atmospheric input of nutrient elements from 1953–1983 in kg ha^{-1}, and the input as a percentage of amounts in the tree stand biomass at the end of the period

Experiment	Mean stand volume (m^3ha^{-1})	Atmospheric input					
		N	P	K	Ca	Mg	S
		kg ha^{-1} (%)					
Pinus sylvestris on drained mire	56.8	391 (223)	* –	75 (97)	168 (220)	81 (431)	476 (3579)
Pinus sylvestris on drained mire	174.3	391 (124)	* –	75 (51)	168 (86)	81 (196)	476 (1770)
Picea abies on drained, fertilized mire	149.0	290 (58)	* –	62 (40)	130 (40)	68 (118)	395 (1039)

Source: Braekke 1988a; * = negligible.

The atmospheric input and increased mineralization in surface peat probably eliminate the deficiency of nitrogen in Scots pine stands on drained but unfertilized mires in the long run, except for the most nitrogen-poor sites. The higher amount of nitrogen deposition may, on the other hand, induce deficiencies of other nutrients; for example, phosphorus in the case of ombrotrophic mires and potassium in the case of minerotrophic mires. To correct these nutrient imbalances, appropriate fertilization is needed (see Chap. 5).

The uptake and binding of nutrients by trees on peatland sites are notable when compared with their amounts in peat (Fig. 3.15). The stores of potassium

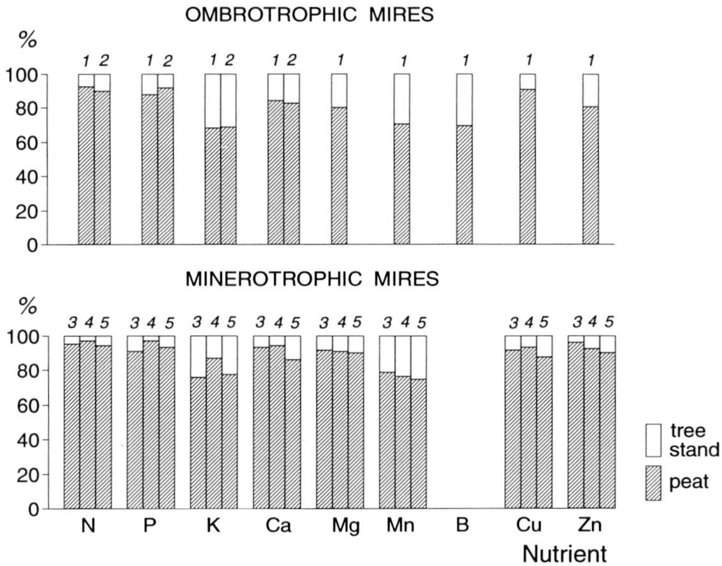

Fig. 3.15. The relative amounts of nutrients bound in the tree stand and surface peat (0–20 cm) for various study sites (see Table 3.9)

Table 3.16. The removal of nutrients (kg ha^{-1}) in stemwood and whole tree harvesting from an ombrotrophic pine bog (stemwood volume 135 m^3 ha^{-1})

Element	Stemwood harvesting	Whole-tree harvesting
N	32.41	165.55
P	4.36	19.52
K	22.29	64.27
Ca	49.31	96.32
Mg	12.36	25.22
Mn	3.19	6.47
B	0.102	0.212
Cu	0.065	0.114
Zn	1.02	1.72

Source: Paavilainen (1980).

and boron, in particular, are often small with respect to stand requirements (Holmen 1964; Mannerkoski 1973; Huikari 1977a; Braekke 1979, 1983b; Kaunisto and Paavilainen 1988; Finér 1989). These requirements can, to a certain extent, be covered by retranslocation within trees as well as by nutrients returning to the soil in the form of litter. It is possible, however, that after repeated harvestings, a deficiency of potassium or some micronutrients may occur. This risk is much higher on nutrient-poor peatlands when using whole tree than stemwood harvesting, as the data in Table 3.16 indicate (site 2, Table 3.9).

Preliminary estimates, based on nutrient balance calculations, indicate that the amount of potassium in the soil decreases as trees grow without fertilization. A shortage of this nutrient is possible 50–80 years after drainage in repeatedly thinned stands without fertilization (Kaunisto and Paavilainen 1988). However, the loss of potassium from the surface layer of peat by uptake is to some extent replaced from originally deeper peat layers as a result of subsidence and compression of peat (Laiho and Laine 1990; Laiho 1991).

Only part of the fertilizer dose is taken up by the trees. For example, the recovery of added nitrogen was < 5–65%, phosphorus 5–16% and potassium 14–54%, according to a study by Finér (1989). The remaining part is bound by ground vegetation, fixed biologically or chemically in the peat and leached away. Part of fertilizer nitrogen may also be volatilized. The leaching losses can to a great extent be avoided by using proper fertilization rates, techniques and times (see Chap. 5).

3.4 Mire Classification

3.4.1 Virgin Mires

Classification provides a framework for organizing and retrieving information. In forestry, the target of site classification is to help foresters decide how to manage any specific piece of land according to the experience and knowledge already gained from similar sites (Burger and Pierpoint 1990). In this chapter, **mire classification** systems developed and applied for forestry purposes in different regions and countries will be discussed. In forest amelioration of peatlands, the main emphasis is on evaluating sites for forest growth after drainage. On drained peatlands, on the other hand, site classification is the basis for silvicultural measures (e.g. selection of the appropriate tree species, tree stand density, regeneration method, need for fertilization, etc.).

In **Finland,** the land used for forestry (forestry land) – on both mineral and peat soil – is divided into three classes according to site productivity (Yearbook . . . 1992; see also Sect. 7.1):

- **forest land**: mean annual increment > 1 $m^3 ha^{-1}$
- **scrub land**: mainly exposed bedrock and scree or mires, mean annual increment < 1 $m^3 ha^{-1}$ but > 0.1 $m^3 ha^{-1}$
- **waste land**: if not naturally treeless, it has a mean annual increment of < 0.1 $m^3 ha^{-1}$.

Forest and mire site classification in Finland is based on Cajander's (1909, 1913) works concerning mineral soil site types (**forest types**) and wetland site types (**mire site types**). Sites are primarily classified into vegetational communities. According to Cajander's theory (Cajander 1949), a vegetation unit characteristic of a particular quality of the site is the result of: (1) the specialized requirements of the plants with regard to site; (2) the competition between the plants; and (3) the favouring of one species by another. Cajander considered that his site quality classes – forest types and mire site types – based on such "a natural, biological" (i.e. ecological) classification would reflect soil fertility (Starr 1986). Thus, habitats which are ecologically similar can be expected to support similar vegetation (Eurola et al. 1984). The vegetational communities (site types) have been categorized into trophic classes, usually based upon their vegetational and floristic relationships (see review by Jeglum 1991b). The term **ecosystem** had not yet been proposed in the beginning of this century when Ca-

Table 3.17. List of Finnish mire site type abbreviations (see Fig. 3.16) used in practical forestry and suggested English names

Genuine, forested types	
LhK	Eutrophic paludified hardwood-spruce forest
RhK	Herb-rich hardwood-spruce swamp
KgK	Paludified *V. myrtillus* spruce forest
MK	*V. myrtillus* spruce swamp
PK	*V. vitis-idaea* spruce swamp
PsK	*C. globularis* spruce swamp
KR	Spruce-pine swamp
KgR	Paludified pine forest
IR	Dwarf-shrub pine bog
RaR	*S. fuscum* pine bog
Treeless and composite types	
VLK	Eutrophic hardwood-spruce fen
KoLK	Eutrophic birch fen
VLR	Eutrophic pine fen
RaLR	Eutrophic *S. fuscum*-rich pine fen
VL	Eutrophic fen
RiL	Eutrophic flark fen
RhSK	Herb-rich sedge hardwood-spruce fen
RhSR	Herb-rich sedge birch-pine fen
RhSN	Herb-rich sedge fen
RhRiN	Herb-rich flark fen
VSK	Tall-sedge hardwood-spruce fen
VSR	Tall-sedge pine fen
VSN	Tall-sedge fen
VRiN	Flark fen
TSR	Cottongrass-sedge pine fen
LkR	Low-sedge *S. papillosum* pine fen
LkKaN	Low-sedge *S. papillosum* fen
TR	Cottongrass pine bog
KeR	Hummock-hollow pine bog
LkN	Low-sedge bog
RaR	*S. fuscum* bog

jander was developing his theory of forest and mire site types. However, Cajander's concept of a forest type or mire site type (Cajander 1909, 1913), which can be used to mean a certain type of vegetation, the corresponding habitat (site), or both, is clearly similar to that of an ecosystem (Mikola 1982).

Heikurainen (1979) explained that mire vegetation could be considered as a continuum based on fertility (trophic status), water regime, and stage of mire development. Thus, **mire site types** are vegetation community accumulations in a continuum. Laine and Vasander (1990) described a mire site type as an average, abstract description of vegetation communities.

For practical forestry purposes, about 30 mire site types have been described (Heikurainen 1986; Laine and Vasander 1990). The most extensive mire site classification was proposed by Huikari (1952, 1974) and is described later in this chapter. The 40 mire site types listed by Heikurainen and Pakarinen (1982) are more in accordance with a plant sociological approach. The strict phytosociological classification of mire vegetation describes more than 80 mire site types (e.g. Eurola et al. 1984).

The translation of the Finnish mire site types into the English language has varied from one publication to another. The names suggested by Laine et al. (1986) and Laine and Vasander (1990) will be followed here (Table 3.17). For minerotrophic, genuine forested site types, the term **swamp** is used, whereas for ombrotrophic site types in this group the term **bog** is used. Sites with a shallow peat layer (< 30 cm) are also classified as mires in Finland. These types are, however, referred to as **paludified forests**. In the treeless and composite site type groups, the term **fen** is applied to the minerotrophic site types and **bog** to the ombrotrophic ones. In the case of sparsely forested composite site types, the use of the term **fen** stresses the close eco-genetic relationship of these sites with the treeless mires (Laine et al. 1986). For a description of the individual mire site types, see Heikurainen (1979) and Laine et al. (1986).

The Finnish mire site types used in practical forestry have been plotted against two environmental gradients, trophic status and degree of wetness, using abbreviations derived from the Finnish names (Fig. 3.16). The ecological distances between the site types are to be considered both subjective and relative, rather than objective.

The site types are grouped into three main categories (see Laine et al. 1986; Fig. 3.16):

1. **Genuine, forested mire site types**, which have a natural tree stand and consist of relatively uniform hummock or lawn-level surface vegetation.
2. **Sparsely forested composite mire site types** characterisized by a mosaic of hummock and lawn hollow vegetation. They can be considered as mixtures of forested and treeless site types. These kinds of sites have been referred to as treed fens and treed bogs in Canada (see Jeglum 1991b).
3. **Treeless mire site types** without a natural tree stand in commercial terms.

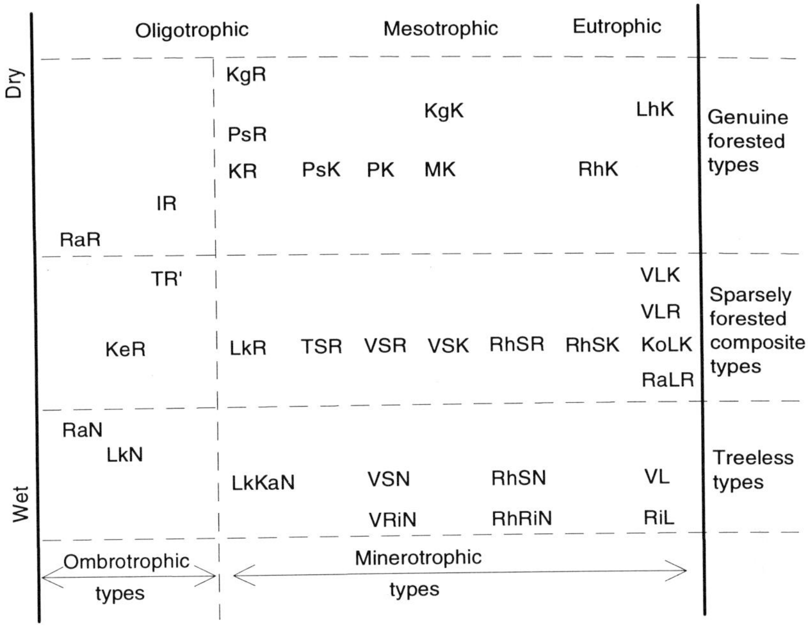

Fig. 3.16. The Finnish mire site types in forestry use schematically plotted against two environmental gradients, nutrient status and degree of wetness; for site types, see Table 3.17. (Redrawn from Laine et al. 1986)

In the **extensive site quality classification of peatlands** proposed by Huikari (1952, 1974), virgin mire sites have been grouped into six site type classes according to nutrient status and estimated post-drainage tree stand productivity:

I Eutrophic mires
II Herb-rich mires
III *Vaccinium myrtillus* and tall-sedge mires
IV *V. vitis-idaea* and small-sedge mires
V Cottongrass and dwarf-shrub mires
VI *Sphagnum fuscum* mires

These site quality classes have been used in National Forest Inventories since 1960. This classification is very useful especially in teaching and learning the bases of mire site classification, but needs sometimes additional explanations (Heikurainen 1964; Mattila and Penttilä 1987).

Up to the early 1970s the **suitability of peatlands for forest drainage** had been evaluated mainly on an ecological basis. Later, however, Heikurainen (1973b) took into consideration not only site type, climatic region (temperature sum) and increase in tree growth due to drainage, but also gross profit and costs. The method includes a profitability coefficient which expresses the ratio between the returns in terms of increased tree growth and the costs of drainage. For practical use, the limiting temperature sum at which the profitability coefficient equals 1.0 has been calculated for each mire site type. With a temperature sum of this figure or higher, forest drainage is considered profitable at the site in question (Heikurainen 1973a).

In **Sweden,** the terrain has to have a peat layer of 30 cm or more in order to be considered **peatland** (mire). If the peat layer is shallower, the site is called **wet mineral soil. Peat-covered wetlands** include both peatlands and wet mineral soils (Hånell 1991a).

Malmström (1928) roughly described the forest production of wetlands after drainage and developed a classification into six major types. Holmen (1968, 1969) evaluated the classification later on and formed a preliminary site quality table for Swedish peatlands. This partly modified the scheme prepared by Huikari (1952, 1974) in Finland.

However, it was not until the 1980s that Hånell (1984, 1991a) based the **peatland site classification on the post-drainage productivity**. The classification is carried out in three steps (see Hånell 1991a):

1. The site is classified into one of eight site types based on the presence and abundance of a number of indicator plants in the field layer (Fig. 3.17):
   a) Tall herb type
   b) Low herb type
   c) Bilberry-horsetail type
   d) Tall sedge type
   e) Dwarf shrub type
   f) *Carex globularis* type
   g) Low sedge type
   h) *Andromeda*-cranberry type
2. The effect of climate is taken into consideration by determining the temperature sum for the area in question.
3. The post-drainage productivity ($m^3\ ha^{-1}\ a^{-1}$) is given in the form of a table where the variables are the virgin site type and temperature sum. From low herb type to low sedge type two values for forest productivity are given depending on the dominant tree species. Thus, the final number of mire site types is increased up to 14.

The difference between the Finnish (Heikurainen 1973b) and the Swedish (Hånell 1984, 1991a) approaches is that in the latter, the mire site types are described more extensively – the ground layer vegetation is omitted. In the Swedish system, the actual post-drainage productivity is given, whereas in the Finnish system the post-drainage productivity is given as a site quality index.

In **Norway**, the classification system used in practical forestry is that proposed by Thurmann-Moe (1941a). According to this system, both wet mineral soils and forested mires are divided into the following site quality classes (Jerven and Wisth 1967):

Follow the question scheme. Start from above and proceed downwards. Check if the abundance of site type species fulfils the site type requirements. If not, proceed downwards and add the site type species you found, as valid type species for the next site type. Decide again whether site type requirements are fulfilled. If so, the site type is defined.

Remember
- It is important that your decision is not based on a too small area.
- The type species must be evenly spread over the area that is classified.
- When the cover of the type species is estimated, imagine thet the ground is seen straight from above.

Site type species:	Question scheme:	Yes →
Cirsium heterophyllum *Tall Dryopteris-species* *Crepis paludosa* *Cirsium palustre* *Paris quadrifolia* *Aconitum septentrionale* *Angelica silvestris* *Filipendula ulmaria* *Rumex acetosa*	The site type is characterized by tall herb species (No ↓)	*Aconitum-Filipendula /* tall herb type
Potentilla erecta *Broad-leafed grass species* *Gymnocarpium dryopteris* *Maianthemum bifolium* *Oxalis acetocella* *Geum rivale* *Equisetum palustre* *Orchis-species* *Viola-species*	The site type is characterized by low herb species (No ↓)	*Maianthe-mum-Viola /* low herb type
Vaccinium myrtillus *Comarum palustre* *Equisetum silvaticum* *Trientalis europaea*	The site type is characterized by bilberry-horse-tail species (No ↓)	*Vaccinium myrtillus /* bilberry-horsetail type
Carex rostrata *Carex chordorrhiza* *Carex lasiocarpa* *Menyanthes trifoliata* *Eriophorum angustifolium*	The site type is characterized by tall sedge species (No ↓)	*Carex rostrata /* tall sedge type
Vaccinium vitis idaea *Vaccinium uliginosum* *Ledum palustre* *Betula nana*	The site type is characterized by dwarf shrub species (No ↓)	*Ledum palustre /* dwarf shrub type
Carex globularis	The site type is characterized by Carex globularis (No ↓)	*Carex globularis* type
Carex limosa *Carex pauciflora* *Scirpus caespitosus* *Eriophorum vaginatum*	The site type is characterized by low-sedge species (No ↓)	*Eriophorum vaginatum /* low-sedge type
Empetrum nigrum *Calluna vulgaris* *Andromeda polyfolia* *Vaccinium oxycoccus*	The site type is characterized by Marsh Andromeda-cranberry species (No ↓)	*Calluna vulgaris /* Marsh Andromeda-cranberry type
	Is that so? Try again from the beginning	

Site type requirements:

For herb types, one of the following requirements should be fulfilled.

1. At least three species.
2. At least two species covering 1/16 of existing field layer.
3. One species covering 1/8 of existing field layer.

For other site types (except herb types):

Site type species must cover at least 1/4 of existing field layer. At least three species.

Fig. 3.17. Key to the Swedish mire site type classification for forestry use. (Hånell 1991a)

Site quality class	Average annual post-drainage productivity, $m^3 ha^{-1}$
A Extremely good	8
B Good	5–6
C Fairly good	3–3.5
D Fairly poor	2.5
E Poor	Not suitable for forestry without fertilization

The mire classification for forestry has also been expanded to open peatlands. It is based mainly on the vegetation indicating fertility of the peat (see Haveraaen 1966):

A Eutrophic open mires
B Herb-rich open mires
C Sedge-rich open mires
D Low sedge open mires
C *Sph. fuscum* open mires

The Norwegian wetland site type classification is thus divided into wet mineral soils, forested peatlands and open, treeless peatlands.

Masing (1975) proposed a detailed mire classification system for **Estonia** (see also Zobel 1992). The first division is into four types of minerotrophic sites (soligenous, topogenous, limnogenous and transitional) and two types of ombrotrophic sites (moor sites, i.e. shallow peat sites, and bog sites). These are further divided on the basis of floristic composition. In total, there are 36 plant community types.

Bush (1961) developed a mire classification system for **Latvia**. Sites are divided into trophic classes, which are described separately for upland sites, paludified sites (peat layer < 30 cm), forested swamps, treed fens and open bogs (see also Heikurainen 1964; Vompersky and Ivanov 1993).

In **Russia**, peatlands and paludified forests have traditionally been classified as forests on peat soil. Open, treeless mires and sparsely forested mires (i.e. treed bogs and fens according to Canadian concepts, Jeglum 1991b) have been considered as "genuine" mires. Therefore, in Russia, the term **forest type** is used instead of **mire type** (Vompersky and Ivanov 1993). Furthermore, in the Russian forest site classification, the dominant tree species is used as one of the indicators of the site type and is thus included in the name of the forest type. Since the 1950s, the classification of paludified forests and forested peatlands suggested by Pyavchenko (1956; ref. Vompersky and Ivanov 1993) has been widely used. In this system, the sites are classified into three ecological series (trophic levels) according to the origin of water and mineral nutrition: (1) eutrophic (flowing water), (2) mesotrophic (semi-flowing water), and (3) oligotrophic (stagnant water). These trophic levels are further divided into 17 groups of forest types, and the total number of forest types (site types of paludified forests and forested mires) is 31. In this way, formed site types could be considered **forested mire site types**.

In **Great Britain**, Toleman (1973) described the mire classification system developed by Forestry Commission for forestry purposes. In this system, there are four mire groups, and a total of 14 mire site types. Recognition of the site types is based on the vegetation.

In **Canada**, wetland classification includes a number of criteria, including vegetation composition, nutrient regime, hydrology, soil materials and developmental trends (Zoltai and Pollett 1983). The broadest subdivision, **Wetland Classes**, is based on the physiognomy of wetlands: bogs, fens, marshes, swamps and shallow open waters (Zoltai et al. 1973). The second level in the classification is **Wetland Form** and it is primarily based on the surface morphology of wetlands, including the distribution of surface waters and the morphology of the underlying mineral terrain and enclosing basin. The third level is **Wetland Type** and it is based on the dominant aspect of the vegetation, where the various physiognomic forms are within predetermined limits (Jeglum et al. 1974). **Wetland Varieties** are the most detailed level of this classification (Zoltai and Pollett 1983).

Because of the huge area of Canada, it is logical that in some provinces the peatlands are better studied than in others. For instance, for Newfoundland, Heikurainen (1968) suggested the Finnish approach of mire site type classification. Pollett and Bridgewater (1973) developed a mire classification system for the open peatlands of central Newfoundland, leading to 11 plant associations.

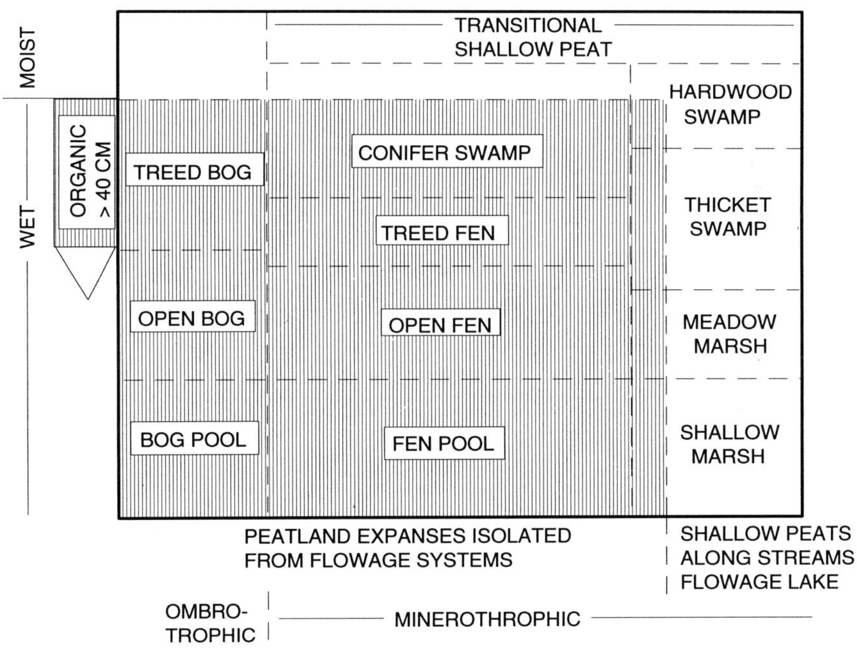

Fig. 3.18. A conceptual edaphic model of the main units of wetlands in Ontario. (Redrawn from Jeglum 1985)

Jeglum (1985) presented a conceptual edaphic model for Canadian wetlands (mainly mires), in which the main environmental gradients are moisture regime and a complex factor gradient related to the trophic status and the influence of flooding (Fig. 3.18). The scheme resembles that proposed for the Finnish mire site types by Heikurainen and Pakarinen (1982) and developed further by Reinikainen (1983, 1988a).

For forestry purposes, the most important classification attempt in Canada is the Forest Ecosystem Classification (FEC) for the Clay Belt area in the mid-boreal zone of eastern Ontario (Jones et al. 1983; Jeglum 1985). This is proposed for practical forest site classification for forest management. The classification encompasses both uplands and forested peatlands.

Both in connection with the above-mentioned project and in other studies involving mire site type classification, a variety of sophisticated methods (DECORANA, TWINSPAN, MDS) have been applied to delimit virgin site types by means of vegetation (Jeglum et al. 1971; Jones et al. 1983; Starr 1984; Jeglum 1991b). With ordination techniques, not only the ecological gradients relating to moisture and trophic status, but also the gradient from the mire margin to the mire centre plant communities have become apparent (Pakarinen and Ruuhijärvi 1978). In large data sets, the delimited species groups agree reasonably well with conventional mire site types (Pakarinen 1979; Hotanen and Nousiainen 1990).

3.4.2 Mires Drained for Forestry

The classification of drained mires has received little attention worldwide. However, because of the long history of forest drainage activity on an operational scale in Finland, the classification of drained sites has been an important field of research for several decades.

After drainage the effect of one of the main ecological factors – moisture – is diminished and nutrient status is increased. This effect was already evident in the early works of Fennoscandinavian scientists (Tanttu 1915; Melin 1917; Cajander 1926; Lukkala 1929), and led to the preliminary classification of drained peatlands for practical forestry purposes (Lukkala 1939; Lukkala and Kotilainen 1945).

It was assumed earlier (e.g. Cajander 1926) that the succession after drainage eventually resulted in more or less stable climax communities equivalent to mineral soil site types. Later studies (Sarasto 1957, 1961a,b) did not support this assumption. The succession of plant communities after drainage led to communities which resembled the forest types on mineral soils of the same trophic level, but retaining special characteristics due to their mire origin and the peat substrate (Reinikainen 1990).

Peatlands drained for forestry have traditionally been classified into three main categories, primarily depending on the extent to which the vegetation has changed (Sarasto 1961a,b; Heikurainen and Pakarinen 1982; Mikola 1982):

Recently drained mires are ditched sites where the ground vegetation still closely resembles that of the original site type in its virgin state. This is due to the short time since drainage or ineffective drainage. Tree growth has not yet responded to drainage, or afforestation of an open mire is incomplete.

Transforming drained mires are sites where the succession of the ground vegetation has gone further and the tree stand has increased its growth or originally treeless mires have become afforested. Coverage of *Sphagnum* mosses (75–25%) usually decreases significantly in this phase, while certain dwarf shrubs (e.g. *Betula nana*) may increase in abundance. Both of these two categories are classified in more detail on the basis of the original mire site type.

Transformed drained mires represent the final stage of succession; the ground vegetation has developed into a relatively stable form, which clearly differs from virgin mire vegetation, and resembles more the vegetation of mineral forest soils. The coverage of *Sphagnum* mosses on the lawn level surface is less than 25%. The growth of the tree stand corresponds to that of the upland forest stands at a parallel level of fertility. The structure of the tree stand may vary depending on the development class and cuttings that have been carried out. Four transformed drained mire site types have been described (Sarasto 1961b; Heikurainen and Pakarinen 1982).

It has been suggested that the classification system could be made simpler and more operational by allocating the drained mire sites during the successional phases to corresponding **drained peatland forest site types**. Seven drained peatland forest site types have been suggested by Laine (1989):

1. Herb-rich type develops from the most fertile spruce mires. Ground vegetation is characterized by tall ferns (e.g. *Athyrium filix-femina*) and herb species. *Oxalis acetosella* is a typical species in southern Finland.
2. *Vaccinium myrtillus* type I develops from genuine forested spruce mires. *V. myrtillus* and *V. vitis-idaea* dominate the field layer. Herbs typical of mesic upland sites (e.g. *Trientalis europaea* and *Dryopteris carthusiana*) are indicator species.
3. *V. myrtillus* type II develops from mesotrophic treeless and composite pine or spruce mires. Indicator species are largely the same as for type I, but tree stand and peat characteristics differ.
4. *Vaccinium vitis-idaea* type I develops from less fertile spruce mires and minerotrophic genuine pine mires. Dwarf shrubs typical of pine mires (*Ledum palustre, Vaccinium uliginosum*) grow scattered amongst the *V. myrtillus* and *V. vitis-idaea* which dominate the community.
5. *V. vitis-idaea* type II develops from treeless and composite types of oligotrophic tall-sedge mires. *Betula nana* usually dominates in the field layer of younger drainage areas. The more stabilized communities are rather similar to those in type I; the major differences are found in the peat properties.
6. Dwarf-shrub type develops mainly from ombrotrophic pine mires. The pine mire dwarf shrubs (*Ledum palustre, V. uliginosum*) usually dominate in the field layer, even in stabilized plant communities of older drainage areas.

7. *Cladina* type develops from the most nutrient poor mires. *Sphagnum fuscum*, along with lichens, usually dominates the bottom layer. The most common field layer species are *Calluna vulgaris, Empetrum nigrum* and *Eriophorum vaginatum.*

Volume increment values of tree stands of different drained peatland forest site types at the thinning stage are similar to those for upland forest stands of parallel fertility (Laine 1989). The same has been found for older forest drainage areas with mature tree stands, but only if the ditches are in good condition and the drainage has been effective enough (Seppälä 1969).

In **Sweden**, in spite of the large area of peatlands that has been drained – 1.4 million ha – the classification of peatlands drained for forestry is not especially advanced. It has been suggested that the drained sites should be divided into three productivity classes (Hånell 1988b; for corresponding virgin sites, see Hånell 1991a):

1. High productive sites
   - Tall-herb type
   - Low-herb type
   - Better variant of bilberry-horsetail type
2. Medium productive sites
   - Poorer variant of bilberry-horsetail type
   - Tall-sedge type
   - Dwarf-shrub type
3. Low productive sites
   - *Carex globularis* type
   - Low-sedge type
   - *Andromeda*-cranberry type.

In **Estonia**, Lohmus (1981) distinguished four mire site types (pine bog, transitional pine swamp, birch swamp and alder fen) and the corresponding drained sites in the successional stage. After a long period of drainage, these sites converge into two site types – "decayed" *Vaccinium myrtillus* site type (mainly oligotrophic sites with pine) and "decayed" *Oxalis acetosella* site type (mainly eutrophic and meso-eutrophic sites with predominating spruce). In this stage of the succession, coverage by hygrophilous mire plants is less than 20% (Lohmus 1981; Zobel 1992). It is expected that under Estonian climatic conditions the final stage of succession, a peaty *Oxalis* spruce forest, will only be reached after two tree stand generations. This final stage should be considered as a separate, relatively stabilized forest site type on peat (Lohmus 1981).

It is evident that the total variety of the mire vegetation decreases after drainage as the water regime in the substrate becomes less limiting and more uniform (Hotanen and Vasander 1992). The post-drainage vegetation succession is the most rapid in the more fertile site types, and the distinction between the site classes seems to be better in the nutrient-rich part of the trophic series (Laine and Vanha-Majamaa 1992). However, even in Finland the succession of the plant communities after forest drainage has been studied mainly in the southern part of the country. There is a need to improve the site classification system of drained mires (Mannerkoski 1979; Reinikainen 1988a).

3.5 Hydrology of Peatlands

3.5.1 Hydrology of Natural Peatlands

3.5.1.1 Concepts

Hydrology is the science that deals with the waters of the Earth – their occurrence, circulation and distribution, their chemical and physical properties, and their interaction with the environment, including their relation to living things (Johansson 1984).The hydrological quantities (P = **precipitation**, R = **runoff**, ET = **evapotranspiration**, ΔS = **change in water storage**, all in mm) for an area and period of time may be expressed as a **water balance**:

$$P = R + ET + \Delta S.$$

In a forested area, only part of the **gross precipitation** reaches the ground as **throughfall** or **stemflow**. **Interception loss** is the amount of water evaporated from plant surfaces which have been wetted by precipitation, dew or fog (Fig. 3.19).

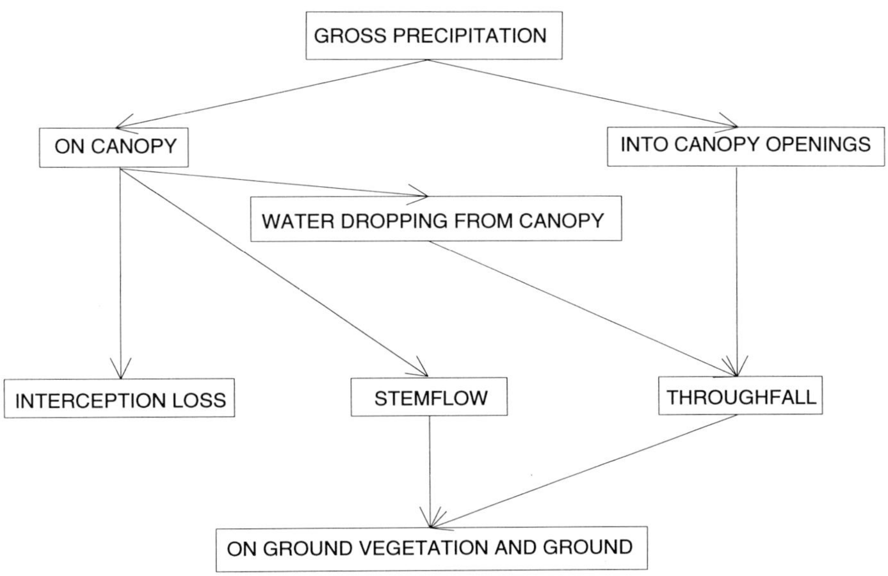

Fig. 3.19. Distribution of precipitation in a stand. (Redrawn from Päivänen 1966)

If the water leaving the area in question is measured as the quantity of water passing through a cross section of the outlet stream in unit time (e.g. $m^3 s^{-1}$), it is termed **discharge** or **water flow**. **Runoff** usually refers to discharge per unit area.

Evaporation refers to the process whereby the state of water changes from liquid to vapour. **Transpiration** means the evaporation by plants, and **direct evaporation** is from the non-living soil surface; together they are called **evapotranspiration (ET)**. **Potential evapotranspiration (PET)** refers to the potential maximum of evapotranspiration, i.e. the evapotranspiration that would occur if the amount of available water was not limited. It is defined as the evapotranspiration from short, physiologically active, green vegetation, completely cover-

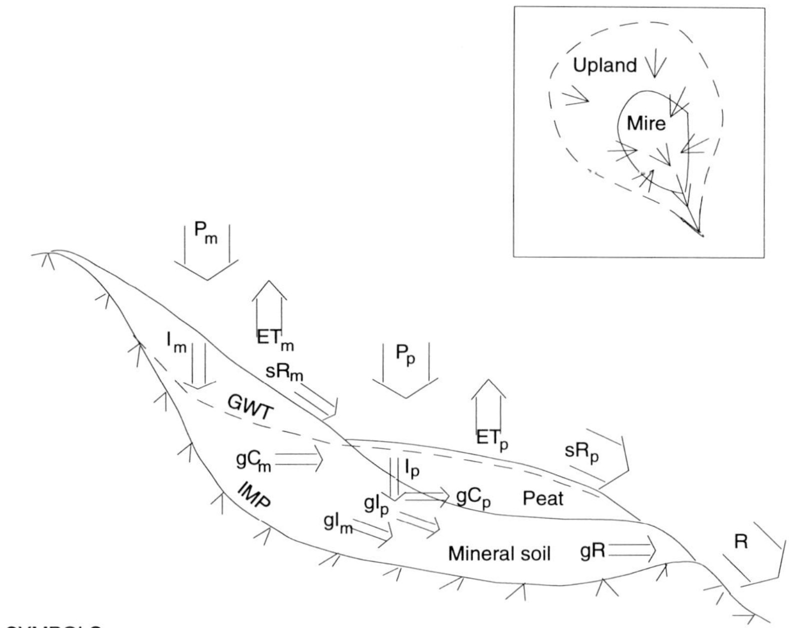

SYMBOLS:

P_m	= Precipitation on mineral soil		gC_p	= Groundwater flow from peatland to mineral soil and back to peatland
P_p	= Precipitation on peatland		gl_m	= Groundwater flow from mineral soil
ET_m	= Evapotranspitation from mineral soil		gl_p	= Groundwater flow originating from peatland
ET_p	= Evapotranspitation from peatland		gR	= Groundwater runoff
I_m	= Infiltration from mineral soil		R	= Total runoff
I_p	= Infiltration from peatland		GWT	= Groundwater table
sR_m	= Surface flow and interflow from mineral soil to peatland		IMP	= Impermeable layer
sR_p	= Surface flow and interflow from peatland			
gC_m	= Groundwater flow from mineral soil to peatland			

Fig. 3.20. Water balance of a minerogenic mire complex, schematic presentation. (Redrawn from Sallantaus 1988)

ing an extensive area of ground kept sufficiently well moistened to prevent moisture stress. PET is often estimated from climatological data alone. For a further discussion of these concepts, see Penman (1956), Ingram (1983) and Johansson (1984).

Peatland or **mire hydrology** deals with hydrologic processes of a mire ecosystem with its surroundings. As Ingram (1983, p. 69) points out, "it is helpful to consider the system as comprising four compartments, namely the atmosphere, the mire itself, the mineral soils and their parent rock adjacent to the mire, and the local system of surface streams". If a mire is considered as an ecosystem isolated from its surroundings, we may apply the general water balance to it: any difference in the rate at which water enters and leaves the system must appear as an increase or decrease in the amount stored within it (Ingram 1992). A schematic presentation of the water balance for a minerogenic mire is shown in Fig. 3.20 (see also Sallantaus 1988). The term **ecohydrology** has been suggested to demonstrate that the ecosystem of a mire, or any part of a mire, cannot be definitively evaluated unless account is also taken of its water balance (Eggelsmann et al. 1993; Heathwaite et al. 1993).

In this chapter, mire hydrology is discussed from the point of view of water quantity and temporal processes rather than water quality. The water quality questions are dealt with later in connection with environmental issues (Sect. 8.2).

3.5.1.2 Soil Water and Groundwater

Water is a prerequisite for peat formation (see also Sect. 1.2.1), and topography or climate exerts a retarding influence on water movement. Moisture conditions that are sufficient for peat formation prevail in regions where precipitation exceeds evapotranspiration. Depending on topography, peat deposits may also occur in places where the area of the forming mire is small in comparison with the area of the catchment. In such cases, the runoff from the surroundings, together with precipitation, exceeds evapotranspiration. This makes peat formation possible.

The vegetation cover of mires (the mire site type) is determined by the origin of the incoming waters, whereas the formation of **mire complex types** is determined by climate and topography (e.g. Damman 1986; Damman and French 1987; Verry 1988). **Aapa mires** are characterized by flarks – long depressions dammed by *Sphagnum fuscum* hummocks. These flarks are filled by snowmelt water which escapes mainly through evaporation during the course of the summer. High air humidity and spring floods are prerequisites for the formation of aapa mires (Solantie 1986). **Raised bogs**, on the other hand, are totally dependent on precipitation. The height of a mire complex in relation to its extent in area is mainly determined by the annual precipitation. The lateral extension of raised bogs seems to slow down markedly once they reach the steeper slopes at the edges (Korhola 1992).

The groundwater in mires lies close to the soil surface. In general, **groundwater** is water stored in the saturated zone (Ingram 1983), having a hydrostatic pressure equal to or higher than atmospheric pressure (Johansson 1984). Groundwater is partly formed by **free water**, which can be displaced through gravitation, and **bound water**, which remains immobile and does not seep out when a relatively small pressure gradient is applied to it (Ivanov 1981). The boundary level of groundwater is called **groundwater table** or **groundwater level**. It is the elevation of the phreatic surface of an aquifer. Water bound in the soil matrix above the groundwater table is called **soil water** or **soil moisture** (Johansson 1984).

The primary factors which influence mire water table fluctuations are vegetation, local hydrogeology, climate, and type of peat material (Bay 1968). In forested mires – both virgin and drained – the tree stand has a strong effect on the water table level through interception and evapotranspiration. An increase in the stand volume lowers the groundwater table and intermediate cuttings and clear-cuttings have the opposite effect (see Sect. 3.5.2 and 3.5.3).

The groundwater table is usually measured as the distance from the soil surface in unlined or lined holes called **groundwater wells**. For hydraulic analyses involving comparisons of **water table elevation** at several points, the elevations are computed with respect to a general datum plane, for instance to sea level (Heikurainen 1971a; Ingram 1983). Because the level of the surface of mires changes through the influence of a great number of different factors (see Sect. 3.2.1), the distance (or depth) to the groundwater table and the water table elevation are not exactly parallel concepts.

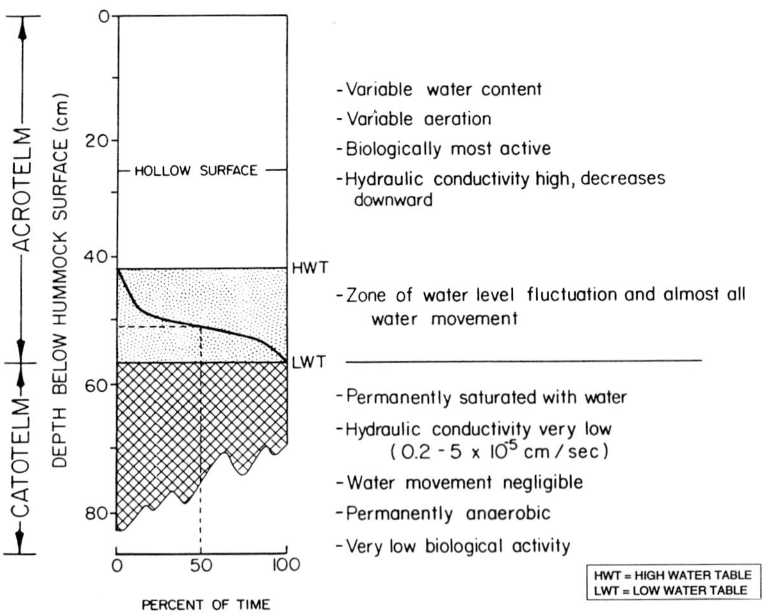

Fig. 3.21. Properties of acrotelm and catotelm of a bog. The *heavy black line in the stippled horizon* is the water level duration curve, indicating the percentage of time that the water table is at or above that level. (Redrawn from Damman and French 1987)

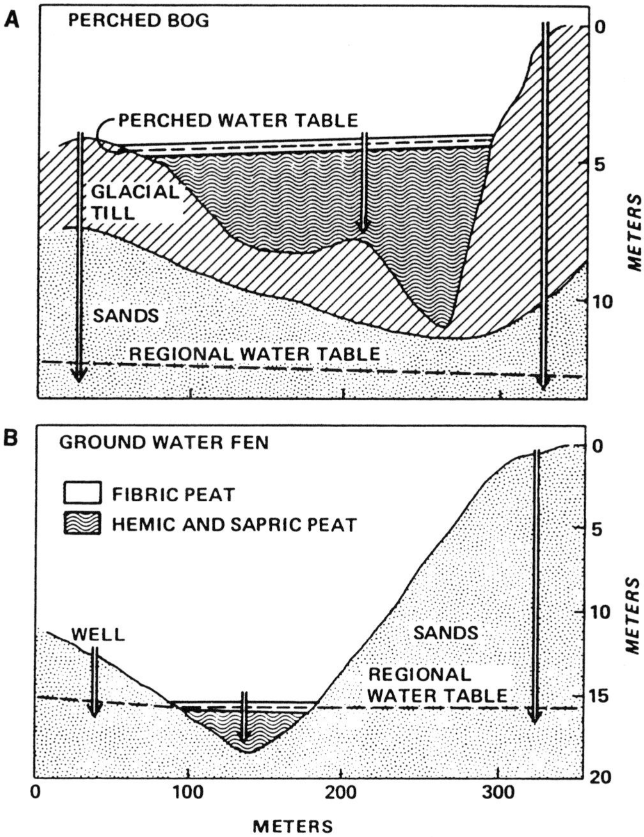

Fig. 3.22 A, B. Division of peat deposits according to their relation to the regional groundwater table. **A** Perched bog; **B** groundwater fen. (Redrawn from Boelter and Verry 1977)

In the case of natural mires, it has been suggested to distinguish between two horizons in the surface peat: the acrotelm and catotelm (Ivanov 1981; Ingram 1983; Damman and French 1987). The **acrotelm** is the surface peat layer above the low-water table, and **catotelm** is the permanently anaerobic peat layer below this level (Fig 3.21). Many hydrologically and ecologically important properties of the peat substrate are associated with this subdivision. However, the concepts may be more useful for ombrotrophic than minerotrophic conditions (Ingram 1992).

A mire complex may have developed in a basin that is separated from the regional groundwater system because of very slowly permeable peat or lacustrine deposits. This type of mire complex has been called a **perched bog** (Fig. 3.22A) because its water table is perched above the regional groundwater table. There may be an unsaturated zone between the bottom of the peat deposit and the regional groundwater table, and there is essentially no mixing of their waters. In contrast, **groundwater fens** (Fig. 3.22B), are minerotrophic mires in which the water table is simply an extension of the regional groundwater table (Bay 1967; Boelter and Verry 1977).

3.5.1.3 Higher Plants and Soil Water

The difference in the soil water content between field capacity and permanent wilting point is usually considered as being available to higher plants (e.g. Kramer 1949; Aslyng 1952; Marshall 1959a). However, the **field capacity** (for the concept, see Veihmeyer and Hendrickson 1949) can only be applied to horizons where natural drainage is more or less vertical and free from waterlogging or lateral water input (Baize 1993) and thus it cannot be applied to peat soils, where the groundwater level is close to the soil surface and the free draining of soil is impossible (Päivänen 1973b).

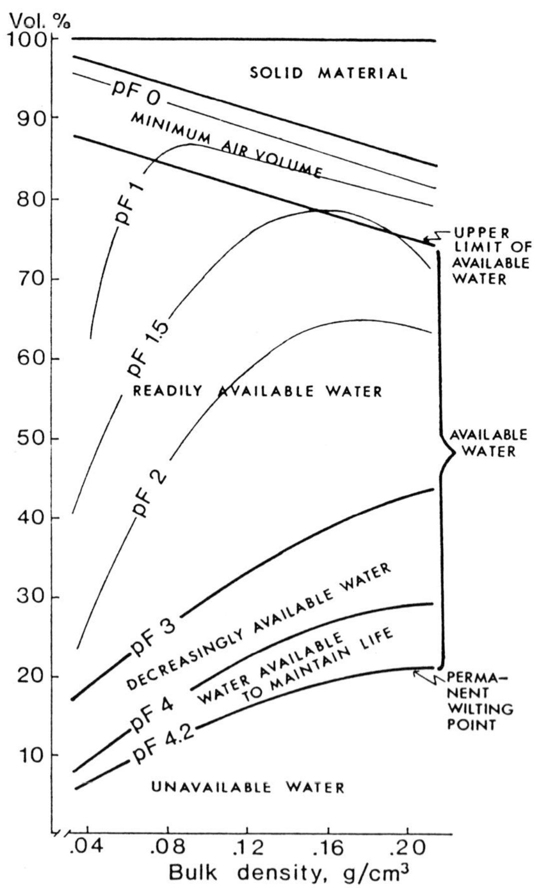

Fig. 3.23. Volumes of solid material and minimum air space as well as of water available and unavailable to plants in peats with different bulk densities. The *thin lines* show the quantities of water retained in the peat at pF 0, 1, 1.5 and 2. (Redrawn from Päivänen 1973b)

All the water which is present in the soil at water contents between **saturation** and the **permanent wilting point** is available to plants (Fig. 3.23), at least in the sense that its binding does not prevent it from being taken up by plants (Andersson and Wiklert 1970). Excess water in the soil may, however, impair aeration of the soil to such an extent that the low oxygen supply and lack of carbon dioxide removal inhibit growth of higher plants. Such conditions persist in virgin mires if the water in the soil does not move fast enough. This creates anaerobic conditions in the soil (Lähde 1969). Under such conditions the limited air void volume of the soil becomes a growth-limiting factor.

Flooding during the growing season is much more harmful for the trees than flooding during the dormant season or early summer (Pelkonen 1975; Päivänen 1984b). This is because of the high oxygen requirement of actively growing roots with high respiration rates (Kozlowski 1982).

Oxygen and carbon dioxide move in the soil either dissolved in the water or by diffusion in the air space of the soil, the latter being the more effective way. Thus, **soil aeration** depends first and foremost on the volume of the air space of the soil (Hillel 1971). The rate of gas diffusion is primarily determined by the total air space of the soil and only secondarily by the size distribution of air-filled pores (Marshall 1959b). The development of toxic concentrations of ferrous iron, sulphides and manganese under anaerobic conditions also limit root development (Robinson 1930). However, it is unclear whether oxygen deficiency, excess carbon dioxide, toxic compounds produced under anaerobic conditions, or hormonal imbalance are most important in inducing symptoms of flooding injury in trees (see Kozlowski 1982, 1986).

The percentage of soil air void volume generally used to differentiate between well and poorly aerated conditions is from 10 to 15% (Kramer 1949; Taylor 1949; Glinski and Stepniewski 1985; Mannerkoski and Möttönen 1990). However, there is very little information available on the **minimum air space** required by the roots of trees growing on peat. Paavilainen (1967) found that the roots of Scots pine (*Pinus sylvestris*) growing on pine mire did not penetrate the soil down to a depth at which the air space dropped below 10%.

The distance from the soil surface to the groundwater level reflects the soil water content in the rooting zone (Heikurainen et al. 1964; Juusela et al. 1969), the anaerobic limit (Lähde 1966; Mannerkoski 1985), and the matric suction in the unsaturated zone (Ahti 1971, 1978; Laine and Mannerkoski 1975), as well as the tree root penetration depth (Paavilainen 1966a; Mannerkoski 1991).

3.5.1.4 Evapotranspiration

Evapotranspiration can be directly estimated from diurnal water table fluctuations if the water table is sufficiently near to the mire surface and if the groundwater coefficient has been determined (Heikurainen 1963; Laine 1984). The **groundwater coefficient** is the ratio of the fall in the groundwater table to the amount of water lost. The coefficients have been determined empirically using large undisturbed peat samples and removing (or adding) a known quantity of

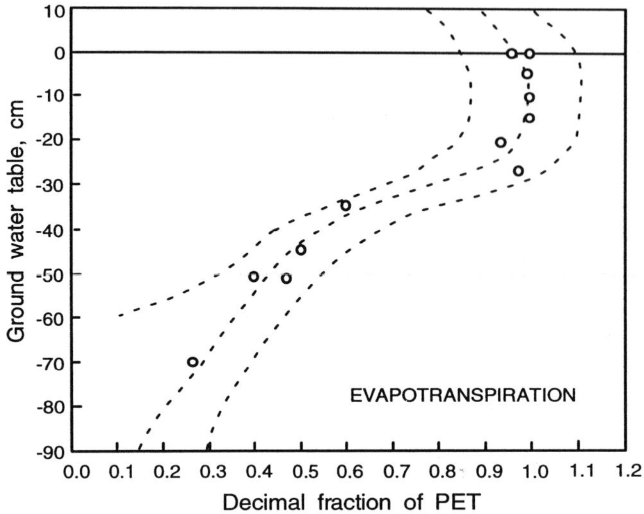

Fig. 3.24. Evapotranspiration as a decimal fraction of potential evapotranspiration (PET) in relation to groundwater table with theorized bounds of variation parallel to the central trend. (Redrawn from Verry 1988)

water and measuring the corresponding fall (or rise) in the water table. The coefficients have been obtained by the formula (Heikurainen 1963; Päivänen 1964; Laine 1984):

$$C = G/W,$$

where C = groundwater coefficient, G = fall or rise of the water table, and W = the amount of water causing the change in the water table.

Water from virgin mires is mainly lost through evapotranspiration. The small proportion that leaves as runoff is due to small gradients, poor hydraulic conductivity of peat and scarcity of natural channels. Thus, we may state that the **runoff threshold** of virgin mires is high. Evapotranspiration from natural mires almost equals the potential evapotranspiration (PET) because the groundwater table is usually rather near the peat surface (Virta 1966; Verry 1988). Moreover, the unevenness of the micro-relief of a peat surface offers a large evaporating surface and, particularly under the conditions prevailing in treeless mires, wind conditions that are favourable for evapotranspiration. According to several studies, transpiration from mire vegetation exceeds evaporation from free water surfaces (e.g. Romanov 1968; Sturges 1968; Nichols and Brown 1980). Evapotranspiration from natural mires is of a similar magnitude as PET as long as the escaping quantity of water is replaced by capillary forces from the groundwater table (Fig. 3.24). With an increase in the distance between the groundwater table and the peat surface, evaporation decreases considerably (e.g. Virta 1966; Laine 1984; Verry 1988).

3.5.1.5 Runoff

According to a popular belief, virgin mires are considered to act as hydrological buffers - floodwaters percolate into the peatland and then gradually flow out into creeks and rivers during dry periods. However, this is an oversimplification at best and can lead to the misunderstanding of a complex natural system (Heikurainen 1976; Boelter and Verry 1977; Verry 1988). Several studies have shown this so-called sponge theory to be wrong. For instance, in central Europe the runoff of flood water from peatlands under natural conditions has been observed to be fast and strong with scarcely any runoff during dry periods (Baden and Eggelsmann 1964; Eggelsmann 1972; Uhden 1972). Similar results have also been published for mires in the boreal zone (Heikurainen 1976, 1980a; Boelter and Verry 1977) – a saturated sponge without a continuous supply of water will quickly dry out by evapotranspiration and little water will flow from it. Thus, the very thinness of the acrotelm (in which the water table is confined) means that peat-covered catchments are not only poor suppliers of base flow during the summer recession, but also have only a limited capacity to store inputs of precipitation (Burt et al. 1990).

Different mire complex types differ in their runoff characteristics. Mire complexes with a **perched groundwater table** receive water mainly from precipitation. Summer runoff from such mire areas is slight because conditions favour evapotranspiration and most of the water received in the form of rain evaporates. In winter, runoff is negligible and may even cease altogether because the precipitation is received in the form of snow and accumulates into the snow cover (Bay 1969; Verry and Boelter 1972; Boelter and Verry 1977). Thus, most of the runoff from mires with a perched groundwater table takes place in the spring when the snow melts away (Fig.3.25).

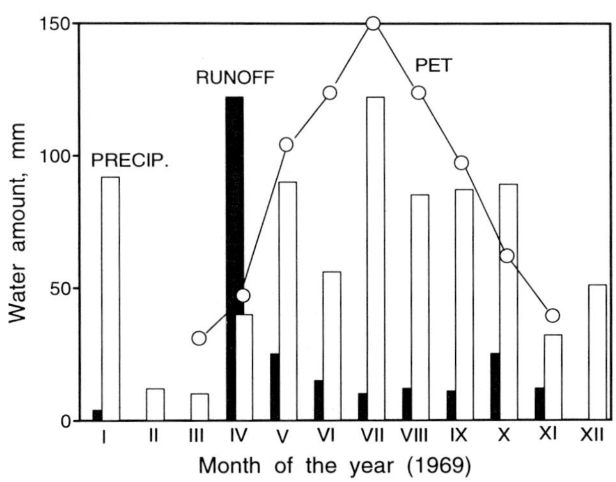

Fig. 3.25. Monthly precipitation, runoff, and potential evapotranspiration for a perched bog watershed in Minnesota. (Redrawn from Boelter and Verry 1977)

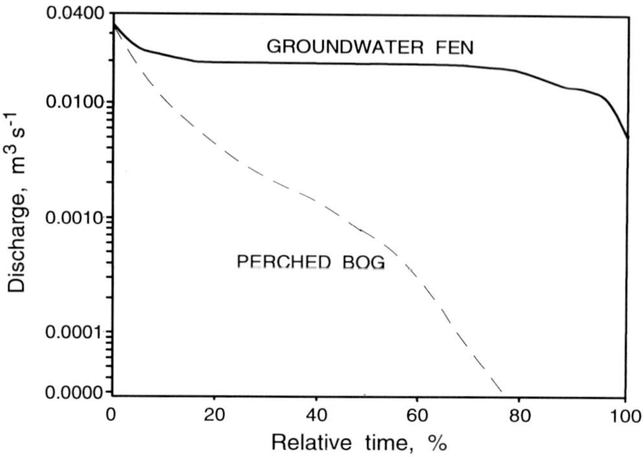

Fig. 3.26. Discharge-duration curves for a groundwater fen and perched bog watershed. (Redrawn from Verry 1988)

The runoff pattern of **groundwater mires** is different. Here, too, water discharge reaches its maximum when the snow melts. However, as the water supply of groundwater mires is supplemented by groundwater from surrounding mineral soils, runoff conditions are more stable than in mires with a perched groundwater table (Boelter and Verry 1977). The differences in the behaviour of runoff between groundwater mires and mires with a perched groundwater table are particularly pronounced in the duration of the discharge (Fig. 3.26).

From a groundwater fen fed with an annually constant groundwater flux, the summer storm-flow volume produced by a rainfall event may be small, but the response and recession are rapid. The storm-flow volume is also directly proportional to the volume of rainfall. This indicates that the size of the saturated zone is constant and not influenced by storm rainfall (Roulet 1991). The groundwater fen is thus not able to reduce storm-flow volumes because storage capacity is limited.

Due to the general evenness of their surfaces and to water stagnation in depressions, virgin mires may cut off discharge peaks (Hyvärinen and Vehviläinen 1980). The damming effect of hummocks has been demonstrated with surface water and groundwater level measurements on an open low-sedge bog (Päivänen 1968). However, as the runoff threshold of natural mires is high (groundwater table is near the peat surface), the free space for storing melting water or heavy rain may be filled up very quickly (Burt et al. 1990). This, in turn, may lead to high rates of discharge. Thus, the maximum runoff is determined primarily by meteorological factors (the water equivalent of snow, air temperature, snowmelt rate, precipitation), and then by the properties of the peat and the vegetation (including the tree stand) (Starr and Päivänen 1981). Also of importance are the ratio between the area of the mire and that of the entire catchment area, and the length of the mainstream of the catchment area (Brandesten 1987).

3.5.2 Effects of Drainage on Peatland Hydrology

3.5.2.1 Methods of Study

The study of minor and long-term effects on watercourses of forestry measures (forest drainage, timber harvesting, forest fertilization) is usually carried out on only a small area of a catchment. The hydrological effects also depend to an essential degree on the size of the catchment area and the location of the point of measurement in relation to that of the area where measures have been taken.

In large catchments **trend** or **frequency analyses** have usually been used to study possible changes in runoff characteristics (e.g. Hyvärinen and Vehviläinen 1980). This means that the results also include the possible effects of changes in weather conditions. These sources of error may be eliminated to some extent by means of hydrological models. An additional problem in large catchments is that several factors affecting the hydrology of watercourses may change at the same time (Hyvärinen and Vehviläinen 1980).

In small catchments, the **reference-area method** is commonly used. With this method two or more adjacent areas that differ – or should differ – from each other only with respect to the measure to be studied are compared (e.g. Multamäki 1962; Burke 1972; Heikurainen 1974, 1976). The study areas used have either been located in natural catchments, which usually also include surrounding mineral soils, or artificial catchment areas laid out on mires and delimited by ditches. With artificial catchments it is possible to vary the intensity of the measure to be studied in parallel plots. In studies of the hydrological effects of drainage, this method has been used in the form of so-called ditch spacing series (Huikari et al. 1966). In this approach, it has been supposed that conditions prevailing in the widest strip between two ditches show the smallest differences as compared with virgin mires (Ahti 1980). In the ditch spacing series, each catchment area is of the same size. An advantage of the reference-area method is that results may be obtained in a rather short time. The greatest weakness of this method is that we have no certain information about the similarity of the areas to be compared before treatment. This is a great draw-back and may explain some of the contradictory results obtained from studies using the reference-area method (see e.g. Berry 1991).

The best way to study the effects of forestry measures is to calibrate the areas before treatment – the **calibration-time/control-area method**. The principle of the method is as follows: The research basins are kept in their natural state during a calibration period. One of the basins is kept in its natural state as a control throughout the study period. Regression equations are computed for the hydrologic parameters of the control and treatment (drainage/thinning/clear-cutting) basins for the calibration period. The equations are used after the treatments to calculate what the parameters would have been if the measures had not been performed. The difference between the calculated and observed hydrologic parameters indicate the change caused by the measure concerned, provided that changes in other factors have not taken place in the study area (e.g. Mustonen and Seuna 1972; Seuna 1980; Lundin 1984; Päivänen and Ahti 1988). This method eliminates the effects caused by differences in topography, soil types, stands etc. It also implies that the climatic conditions are more or less similar (on average) during the calibration period and the period following the treatments.

3.5.2.2 Immediate Effects of Drainage

The immediate effects of draining on peatland hydrology are a drop in the groundwater table and faster runoff. The drop in the groundwater table due to ditching means that the quantity of water stored in the peat soil diminishes and that the potential storing space of the peat consequently increases. For a given distance from a ditch, the distance to the groundwater table increases as ditch spacing decreases (Päivänen 1974c; Berry and Jeglum 1991a; Hillman 1992). As the distance between the peat surface and the groundwater table increases, the evaporation from the peat surface decreases. Draining also changes the transpiration from vegetation and the temperature conditions of the mire.

Increasing runoff decreases the water store and lowers the water table, whereafter direct evaporation from soil surface will decrease. The water store decreases particularly during the first 2-year period after ditching. The increase in runoff due to a decrease in evaporation is of a more lasting character. In most cases, the change in transpiration seems first to lead to an increase in runoff as the former mire vegetation is partly destroyed. However, as the vegetation later becomes adapted to the new growing conditions, and particularly when tree growth recovers, transpiration increases. The increase in the space for storing water cuts off small runoff peaks, which in practice obviously levels out floods only in their initial phase and at small high runoffs. Thus, it is evident that these factors jointly – at least during the first years after ditching – lead to increasing runoff from drainage areas (e.g. Mustonen and Seuna 1972).

Only few studies based on a calibration period and a control area have been carried out covering the whole annual water balance. A study in catchment areas in south-eastern Finland (Huhtisuo and Latosuo) showed a 33% increase in the annual runoff during the first 10 years when 40% of the total catchment area had been drained using 50-m ditch spacing (Mustonen and Seuna 1972; Seuna 1981). In north-eastern Finland (Koivupuro) the increase in annual runoff was 13% when 26% of the catchment had been drained and 6% of it had been simultaneously clear-cut (Seuna 1988).

3.5.2.3 Peak Flows

Concerning the **effects of drainage on peak flows**, contradictory results, which may be simplified into three models, have been presented (see McDonald 1973; Starr and Päivänen 1981; Fig. 3.27). Through draining, the groundwater table is

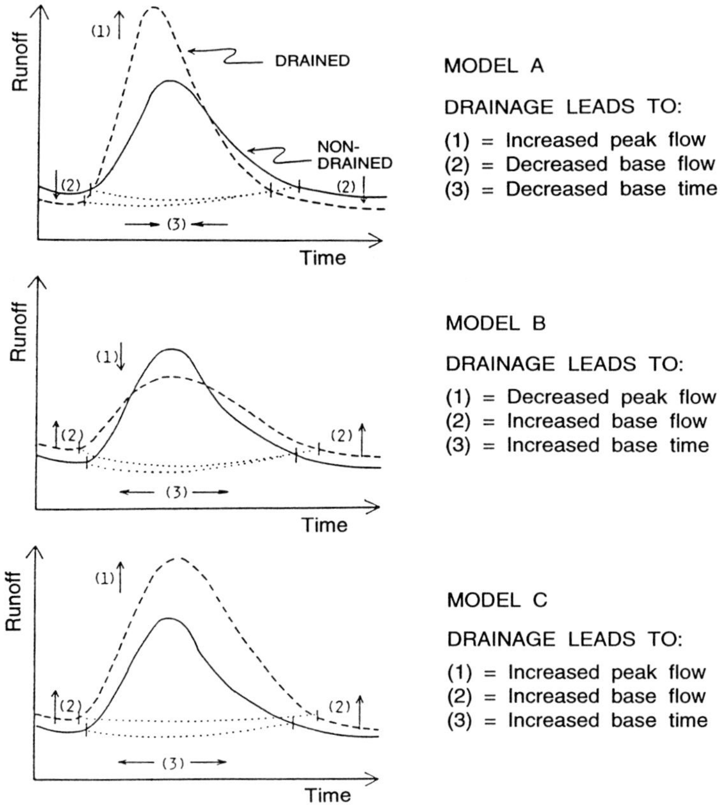

Fig. 3.27. Hydrograph models for the possible effects of forest drainage on peak flow. (Redrawn from McDonald 1973; Starr and Päivänen 1981)

lowered, resulting in a potential increase in the capacity of the peat layer to store rainwater. Simultaneously, a drop in the runoff threshold makes an increase in the base flow possible. This, in principle, would lead to a levelling out of the runoff (Multamäki 1962; Burke 1972; Heikurainen 1976, 1980a)(Model B). On the other hand, the ditch network allows a fast discharge of rainwater from the mire area concerned and hinders water from being stored in the surface depressions. These factors, in turn, tend to increase peak flows (Mustonen and Seuna 1972; Ahti 1980)(Models A and C). In the aforementioned Huhtisuo catchment area, the spring peak flow increased as a consequence of drainage by an annual average of 31% during the first 9 years after drainage. In the Koivupuro area, the average increase was 8% during the first 4 years after drainage (Seuna 1986, 1988) (Model A).

A peatland basin may show runoff patterns of all models, depending on the weather conditions during the snowmelt. In Sweden, Lundin (1984, 1990) found that the snowmelt peak flow was greater from the undrained area in spring – thus supporting Heikurainen's (1976, 1980a) findings – when precipitation was low (Model B), but when precipitation was high in spring, the peak flow from the drained area was greater (Model A).

As the groundwater table in virgin peatlands is near the peat surface, the water-storing capacity of the unsaturated zone is exceeded more quickly than in drained peatlands, where drainage has increased the water-storing capacity of the peat by lowering the groundwater table. However, after particularly heavy rains or if the snow melts during a very short period, the water-storing capacity of peat is exceeded in both virgin and drained peatlands. In such situations, the ditch network of drained peatlands immediately leads the excess water into a watercourse; the faster, the more effective the drainage. Obviously, one of the immediate effects of forest drainage is that peak flows are increased even if there is some variation from case to case. The actual changes in peak flows, i.e. what model the hydrology of a drained peatland area corresponds to, depend to an essential degree on the character of the rainfall (or snowmelt), the effectiveness of drainage, and the vegetation (particularly the tree stand) of the area concerned (Starr and Päivänen 1981).

3.5.2.4 Drainage Intensity

The **intensity of drainage** (**ditch depth** and **spacing**) obviously has an effect on the magnitude of the hydrological changes. A theoretical analysis of the effect of drainage intensity on runoff has been presented by Ahti (1977). Based on ditch-spacing experiments on treeless mires, Ahti (1980) subsequently showed that summer peak flows were, indeed, inversely proportional to ditch spacing. For monthly runoff (Ahti 1980) and for the total summer runoff (Huikari et al. 1966), the influence of ditch spacing showed the same tendency. However, whole catchment-scale studies covering the entire year have not yet been carried out.

3.5.2.5 Types of Ditches

The drainage effect as well as the quantity and timing of the water discharging from a peatland area are to a certain extent related to the **type of drainage ditches** used.

The bulk of forest ditching has been done using plowing or excavating techniques with **open ditches** (see Sect.4.1.3). In Finland, the cross section of ditches conforms to the standard forest ditch measurement recommendations (Fig. 3.28A). The effects of forest drainage on peatland hydrology discussed above are associated with this type of drainage. In areas drained with open ditches, part of the rain falls into the ditches. Surface and subsurface runoff make it possible for the water to reach a ditch in a short time.

In areas with **covered drains** (subsurface drains), the water has to penetrate an unsaturated layer in order to reach the groundwater table, move into the covered drain in the direction of the gradient through a saturated layer, penetrate the

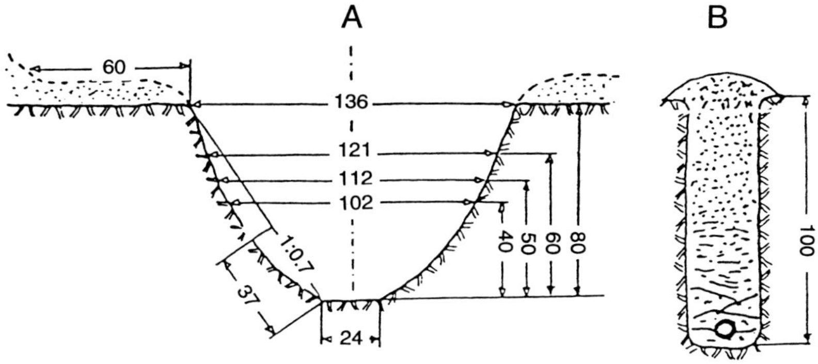

Fig. 3.28 A, B. Schematic presentation of the standard measurements of a forest ditch **(A)** and a plastic pipe drain **(B)** used in Finland. (Redrawn from Aitolahti 1972; Niskanen 1980)

wall material of the covered ditch and flow in a space limited by the material of the covered drain into an open ditch (Päivänen 1976; Verry 1988). Contradictory opinions have been presented concerning the functioning of covered drains in winter when the surface layer of the peat soil is frozen (Aitolahti 1972). It is possible, however, that the functioning of covered drains is limited, if the outlets – open ditches – are frozen. This is often the situation if the quantities of water concerned are small.

Some covered drains were made in the times of manual ditching. These ditches were filled with branches and poles and retained their functioning capacity for many decades (Aitolahti 1972). **Unfilled, covered drains** (e.g. **mole drains**) have been widely used in agriculture in central Europe (e.g. Eggelsmann 1973; Scholz 1986) and experimentally in peat-harvesting areas in Finland (Klemetti et al. 1988). However, they are not suitable for forest drainage.

In the mid-1960s, experiments using **plastic pipe-covered drains** were carried out in Finnish peatlands (Fig. 3.28B). However, the results were not promising. The following conclusions were drawn from an experiment applying the reference area method (Päivänen 1976):

The water-carrying capacity of open ditches was superior to that of plastic pipe-covered drains, particularly during wet periods. During dry spells, runoff from areas drained using covered drains continued for a longer time than from areas drained with open ditches, in the case of which it sometimes even ceased altogether. During the growing season, the groundwater table was, on average, 10 cm closer to the soil surface in the areas with covered drains than in open-ditched areas. Even if the groundwater table was at the same depth in cover-ditched and open-ditched areas, runoff was considerably higher from the open-ditched areas. Furthermore, the water-conducting capacity of plastic pipe-covered drains grew poorer as time elapsed from drainage.

These observations about covered ditches are based on the techniques and materials used in the 1960s. Drainage with modern plastic pipe drains of a larger diameter and embedded in a filtering material could be more effective, especially if sufficient regard is paid to both ditch digging and placing of the pipe. The costs of a draining operation would probably increase to such a level, however, that the benefits obtained would not cover the investment.

3.5.2.6 Location of the Area to be Drained

The **location of the drainage area** within the whole basin is obviously of great importance – especially in terms of maximum discharge downstream (Fig. 3.29). If a drainage operation is carried out in the lower parts of a basin, the maximum discharge becomes lower. At the same time, however, the high flow period is extended. In contrast, if the drainage operation is carried out in higher parts of a watercourse, the maximum discharge is greater and the high flow period shortened (Mustonen and Seuna 1971).

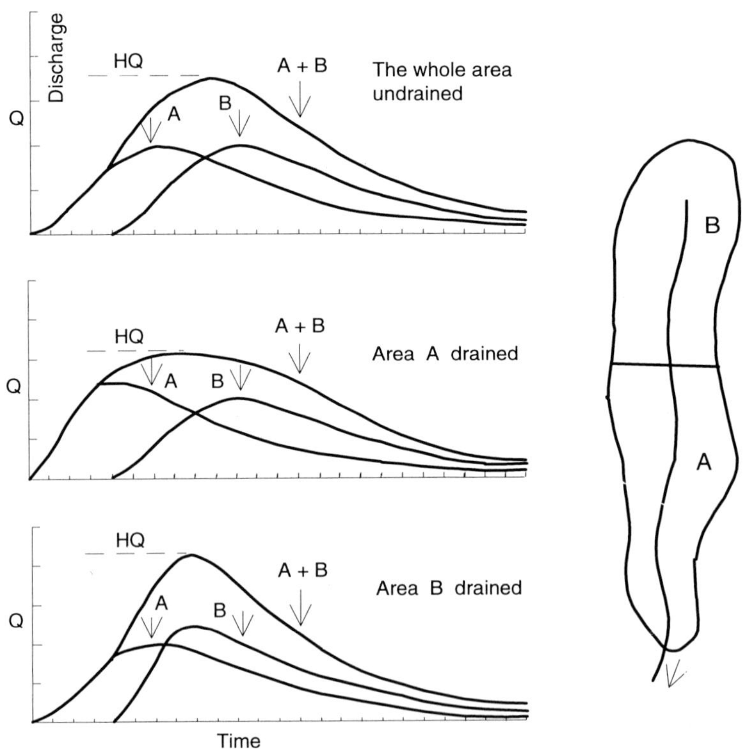

Fig. 3.29. Effect of the location of the drained area on the maximum discharge *(HQ)* of a watercourse. Time scale is relative. (Redrawn from Mustonen and Seuna 1971)

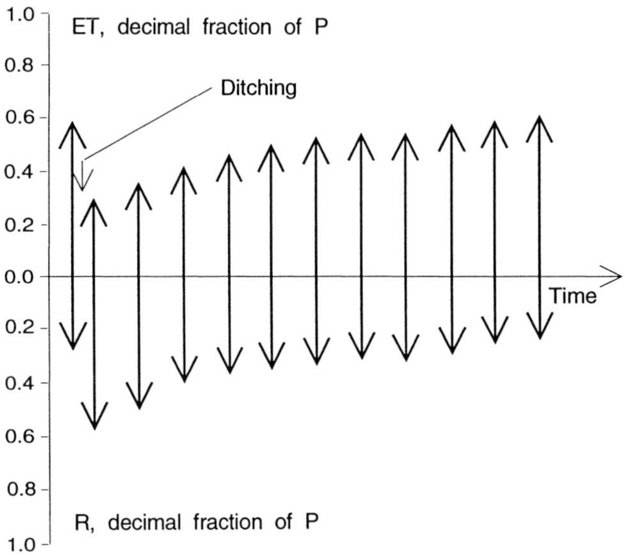

Fig. 3.30. The effect of time since drainage on the share of evapotranspiration *(ET)* and runoff *(R)* of precipitation *(P)*. (Redrawn from Ahti 1988)

3.5.2.7 Time Since Drainage

The hydrological changes due to drainage, especially in runoff, decrease with time (Braekke 1970; Ahti 1988). The increase in interception and evapotranspiration as a consequence of tree stand development is one of the most important factors accounting for this decrease in runoff (Fig. 3.30). As the tree stand growing in a drainage area becomes closed, the average rate of snowmelt is reduced and the flood peaks in the spring level out. Peak flows and total runoff may resume their predrainage levels within 10–20 years of drainage (e.g. Seuna 1980). However, low runoff may remain higher from drained than from undrained peatlands for several decades, which tends to even out hydrological conditions (Heikurainen 1980a; Seuna 1980).

3.5.2.8 Drainage Effect of the Tree Stand

There is an interdependence between drainage techniques, the drainage depth achieved and stand characteristics. The drainage effect of the developing tree stand is heavily dependent on the tree species in question (see King et al. 1986). The drainage effect for Scots pine (*Pinus sylvestris*) stands on peat is described by Laine (1986) as follows (Fig. 3.31).

At the time of ditching, the ditch spacing and depth (characteristics determined by the drainage techniques) have a dominating influence on the drainage depth, i.e. the average distance between the surface of the peat layer and the groundwater table. Of the substrate properties, the degree of decomposition of

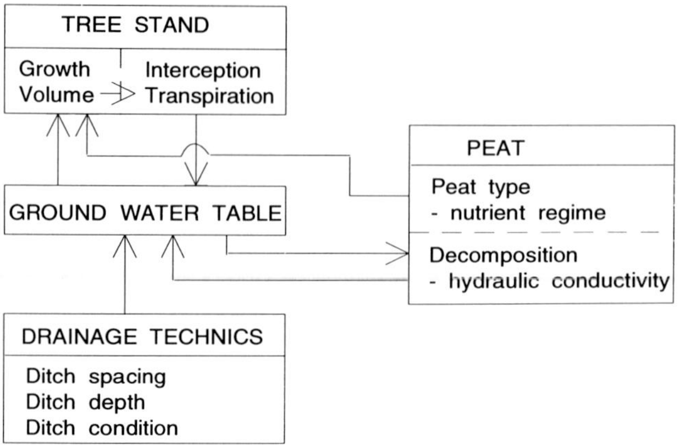

Fig. 3.31. Factors affecting the drainage depth (groundwater table) achieved. Climatic conditions are assumed to be constant. (Redrawn from Laine 1986)

the peat in particular but also the peat type affect the drainage depth through their effects on the hydraulic conductivity of the peat. As time elapses, the condition of the ditches also affects the drainage depth.

Tree stand growth and the level of the water table are interdependent through time. First, the drop in the groundwater table due to ditching increases tree growth and timber volume (Laine 1986). The greater stand volume results in greater interception and transpiration (biological drainage) which lowers the groundwater table further or, at least, maintains a drainage effect that is sufficient for the continued growth of the trees. The quality of the substrate, particularly the nutrient content of the peat, is also important to the revival and initial development of the tree stand. Thus, the negative effects of ditch deterioration tend to be compensated for on sites with a faster initial stand development and on sites with older stands and large volumes; the more so, the richer the nutrient status of the mire site.

Of the technical drainage characteristics, the **effective ditch depth** (distance to the ditch bottom or to the water level in the ditch from the average soil surface) best describes the average depth of the groundwater table. However, clearly the best explanator of the distance to the groundwater table (drainage depth achieved) is the volume of the tree stand (*Pinus sylvestris*), which alone can explain more than 50% of the variation in drainage depth (Laine 1986). An increase in the timber volume by 100 $m^3 ha^{-1}$ equals a lowering of the groundwater table by an average of 20 cm (Fig. 3.32). Apart from stand characteristics, the level of the groundwater table is largely determined by the physical properties of the peat profile.

The lowering of the groundwater table, resulting from an increase in timber volume, is of a similar magnitude to the rise in the groundwater table resulting from cutting a corresponding volume (Sect. 3.5.3). One can therefore draw the

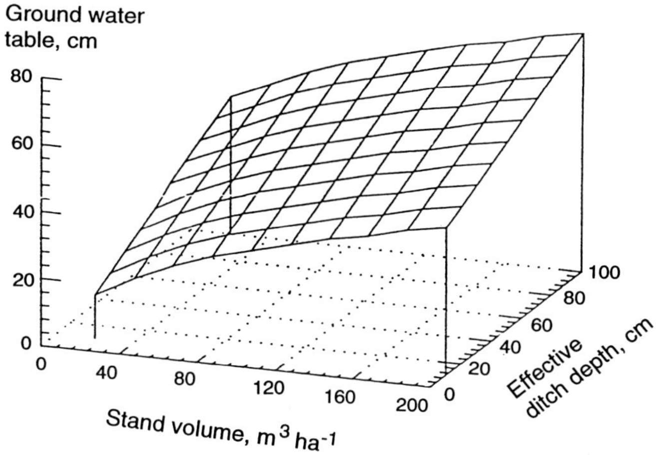

Fig. 3.32. The effect of the tree stand (*Pinus sylvestris*) volume and the effective ditch depth on the average drainage depth achieved (groundwater table), when ditch spacing is assumed to be 50 m. (Redrawn from Lukin 1988)

conclusion that the water regime of heavily stocked drainage areas, to a large extent, is almost independent of the technical drainage effects of the ditch network and that the tree stand itself maintains a favourable water regime. A decrease in the timber volume (by cutting) will result in the need to improve the "physical" drainage of the area.

3.5.2.9 *Effects of Improvement Drainage*

There is little information available about the effects of improvement drainage (see also Sect. 4.2) on water discharge from peatlands. The effects are, however, probably smaller than those of first ditching because the original drainage (performed more than 20 years earlier) will have lowered the runoff threshold and there is an artificial net of channels – even if it is deteriorated. Usually, the tree stand growing on a drained area is closed, causing the snow to melt slower than on a virgin mire. On the basis of ditch spacing experiments, it has been assumed that neither making the ditches shallower nor ditch cleaning affects the high runoff in summer. Complementary ditching or narrowing the ditch spacing, on the other hand, could possibly increase the runoff peaks caused by summer rains (Ahti 1987).

3.5.3 Effects of Cuttings and Forest Fertilization on Peatland Hydrology

3.5.3.1 Cuttings in Virgin Mires

Silvicultural operations affect the quantity of water infiltrating into the soil, the distance between the groundwater table and the peat surface, evapotranspiration, and thus indirectly, runoff.

In virgin mires, the quantity of harvestable timber – considered from the Fennoscandinavian silvicultural point of view – is usually so small that a possible selective cutting would have no effect on the hydrology of the mire concerned. However, on mineral soils in a state of paludification and in paludified *Vaccinium myrtillus* spruce forests, regeneration cutting may revive the paludification process if the regeneration area is not also drained.

Study results concerning the hydrological effects of tree harvesting on mires in North America are not consistent. In Minnesota, black spruce (*Picea mariana*) strip-cutting and clear-cutting on undrained mires did not change the annual amount of water leaving the peatland (Verry 1980, 1986). Strip-cutting did not change the water table elevation in the harvested strips compared to the forested strips either. Clear-cutting an entire undrained peatland did not change the average water table elevation, but it increased the range of water table fluctuation. However, in Ontario tree harvesting on black spruce mires has been observed to cause a significant rise in the groundwater level, both in undrained and drained areas (Berry and Jeglum 1991a).

3.5.3.2 Cuttings in Drained Peatlands

Even in drained peatlands the groundwater level is rather near the ground suface. Thus, the effect of tree stands on the water regime of the peat has often been studied by measuring the distance to the groundwater table in various forest environments (Thurmann-Moe 1941b; Lukkala 1946; Heikurainen 1967a,b). In drained peatlands the effects of timber harvesting may become accentuated due to the presence of the ditch network and an unsaturated soil profile above the groundwater level (Verry 1988). The following discussion of the hydrological effects of tree cutting refers mainly to drained peatlands.

It has been established that during the growing season **interception loss** in percent of gross precipitation is of considerable magnitude in mature tree stands growing in drained peatland areas (Päivänen 1966):

Picea abies stands	23–35%
Pinus sylvestris stands	20–25%
Betula pubescens stands	about 20%

Tree cutting clearly increases the portion of precipitation reaching the ground. In a Scots pine stand (*Pinus sylvestris*) on drained peatland where the volume of the original tree crop averaged 90 m^3 ha^{-1}, the portion of precipitation reaching the ground increased as follows (Heikurainen and Päivänen 1970):

Percentage of original stand volume removed in thinning	Change in throughfall (%)
20	+ 7
40	+ 8
60	+ 12
100	+ 20

During the first growing season after cutting, there was also some interception loss due to the fresh slash.

The thickness and the water equivalent of snow also increase with heavier cutting (Päivänen 1973a). Snow accumulates particularly in openings that have been cut in the forest. In regions where the snow cover is thick, and where consequently the proportion of meltwater in the total annual precipitation is high, the increase in runoff caused by cuttings is concentrated on the maximum runoff in spring (Verry 1986). In addition to a decrease in evapotranspiration and interception, the increase in runoff is affected by an increase in the quantity of snow accumulating in cut areas and by the fact that the snow starts melting earlier and melts faster in such areas (Päivänen 1973a; Heikurainen 1976).

Maximum runoff in spring is also affected by soil frost. If the meltwater from snow can penetrate into the soil, the maximum runoff in spring is smaller. However, if the soil is frozen, the surface runoff tends to increase. Soil freezing is affected by the time when the snow cover comes into existence and by its thickness. The increase in the snow cover, which results from thinning, insulates the soil and may lessen soil freezing (see Sect. 3.2.4).

The effects of thinning and clear-cutting on the level of the groundwater table in drained peatlands have been studied in hydrological experiments on tree stands, based on reference sample plots and calibration periods. These studies have all shown that even light thinnings (20% of the volume is removed) lead to a clear rise in the groundwater table and that the rise is greater the greater the percentage of the stand volume removed (Heikurainen and Päivänen 1970; Päivänen 1974a).

Data from two areas which were drained some 40 years before cutting exemplifies this (Päivänen 1980a,b, 1982a). The original tree stands (*Picea abies* and *Pinus sylvestris*) are described by the following average characteristics:

	Total stand		Commercial timber ($D_{1.3}$[a] > 7 cm)	
	Stem number	Volume ($m^3\,ha^{-1}$)	Stem number	Volume ($m^3\,ha^{-1}$)
Spruce-dominated	1650	165	1400	163
Pine-dominated	2900	114	1400	108

[a] Diameter at 1.3 m-height.

Particularly in the pine-dominated area there were many small trees, which is characteristic of forests on drained areas (see e.g. Hökkä and Laine 1988). The

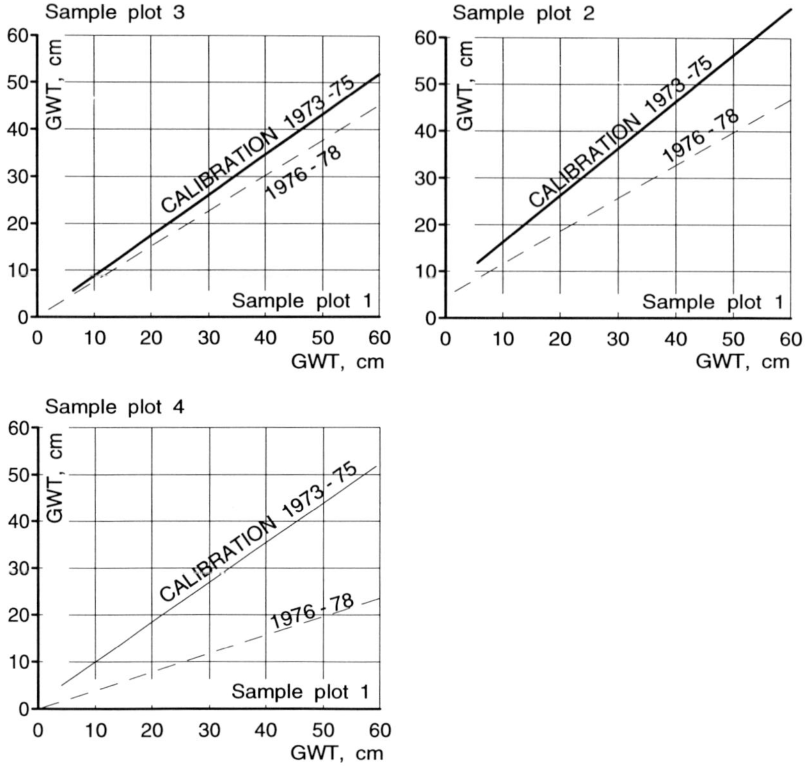

Fig. 3.33. The effect of thinning and clear-cutting on the distance to the groundwater table *(GWT)* in a Norway spruce (*Picea abies*)-dominant area. Thinning intensities 17, 26 and 100% in *sample plots 3*, *2* and *4*, respectively. (Redrawn from Päivänen 1982a)

effects of various degrees of thinning and clear-cutting on the level of the groundwater table are shown in Figs. 3.33 and 3.34.

In addition to silvicultural measures, the changes which take place in the groundwater level are also dependent on the physical properties of the peat profile. When the peat properties are taken into consideration the average increase in soil water storage is strongly related to the volume of the growing stock removed in the cuttings (Fig. 3.35). However, runoff tends to increase with a rise in the level of the groundwater table, which in turn tends to lead to a subsequent decrease in water storage.

In the throughfall study by Heikurainen and Päivänen (1970), the relative runoff from the sample plots was also determined. As the sample plots were not independent watersheds isolated from their surroundings, the runoff values cannot be presented in terms of absolute water quantities. In the Scots pine (*Pinus sylvestris*) stands concerned, in which the volume of the original tree stand was about 90 $m^3 ha^{-1}$, the relative runoff increased due to cutting as follows (Heikurainen and Päivänen 1970):

Percentage of orginal stand volume removed in thinning	Relative change in runoff (%)
20	+ 2.6
40	+ 24.5
60	+ 71.0
100	+ 111.2

The strength of the hydrological effects obviously decreases with time after cutting. In clear-cut areas, the fast-developing ground vegetation and coppice as well as the new tree generation transpire more water than the vegetation of the bottom layer of a drained peatland area where the tree stand has recently been harvested. In the case of thinnings, on the other hand, the remaining tree stand gradually fills up the liberated growing space, which leads to increased interception and evapotranspiration.

Thinnings and the deterioration of ditches exemplify the importance of maintaining drained areas (see Sect. 4.2). The greatest negative effects to the hydrological conditions of drained areas are caused by radical cuttings in which great

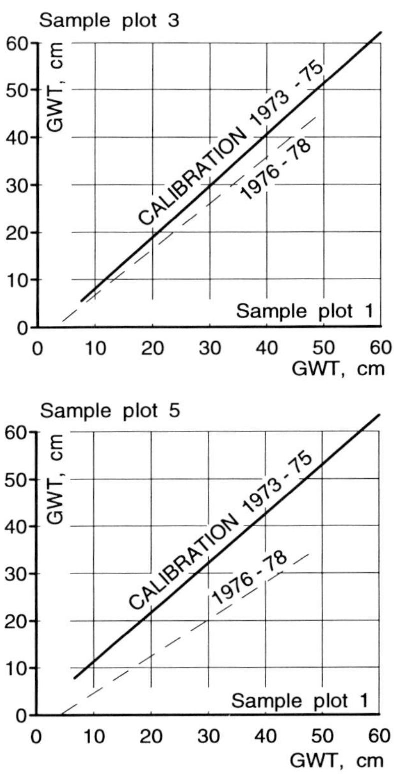

Fig. 3.34. The effect of thinning and clear-cutting on the distance to the groundwater table *(GWT)* in a Scots pine (*Pinus sylvestris*)-dominant area. Thinning intensities 17, 30 and 100% in *sample plots 3*, *4* and *5*, respectively. (Redrawn from Päivänen 1982a)

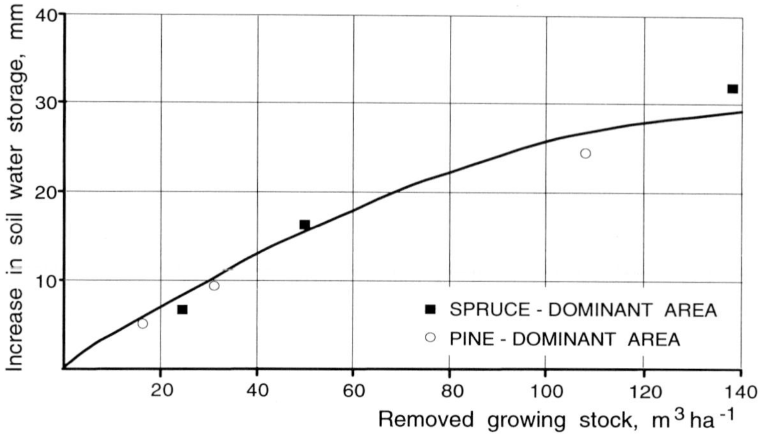

Fig. 3.35. The dependence of the average increase in soil water storage on the volume of the growing stock removed in the cuttings. (Redrawn from Päivänen 1982a)

timber volumes are removed. In such situations, if care is not taken to clean the ditches in a poor state, the paludification process may start again.

In Finland, the effects of cutting on runoff have been studied on a whole catchment scale in only one study. It was observed that 3 to 4 years after partial clear-cutting (56–58%) of a watershed area the annual runoff was 15–20% greater than if left uncut. This is equivalent to an increase in the annual runoff of 5–8 mm for each volume of 10 m^3 ha^{-1} removed (Seuna 1988). These values agree with indirect calculations obtained in Finland from the so-called small catchment area studies (Mustonen 1965). According to these, a decrease in the stand volume by 10 m^3 ha^{-1} resulted in an increase in annual runoff of 7.7 mm.

The increase in maximum discharge in spring caused by cutting has been observed to be 29–37% (Seuna 1988). Apart from an increase in the rate of snowmelt, maximum runoff is also increased by the ditches bordering truck and logging roads in the forest. Cuttings have also increased low runoff in summer.

There has been no long-term monitoring of the changes in runoff after clear-cutting in Finland. On the basis of a number of studies made in other countries, however, it may be concluded that the increase in runoff caused by clear-cutting may last for 20–30 years, decreasing at a logarithmic rate (e.g. Hibbert 1967). Only when the new tree generation closes does the situation return to that prevailing before the cutting (Douglass 1983; Swank et al. 1988). However, it should be remembered that clear–cutting is mostly confined to a small part of the whole watershed at any one time. The regeneration phase is a natural part of the tree stand rotation. Thus, any severe negative hydrological effect of cutting on catchment runoff is reduced by the greater share of stands belonging to the more advanced development classes (Grip 1987; Lundin 1990).

3.5.3.3 Fertilization in Drained Peatlands

It has been found that forest fertilization may have an indirect effect on throughfall and groundwater level in a Scots pine (*Pinus sylvestris*) stand growing on drained peatland (Heikurainen and Päivänen 1970; Päivänen 1972). The increased interception loss due to fertilization reached its maximum during the third growing season after treatment. The first response of trees to fertilization is an increase in the needle length. The increased crown closure and the number of needles attached to the shoots also increased the interception loss. The effect of fertilization was also to lower the groundwater level (Päivänen 1972).

Thus, forest fertilization may increase biological drainage by increasing tree stand growth and thereby interception loss. This leads to a lowering of the groundwater level. Transpiration by tree stand and ground vegetation is probably also intensified. These hydrological changes caused by forest fertilization are favourable from the viewpoint of tree growth under the considerably wet conditions that still prevail in artificially drained peatlands.

4 Forest Drainage

4.1 Ditching

4.1.1 Drainage Requirements

4.1.1.1 Optimum Drainage

The relationship between soil water and the growth of higher plants was partly dealt with in Section 3.5.1. Water is an important growth factor for trees as well as for other plants. Both excess water and drought are limiting factors for tree growth. In the case of virgin mires, excess water in the substrate checks root growth and microbial activity, and may lead to unfavourable biochemical phenomena. One of the most important goals of draining is therefore to adjust the water content of the soil to a level which ensures sufficient aeration (Päivänen 1973b). As discussed by Heikurainen (1964) some scientists speak about removing excess water (Multamäki 1936), some about lowering the groundwater table (Lukkala 1929), and others claim that the purpose of drainage is to increase the movement of water (Huikari 1959b). The purpose of draining may also be to increase the depth of rooting and so reduce the risk of windthrow. This is especially the case in Great Britain (Pyatt 1990a).

A theoretical **optimum drainage** can be implemented for different kinds of peat soils (see e.g. Heikurainen et al. 1964; Heikurainen 1967c). However, the optimum is also dependent on the tree species, the stage of development of the tree stand, and so on. The range of water contents corresponding to favourable tree growth is probably limited, on the one hand, by the water content at the matric suction value of pF 3 and, on the other hand, by the minimum air space (10–15 vol%) required. Studies conducted in Finland seem to indicate that there is no risk of overdrainage in the case of forest drainage under the prevailing humid conditions (Päivänen 1973b).

4.1.1.2 Drainage Depth

The effect of drainage is usually described by the term **drainage depth** or **drainage norm**, which is the mean distance between the soil surface and the water table during the period of June through September (Meshechok 1969). The drainage norm is dependent on such factors as climate, topography and hydraulic conductivity of peat and vegetation. It can be affected or altered by using various ditch depths and spacings (Meshechok 1969; Braekke 1983c).

However, some scientists are willing to separate the achieved and desired drainage depths. **Drainage depth** reflects only the combined effects of natural conditions and ditch drainage, but **drainage norm** takes into consideration the target of forest drainage, which is a "specified depth below soil surface above which the water table should not remain for continuous time intervals longer than the (acceptable) flood duration limit" (Toth and Gillard 1988). Heikurainen (1967c), explained the **optimum drainage depth** with the help of water retention characteristics of different kinds of peats. Vompersky and Eizerman (1990) approached the same problem based on physical soil properties using the term forest **drainage standard**.

Based on different sources (Heikurainen 1964, 1967c; Braekke 1983; Toth and Gillard 1988; Lieffers and Macdonald 1990; Vompersky and Eizerman 1990; Berry and Jeglum 1991a; Hillman 1992), we can make a very generalized synthesis: At the time of ditching, the objective should be to lower the groundwater table to a depth of over 35 cm below the soil surface in ombrotrophic, undecomposed peats and to over 55 cm in minerotrophic, well-decomposed peats. This should be achieved by using a suitable combination of ditch depth and spacing.

4.1.1.3 Ditch Depth

In forest amelioration, open ditches are almost exclusively used for drainage. However, subsurface drains have been used in some experiments (see Sect. 3.5.2).

The appropriate depth of drainage ditches varies between different mire sites. Deeper ditches are necessary in wet, deeper peats because of the expected peat subsidence. The following are recommended ditch depths for tree-covered boreal peatlands (Päivänen 1990):

shallow peat (< 30 cm) soils	70 cm
deep peat (> 30 cm) soils	80 cm
wet, deep peat soils	90 cm

Water furrows, which are plowed on open mires mainly for soil preparation for planting, and only secondarily for local drainage, are shallower than 60 cm (Thompson 1979).

4.1.1.4 Ditch Spacing

Ditch spacing may be described as either geometric or effective (Fig. 4.1). The term **geometric ditch spacing** refers to the perpendicular distance between drainage ditches, whereas the term **effective ditch spacing** refers to the distance between ditches in the direction of the slope of the ground (Meshechok 1960). In

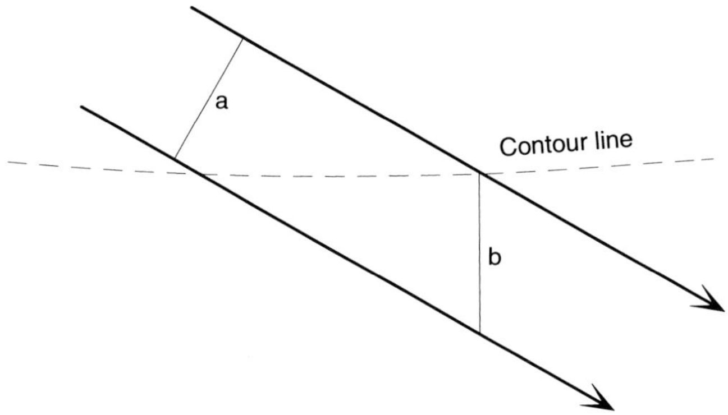

Fig. 4.1. Geometric *(a)* and effective *(b)* ditch spacing

the following discussion, the term ditch spacing refers to geometric ditch spacing.

Ditch spacing may be either **biologically optimal** if it yields the best tree growth, or **economically optimal** if it gives the highest rate of interest on the invested ditching costs.

Several factors affect biologically optimal ditch spacing. Relatively narrow ditch spacing is required on nutrient-poor sites, open peatlands, areas with a plane or a very gently sloping surface, deep peat soils with woody residues absent, and shallow peat soils on impermeable mineral soil (e.g. clay). In contrast, good results may be obtained with relatively wide ditch spacing on naturally tree-covered site types, areas with a good slope, deep woody peat soils and shallow peat soils on permeable mineral soil (e.g. sand) (Päivänen and Wells 1977).

The economically optimal ditch spacing is always wider than the corresponding biological one. It is affected both by the digging costs and by the discounted returns due to drainage (Keltikangas 1971). On originally tree-covered peatlands, the volume of the tree stand per area unit is reduced by the cleared ditch lines. This reduction in tree volume, and thus also in the post-drainage tree growth, increases as the ditch spacing decreases (Seppälä 1972). Although the biologically optimal ditch spacing is narrow for nutrient-poor sites, the economically optimal ditch spacing for such sites is wide.

Operational recommendations for ditch spacing in boreal conditions are as follows (Päivänen 1990, see also Rosen 1989):

shallow peat soils	50 m
forested peatlands	40 m
sparsely treed peatlands with a shallow gradient	30 m

The use of spacings over 80 m has given good results only in the continental parts of Europe (Konstantinov et al. 1983; Sabo 1983; Vompersky 1991). For a more theoretical basis using hydraulic conductivity, precipitation and drainage norm, see Toth and Gillard (1988).

Water furrows are commonly used in connection with afforestation of open peatlands under maritime conditions. The furrows are made with rotary ditchers or plows to a depth of 25–50 cm and with 3–5 m spacing. However, furrowing is considered to be more a measure of site preparation for afforestation than drainage. Consequently, it is recommended in the United Kingdom that, irrespective of the ground preparation method, deep peats have to be drained on sloping ground (> 3°) with 70–m spacing and on flat ground (< 3°) with 20–30 m spacing (Pyatt 1990a). In Norway, on the Atlantic coast a dense system of surface furrows draining into 60–80 cm deep ditches at 10–12 m intervals has proved to give the best afforestation result (Braekke 1978).

4.1.2 Planning

4.1.2.1 General Procedure

Forest drainage planning is discussed in this chapter mainly according to Finnish practice, and partly based on the description given by Päivänen and Wells (1977). Although the Finnish practice has the longest history and widest application in the world, the reader may also like to refer to the following papers and guides for other regions and situations: Sweden (Gustafsson 1980), Norway (Braekke 1978), Russia (Vompersky 1987), Estonia (Sepp 1990), United Kingdom (Pyatt 1990a), and Canada (Haavisto and Päivänen 1987; Hillman 1987; Toth and Gillard 1988; Guide sur le Drainage . . . 1989; Rosen 1989).

The construction of a forest drainage system includes a comprehensive planning stage, involving both office and field research. The area to be drained is selected on the basis of such factors as site type (Sect. 3.4.1) and tree stand (Sect. 6.2). Technical factors include the morphological and physiognomic features of the area. A site may be biologically suitable for drainage but, if it is surrounded by mineral soil ridges or bedrock outcrops, the expense of constructing ditches through such material may be too high. During the preliminary survey, access (roads), underground cables, culverts, and so on must also be considered and the boundaries of the area determined. Land ownership must be ascertained, especially in situations where the ditching operation will involve several landowners.

Aerial photographs and contour maps (e.g. 1:10 000) are essential for drainage planning. A few levelling lines should be run to obtain a general picture of the gradient conditions. All site type patterns observed in the area must be registered on the map. Site type, stand characteristics, and recommendations for silvicultural treatments should be listed.

The essential components of a proper ditch plan are: ditch depth and spacing (Sect. 4.1.1), ditch location and criteria, the inclusion of a sedimentation abatement strategy (Sect. 4.2.3), and the marking of the ditch lines in the field.

4.1.2.2 Main Ditch

The first step in designing the ditch system involves the location of the **main ditch** or **collector ditch** which forms the primary route by which water leaves the drainage area. This ditch usually follows the main gradient in the lowest point of the area where the peat is deepest and subsequent peat subsidence is greatest. The main ditches, with due regard for the settling or screening of sediments, empty into outlet channels such as creeks, rivers or other bodies of water that are at a lower elevation. For factors relating to water protection, see Section 4.2.3.3.

The dimensions of a main ditch mainly dependent on the required capacity to lead water away from the site. The water in the main ditch should not rise above a level that would prevent the drainage ditches from discharging. In bigger drainage areas the dimensions of the main ditch are calculated on the basis of the expected average peak flow (MHq) which is affected by the size, shape, and average gradient of the runoff area, the percentage of open water inside the area, the precipitation, the water equivalent of the snow cover immediately before melting, and the types of land utilization (see Päivänen and Wells 1977).

4.1.2.3 Drainage Ditches

The actual forest drainage is achieved by the **cross-drainage principle**; that is, the bulk of the ditches in the system drain water by collecting it as it moves laterally through the surface and sub-surface soil layers (Fig. 4.2).

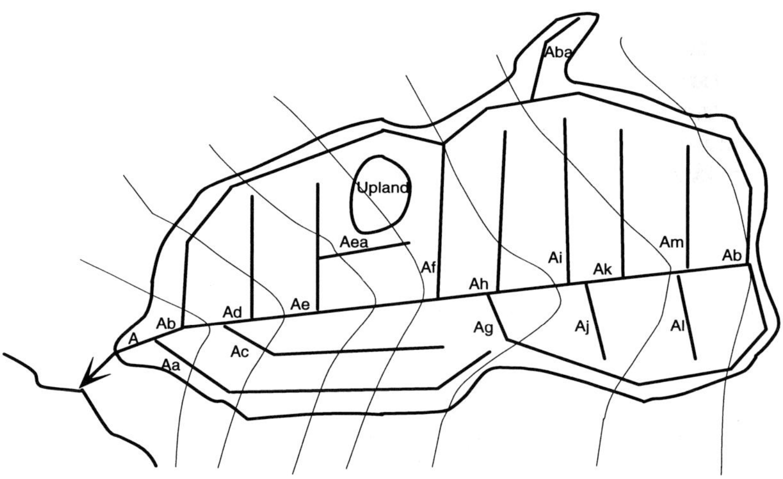

Water - logged mineral soil

Primary ditches - A - Main ditch
Aa, Ab - Trap ditches
Drainage ditches - Ac, Ad, etc. - Lateral or contour ditches

Fig. 4.2. A schematic ditch system showing the location of different ditch types used in forest drainage. (Redrawn from Haavisto and Päivänen 1987)

Several types of **secondary** or **drainage ditches** are used in any ditch system for influencing groundwater level. These are instrumental in gathering water from the area, lowering the groundwater level, and preventing surface and subsurface water from entering the drainage area.

Lateral ditches, sometimes referred to as **contour ditches** (Päivänen and Wells 1977) or **side ditches** (Rosen 1989), collect the drainage water and lead it into the main ditch. They are generally established applying the cross-drainage principle in a systematic network of parallel drains, and are oriented more or less crosswise to the gradient, crossing thc contours at a small angle. Thus, the effective spacing approaches the geometric spacing (Fig. 4.1). The slope of lateral ditches should be around 0.2–0.4%. If lateral ditches are placed on both sides of the main ditch, a herringbone system is formed. If the land slopes in different directions, this has to be taken into consideration in the planning of the drainage ditch system.

Trap ditches (cut-off ditches), sometimes referred to as **surround ditches** (Haavisto and Päivänen 1987), collect the water from surrounding land that would otherwise flow through the area. These ditches are constructed on the border between the peatland and the surrounding lands. Trap ditches do not always flow with the gradient, and so it is sometimes necessary to connect them with the lateral ditches to encourage water movement to the main ditch. **Fork ditches**, **spur ditches** and **protection ditches** are used in special situations (see Päivänen and Wells 1977).

4.1.2.4 Marking the Ditch Lines

The following procedure has been applied in most of the forest drainage undertakings in Finland. The network of ditches first designed on the map is applied to the field. It may be necessary to diverge from this proposed layout, especially if rocky terrain is encountered. The peat depth should be measured and the type of underlying mineral soil determined. If the proposed ditch should encounter bedrock, the location of the ditch must be changed. Whenever possible bedrock blasting should be avoided because of high costs and work safety aspects. Any changes made during pegging of the ditch lines must be indicated on the map.

The ditch lines are marked with stakes, and the beginning and end points of each ditch are marked with a pole with the code of the ditch on it. In Finland, the main ditches are usually marked with capital letters and the drainage ditches are named by attaching a small letter to the letter of the main ditch in order of their occurrence when moving upstream. For example, the first ditch joining the main ditch A is Aa, the second Ab, and so on. The length of each ditch should be measured and indicated on the end pole. A sufficient degree of accuracy may also be obtained by measuring the ditch length on the map.

If the main ditch runs through a mineral soil ridge, the line must be levelled and the peat depth and the type of underlying mineral soil must be determined along the ditch line. This information is required when the profile of the main ditch is finally determined.

Under conditions where bushes prevent the line marker from having an extended line of sight to mark straight lines, a special method – based on bearing compass and tying in to the next-door ditch line – has been developed (see Rosen 1989).

4.1.2.5 Documents

Upon completion of the field assessments, a ditch system map, profiles of the main ditches, site types, stand characteristics and calculation of costs must be documented for each drainage area. Furthermore, the digging contractor requires a map showing the exact layout of the ditch system, with the names of the ditches as marked on poles in the field, and the digging depth at each pole along the lines of the main ditches. The lateral ditches may be dug using a standard ditch depth.

4.1.3 Working Methods

4.1.3.1 Ditching Techniques

Up to the early 1950s, forest ditches were dug **manually** in Finland. Mechanized forest drainage commenced in 1953 with the introduction of **forest ditch plows** (Huikari 1958). The best conditions for plowing were found in the large homogeneous peatland areas of northern and eastern Finland. Forest ditch plows weighed 5 to 6 t and required heavy crawler tractors (14–18 t) with a surface pressure not exceeding 30 kPa to haul them across the terrain. In making the ditch, the plow was pulled by means of a winch, hauling resistance reaching 100 to 120 kN force. A standard-sized lateral ditch was plowed into firm peat at a

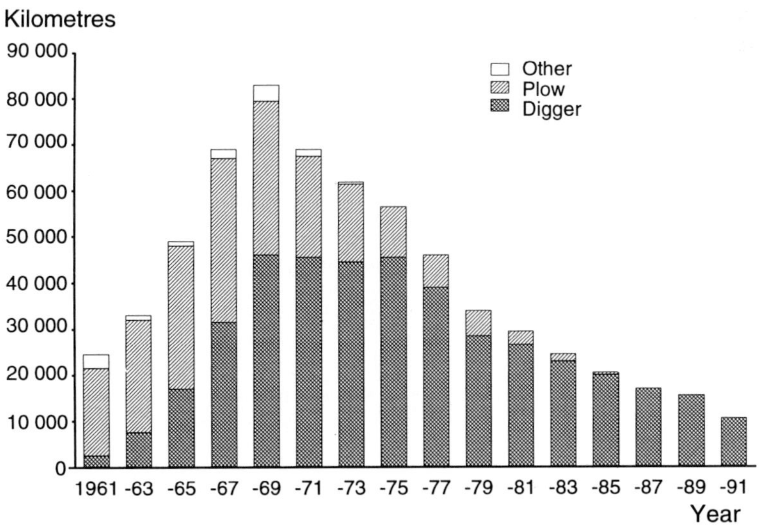

Fig. 4.3. Forest drainage in 1961–1991 by digging methods in Finland. (Data from Aarne 1993)

rate of 500 to 600 m h^{-1} (Niskanen 1980; Päivänen 1984a). The length of ditches made by plowing has, however, decreased since 1969, now being almost negligible (Fig. 4.3).

During the last two decades, the most commonly used machine has been the **tractor digger**. This is a unit consisting of a prime mover and hydraulic excavating equipment (Fig. 4.4). The prime mover is usually a light-track tractor with a 45 to 75 kW engine. The forest ditches are dug longitudinally with a bucket scoop, and the cross-section measurements conform to the standard forest ditch dimensions (Fig. 4.5, see also Sect. 3.5.2.5). The work output of a tractor digger is 70 to 90 m of forest ditch per hour (Niskanen 1980). During recent years, heavier **excavators** with slightly higher work output have also been used for forest drainage.

Fig. 4.4. A tractor digger is a unit consisting of a prime mover (a light-track tractor) and hydraulic excavating equipment with a special bucket scoop. (Photo: Courtesy of Lännen Tehtaat Ltd.)

4.1.3.2 Contracting and Costs

Most drainage has taken place on private land in Finland (Fig. 4.6). Forest amelioration activity on private land is entrusted to the District Forestry Boards, organizations promoting private forestry. Drainage operations on private land have been performed co-operatively and up to 60% of the total drainage costs have been subsidized by the state; partly as grants and partly in the form of low-interest loans.

Fig. 4.5. A tractor digger digging a ditch into stony soil. Observe the bucket scoop conforming to the standard forest ditch dimensions, southern Finland. (Photo: J. Päivänen)

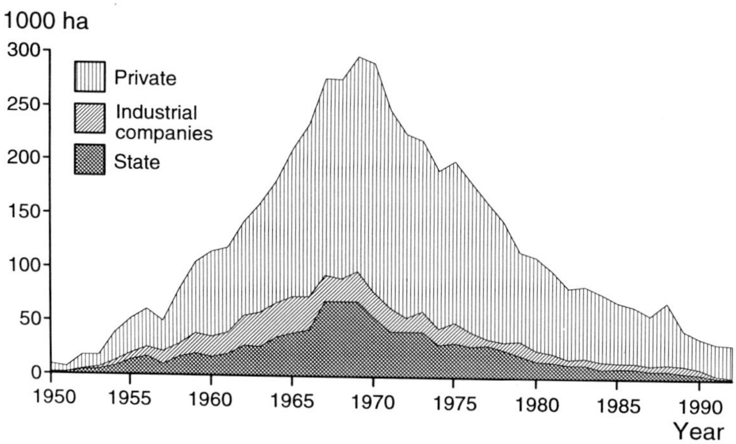

Fig. 4.6. Forest drainage in 1950-1993 by forest owner groups in Finland. (Data from Aarne 1993)

Ditch digging has been performed by private contractors. The prices, based on work studies and cost analysis, have been agreed upon annually between the association of contractors and employers. The payment for the digging work proper has been based on a piece-work system. When a standard-sized drainage ditch ($0.76\ m^3/m$) is dug, the payment is calculated per metre.

The following monetary calculations are in 1984 US dollars. Figures representing Finland are from the Forestry Board Statistics of Tapio, the organization of Finnish private forestry, and those for Canada from the Wally Creek drainage project (Rosen 1987):

	Finland	Canada
	US\$ ha^{-1}	
Planning	15	31
Line clearing	–	53
Digging	92	124
Total	107	208

Fig. 4.7. A tractor digger digging a ditch in the Wally Creek drainage project, Ontario, Canada. (Photo: Courtesy of Ilkka Koivisto)

These figures are, however, hardly comparable. The Finnish costs are average values for some 60 000 ha planned and drained with advanced techniques. The Canadian costs are derived from an operational scale (375 ha) experiment (Fig. 4.7). The ditch line clearing is the duty of the landowner in Finland and not included in the total drainage costs, whereas in Wally Creek there are dense black spruce stands, and the ditch lines had to be cleared separately. As mentioned by Rosen (1987), the costs of the digging work proper represented about a 30% increase over the fee charged in Finland.

4.2 Maintenance of Drainage Areas

4.2.1 Concepts and Approaches

The term **improvement ditching** is becoming established to mean all actions taken in order to improve the impaired water regime in previously drained peatlands. These actions are usually **ditch cleaning** or/and **complementary ditching**. The aim of improvement ditching is either to improve tree growth or to maintain an achieved level of growth which is superior to that of tree stands growing on peaty sites in the natural state.

In this chapter, improvement ditching is discussed with reference to the Finnish experience. There is now considerable emphasis on improving the water regime in peatlands drained for forestry some decades ago. In the last section (Sect. 4.2.6), improvement ditching in other countries is discussed.

4.2.2 Need for Improvement Ditching

4.2.2.1 Ditch Deterioration

The factors leading to the deterioration of ditches have mainly been studied on ditches dug by hand. Nevertheless, most of the results can probably also be applied to ditches made by machines.

Ditch deterioration may be caused by: (1) subsidence of the peat layer; (2) collapse of ditch walls; (3) blocking up through vegetation; (4) slash and other obstacles; and (5) erosion and silting.

Subsidence and compression of the peat layer take place immediately after the drainage work has been completed. Heikurainen (1957) believes that subsidence of the peat layer is the most important factor causing ditches to become shallow. This is fastest immediately after digging and thereafter the decrease in ditch depth declines, the subsidence depending mainly on peat oxidation (Laine and Laiho 1992). Subsidence is greater the deeper the peat layer and the ditches (Lukkala 1948, 1949).

Overgrowing of ditches by coppice is most common on originally sparsely treed fens rich in nutrients. Mosses, sedges and grasses growing in the ditches also impair the flow of water. Slash and temporary culverts hold back the drainage waters and form sites for colonization by hygrophilic vegetation, leading to a further reduction in the flow. Harvesting machinery may cause the collapse of the ditch walls in places where it crosses the ditches. It has been estimated that need for ditch cleaning after cutting is increased by about one fifth (Rantonen and Päivänen 1989).

Collapse of the ditch walls may occur if the walls are too steep, or as a consequence of soil frost or erosion. Where the ditches have been made by plowing, the peat ridges on both sides of the ditch may be too near the ditch so that peat from the ridges falls into the ditch, perhaps aided by soil frost and rains. On shallow peats the need to keep the ditches in repair depends on the type of soil, the gradient and the quantity of water. Erosion of the ditch bottom leads to silting downstreams.

The digging technique used probably also influences the preservation of the dimensions of ditches. It is generally believed that ditches made by plowing become shallower at a faster rate than those made by diggers or by hand (e.g. Keltikangas et al. 1986). The poor condition of plowed ditches is probably also partly due to the desire to have long, straight ditches at the expense of the technical drainage properties of the peat. In plowed ditches, thresholds which dam up the water occur more frequently than in ditches made by diggers.

4.2.2.2 Quality of Forest Ditches

An inventory of areas drained for forestry in Finland included an assessment of the condition of the forest ditches and the need for ditch cleaning (Keltikangas et al. 1986). The average quality class distribution of all the forest ditches (almost 1.2 million km) throughout the country was as follows:

Ditch condition	%
1. Good	23.6
2. Rather good	28.5
3. Fair	28.4
4. Rather poor	14.1
5. Poor	5.4
	100.0

If there was a need for immediate ditch cleaning, the ditch was classified as class 5. Ditches belonging to classes 4 and 5 should be cleaned within at least 10 years of the time of assessment.

On the basis of the quality class distribution and the total length of ditches in existence by the year 1978, the need for ditch cleaning was estimated as follows (Keltikangas et al. 1986):

Region	Immediately (class 5)	Within 10 years (class 4 + 5)
	km	
Southern Finland	53 800	170 400
Northern Finland	9 000	58 500
The whole country	62 800	229 000

According to the results of the study, the need for cleaning was relatively greatest (nearly 30% of the ditches) in areas that had been drained 20–30 years earlier, that is, during the period 1951–1960. With respect to the quantity of work, however, the need for cleaning was greatest in areas drained in the 1960s, when plowing was the dominant method of digging. The greatest need for cleaning occurred in originally wet mire sites (treeless mires, sedge hardwood-spruce swamps, sedge pine swamps).

According to the 7th National Forest Inventory, the need for ditch cleaning in southern Finland was 213 800 ha (Kuusela and Salminen 1983, p. 73). At the average ditch density used in the 1960s (220 m/ha), this would have corresponded to about 46 000 km of ditches. In the national programme for wood production, the annual target for improvement ditching is set at 150 000 ha (Metsä 2000 –ohjelman . . . 1992).

4.2.2.3 Estimation of the Need for Improvement Drainage in the Field

With respect to its water regime a mire that has been drained for forestry is an unstable ecosystem, and its state of drainage requires continuous follow-up (Heikurainen 1980b). Apart from gradient, soil quality, ditching techniques and time, the state of drainage is affected by the tree stand and its development. In heavily forested drained peatlands, water uptake by the stand is usually capable of compensating for a reduction in the effect of drainage, so that the ditches do not have to be improved by cleaning until the final felling (Laine 1986). However, if the tree stand existing at the time of draining is sparse, as in the case of most pine mires, deterioration of the ditches leads to a decrease in growth and a need for ditch cleaning before the final felling.

The need for drainage improvement is indicated by poor tree growth and poor ditch condition. It is necessary to determine whether tree growth has been poor from the very beginning or whether it has declined some time after drainage. If ditch spacing exceeds 50 m, complementary ditching should be considered as an alternative to ditch cleaning. In areas that have been drained by plowing, the ridges of peat along the ditches should be cut. Plowed ditches are usually shallow, and so cleaning should be done irrespective of the condition of the ditches (see Ahti et al. 1988). How to determine the need for improvement draining is schematically presented in Fig. 4.8.

Poor tree growth or a decline in growth may also be due to factors which are only partly connected with the draining effect of the ditch systems: for example, nutrient deficiency and pests of different kinds.

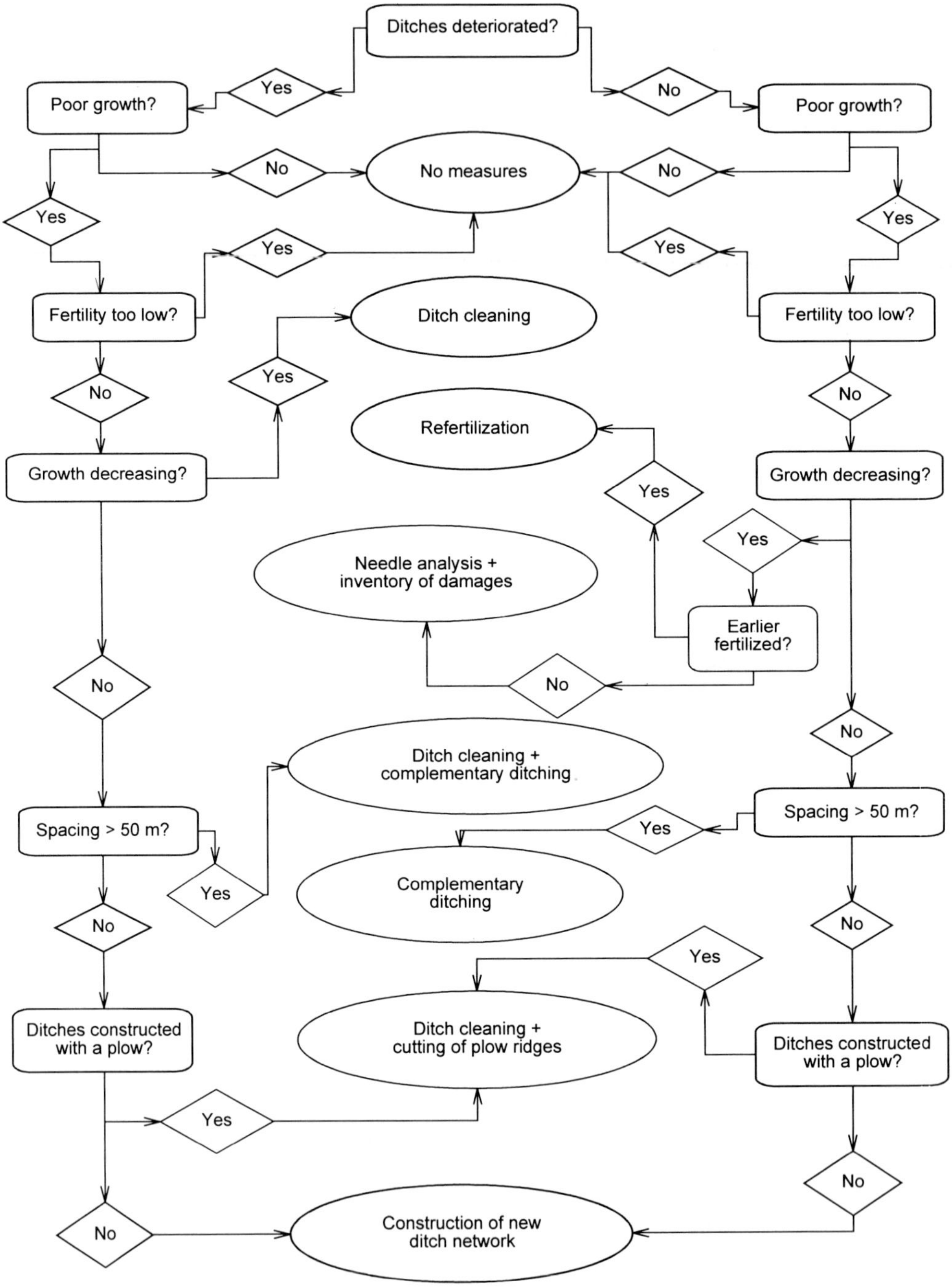

Fig. 4.8. Determination of the need for ditch cleaning and complementary ditching. (Redrawn from Ahti et al. 1988)

A few studies have shown that improvement ditching lowers the groundwater level (e.g. Päivänen and Ahti 1988), but there is very little published information concerning the response of the tree stand to this measure. In an early study in Scotland, Stewart (1958) did not find any growth increase of Sitka spruce *(Picea sitchensis*). This was probably because of the short period of time (4 years) since ditch deepening. The response of Scots pine (*Pinus sylvestris*) to improvement ditching has also been shown to be relatively small during the first 5-year period after treatment (Olkinuora 1990).

4.2.3 Planning of Improvement Ditching

4.2.3.1 Choice of Object

In considering the need for improvement ditching in an area drained for forestry, it must first be ascertained whether the poor stand development is due to insufficient draining or low nutrient status of the site. The latter cannot be improved by increasing the effect of drainage. It is also pointless to carry out improvement ditching in areas where the tree stand has suffered from repeated growth disturbances.

In areas drained for forestry, it is important that the different work phases required be carried out in the right order. Of particular importance are proper allocation of the planning of ditching work and timber harvesting, as well as the execution of these operations. The need for improvement ditching is usually indicated in the forest management plan. The different phases of work on forest drainage areas should be carried out in the following order:

1. Forest improvement planning
   - forest road planning
   - drainage planning
   - other forest improvement planning
2. Road construction
   - forest roads and landings
   - logging roads and culverts
3. Planning of the harvesting work, sales of the timber, harvesting
4. Other action preceding ditching (young stand treatment, soil preparation, etc.)
5. Ditching
6. Fertilization

In planning timber harvesting on drained peatland areas, the first thing to do is to find out if there are existing transport tracks and to establish the possible need for new ones. Forest road construction must be initiated early enough to avoid delay of the other operations.

In connection with the planning of improvement ditching, it is also possible to plan other activities such as young stand treatment, fertilization, soil preparation and regeneration. Young stand treatment should be carried out prior to digging operations. In the choice of regeneration method, the option of using natural regeneration should be considered as this method is usually successful on peatlands (Sect. 6.3.4.2). Which areas require soil preparation, as well as which methods are to be used, should be decided on early in the planning stage, so that ditch digging and soil preparation activities can be synchronized.

The planning of improvement ditching is followed by planning the harvesting, sales of the timber to be felled and the harvesting operation. The ditching operation proper can only be initiated when these other measures have been planned. In improvement ditching the work to be done by landowners, the forestry board, the local forestry society, the timber purchasing and harvesting organizations, the transporter and the ditcher must be coordinated.

4.2.3.2 Planning in the Field

A map at 1:10 000 is good for use in the planning of improvement ditching. The map that was used when the area concerned was originally drained can also be taken into consideration. Likewise, an aerial map might be of help.

The most widely used form of improvement ditching is **ditch cleaning**. In ditch cleaning the same ditch depths should be achieved as in new drainage. **Complementary ditching** should be carried out if it is considered that cleaning of the old ditches will not give the desired drainage effect. If the ditch spacing is less than 50 m, ditch cleaning should not be replaced by complementary ditching (lengthwise splitting of the strips between ditches). Splitting the strips instead of ditch cleaning is not suitable for the following reasons: (1) digging new ditches is more expensive than cleaning old ones; (2) ditches reduce the growing space and so it is not possible to split the strips repeatedly; and (3) ditches make timber harvesting difficult.

Complementary ditching between the original ditches may be considered when ditch spacing is 50 m or more (Ahti et al. 1988; Päivänen and Ahti 1988). Complementary ditching also involves the planning of cut-off ditches if lacking, or their reallocation if they have been dug in wrong places, as well as planning spur ditches into mire slips. Old cut-off ditches may become unnecessary as a result of peat subsidence. In such cases, planning new cut-off ditches in places with a deeper peat layer may be needed.

Stagnant water may occur in the area because of depressions caused by uneven subsidence or because of poor planning of the ditch network. "By-pass" ditches to lead the runoff around the obstacle may be required in order to drain such areas.

The most important ditch in a drainage system is the main ditch. If there is no main ditch, it has to be planned in conjunction with improvement ditching. If the old main ditch is deep enough and its gradient adequate, and if vegetation has stabilized its walls, it may be wiser to refrain from cleaning in order to avoid erosion losses.

For marking of ditch lines in the field orange-yellow fibre flags are recommended. Both ends of a ditch to be cleaned are marked with poles with the name of the ditch concerned, but the ditch does not have to be staked out. Culverts and

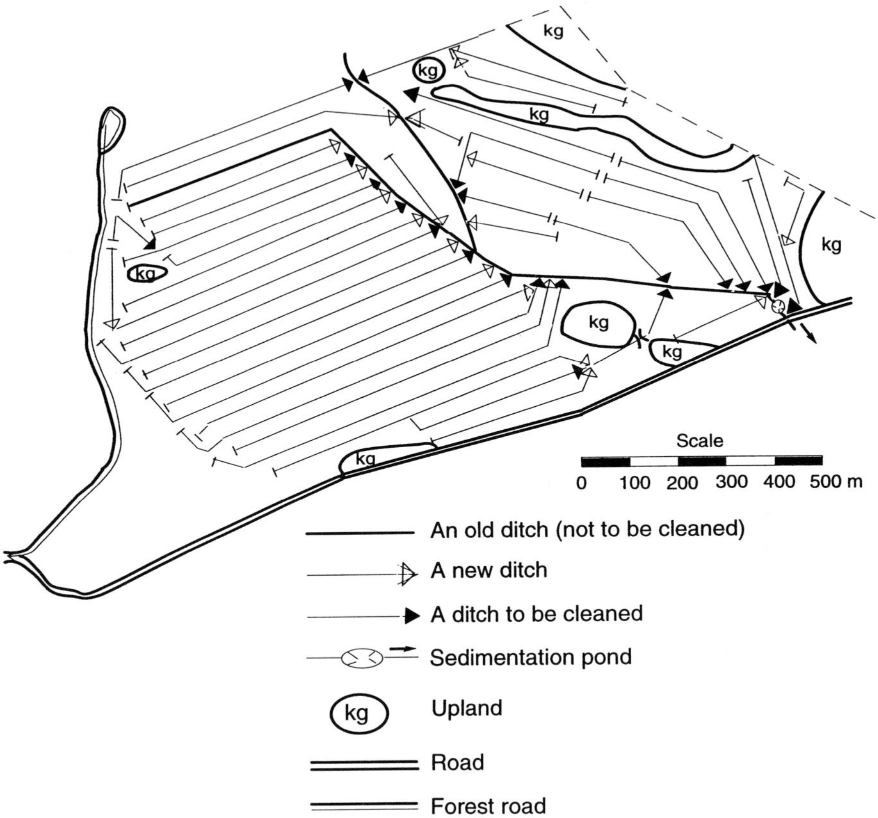

Fig. 4.9. Map symbols in connection with the planning of improvement ditching

other special constructions should, however, be staked out and marked with poles. Complementary ditch lines should be staked out in the field and marked by stakes, poles and fibre strips just as new ditches. A recommendation for standardizing the map markings (Fig. 4.9) for improvement ditching in Finland has been made recently (Päivänen 1990).

4.2.3.3 Factors in Relation to Water Protection

Erosion of ditches is the most frequent cause of the loading of organic matter and nutrients to watercourses after ditching. It is best avoided by using small gradients. Possible erosion should be taken into consideration, particularly in planning the cleaning of the main ditches. Intensive maintenance of main ditches can be a major source of mobile sediment and should be discouraged (Joensuu 1990; Pyatt 1990a).

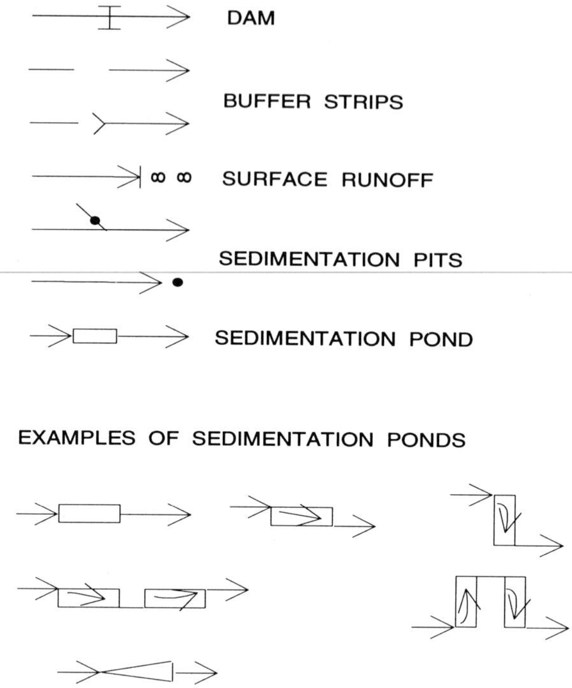

Fig. 4.10. Symbols for different kinds of water protection methods

The water from a drainage area should be led into a natural watercourse by a single main ditch. Preferably, the ditch should not open directly into the watercourse, but should terminate a short distance before so that the discharging water from the main ditch passes through a filter of peat and vegetation before it reaches the watercourse. Alternatively, the water from a drainage area can be led to a mire area unsuitable for drainage, an abandoned ditch net, a boulder field area or a sloping spruce mire narrowing. Ditches can also be broken off if the gradient is steep enough. When necessary, the water can be caught again by fork ditches and led further away. The measures taken with regard to water protection should be indicated using uniform symbols on a map of the area concerned (Fig. 4.10).

The load of suspended solid particles can be limited by choosing the right digging time and work sequence. The load caused during digging can be diminished considerably by digging the upper and lateral ditches first, and only then the lower and main ditches. The main channels leading into the watercourses are made last. The final opening of the main ditches can also be postponed to the driest period of the following summer.

Silt holes are depressions made in the lateral ditches to allow silt from the discharge to settle before the water is led into the collecting ditch. The silt holes are marked on the map and in the list of ditches.

Fig. 4.11. A sedimentation pond, southeastern Finland. (Photo: J. Päivänen)

The term **sedimentation pond** refers to an artificial pond (or ponds) made in the outlet of the main ditch (Fig. 4.11). The water flow in such ponds is slowed down so that solid particles are able to sink to the bottom of the pond (Joensuu 1990). The shape and structure of a sedimentation pond has to be decided upon according to circumstances. The most important thing in this connection is to slow down the running speed of the water to a sufficient degree (Joensuu 1992). Accessibility to the ponds for maintenance must be taken into consideration when determining their location. On the basis of experience in peat production areas, sedimentation ponds in mires drained for forestry should be of the following dimensions:

- area of the pond 2–5 $m^2\ ha^{-1}$
- volume of the pond 2–5 $m^3\ ha^{-1}$
- volume for silting 0.5–2 $m^3\ ha^{-1}$

Along creeks and rivers the ponds must be located upstream so that they are not reached by floods. One of the edges of the pond should be gently sloped so that animals that fall into it can get out by themselves.

4.2.4 Timber Harvesting Considerations

The conditions for tree harvesting in peatland forests are often unfavourable (Eeronheimo 1991). The bearing capacity of the peat soil is poor during the frostless period of the year. Other factors affecting tree harvesting include the ditch network, low timber volume to be harvested, uneven density of stands, superficial root system of the trees as well as the unavailability of forest roads due to high construction costs, resulting in long extraction distances.

Timber harvesting in the improvement ditching area should be initiated as soon as possible after the drainage plan has been completed. The planner of the timber harvesting operation must have a map of the drainage plan, data concerning the compartments in the area and a list of the land owners involved. It should be attempted to centralize timber harvesting in the case where timber sales are divided among several buyers.

Planning the strip road (extraction track) network is the most important stage in planning timber harvesting. In the case of areas to be treated with improvement ditching, the basic strip road system is formed by the lines along which cleaning and complementary ditches are going to be made because the trees have to be removed from the lines anyway. In planning strip road systems, the following principles should be followed (Päivänen 1990):

1. Strip roads should not be made on ditches if there is no intention to clean them after harvesting or if the ditches concerned are situated in places unsuitable for strip roads due to excess wetness, for example.
2. Strip roads should not be located along the very edges of the ditches. Otherwise the ditches may fill up with logging residue and the most productive part of the growing area will be wasted.
3. Later thinnings should be taken into consideration. At that time there may be more ditches in the area hampering the harvesting.

For technical reasons cutting is best performed during the snowless season, but the forest haulage should be carried out in winter, when the soil is frozen. In peatland forests, cutting is mostly done manually with a chainsaw; the productivity in cutting shortwood is about $7\,m^3\,day^{-1}$ in thinnings. With normal distances between extraction tracks (20–30 m), medium-sized forwarders with a carrying capacity of 10 to 12 t and equipped with a long-shafted (10 m) loading device can be used (Eeronheimo 1991). Finally, it must be emphasized again that timber harvesting in drained areas must always be carried out prior to improvement ditching.

4.2.5 Working Methods

4.2.5.1 Ditch Cleaning Techniques

Several methods and machines for mechanized ditch cleaning have been tried. Such machines should fulfill the following requirements (see Vuollekoski 1983):

(1) the ability to move easily in old forest drainage areas; (2) a good working speed; (3) the ability to perform ditch cleaning both on peat and mineral soils; (4) the ability to dig ditches of any size; (5) suitability for digging new ditches as well; and (6) the ability to move along roads.

So far, however, it has proved impossible to develop a ditch cleaning plow or rotary ditcher suitable for the heavily varying Finnish conditions (Niskanen 1980). Three main types of units based on the digging principle have been tested and used for improvement ditching: standard tractor diggers, special diggers for forest ditch cleaning, and ordinary excavators. **Tractor diggers** are units consisting of a tractor and hydraulic excavating equipment (Sect. 4.1.3.1). **Special ditch cleaning diggers** are equipped with a hydraulic power transmission and a special wide track structure which results in a low surface pressure of 21–24 kPa (Vuollekoski 1983; Lauhanen 1992). Nowadays the same scoop is used both for ditch cleaning and the digging of new ditches. Under light working conditions the special ditch cleaning diggers (Lännen S 9 and PATU M 100) reach an output level of about 190 m per operation hour (Vuollekoski 1983; Lauhanen 1992).

Ordinary excavators move on their own base but need more working space in the forest than diggers. The productivity of excavators both in ditch cleaning (220 m per operation hour) and in the digging of new ditches (155 m per operation hour) is higher than that of diggers (82–187 and 45–132 m per operation hour, respectively) under average conditions (Salo H 1988). The excavators have to be transported to the site with trailers.

4.2.5.2 Contracting and Costs

In Finland, most ditch digging – both new ditches and ditch cleaning – is performed by private contractors (see Sect. 4.1.3.2).

For ditch cleaning in Finland, the payment per lineal metre has been around 75% of that paid for new ditch digging.

4.2.6 Volume of Work Carried Out

The annual volumes of improvement ditching for some recent years in **Finland** are as follows (Aarne 1993):

Year	Ditch cleaning		Complementary ditching
		km	
1980	5 700		3 600
1982	5 500		3 000
1984	8 300		2 600
1986	8 800		2 790
1988		10 700[a)]	
1990		19 200	
1992		20 500	

[a)] Ditch cleaning and complementary ditching combined (= improvement ditching) since 1988.

Converting the output figures expressed in lineal kilometres into aerial units involves some uncertainty, because the average ditch density is not known. However, the total area of improvement ditching in 1990 was reported to be 72 000 ha. This indicates a ditch density of about 270 m ha^{-1} (Aarne 1993).

A nationwide field inventory of areas drained for forestry during the period 1930–1978 showed that only 6% of the ditches concerned had been cleaned (Keltikangas et al. 1986). The study also showed that there were surprisingly many areas in which the originally sparse ditch network had been complemented later by additional ditching.

In other countries as well, much attention has recently been given to the need for the maintenance of forest drainage systems. From **Norway**, it has been reported that complementary ditching and ditch cleaning are to be given priority in the near future (e.g. Braekke 1990b). In **Estonia**, it has been estimated that 13% of the drained area requires improvement ditching, mainly ditch cleaning (Valk 1988). In **Russia**, forest drainage efficiency has been found to be low, mainly due to insufficient intensity of the original drainage and lack of ditch maintenance (Konstantinov et al. 1983; Vompersky 1991). Wide ditch spacing is recommended, but only if ditch cleaning is to be considered (Sabo 1983). However, most of the problems involved in the proper mechanization of ditch cleaning have not yet been solved. In **Great Britain**, it has been recommended that, on deep peat soils, ditch depth should be maintained but on mineral soils maintenance should be limited to the clearance of gross blockages, because of the danger of erosion (Pyatt 1990a).

5 Forest Fertilization

5.1 Estimating Fertilization Requirements

5.1.1 Site Type Classification

There are several approaches to estimating the fertilization needs of peatland forests. Instructions for practical forestry are usually based on site type classification and on the results of fertilization experiments. The fertilization recommendations given for the different peatland site types are, however, average values. Therefore it is often useful to employ supplementary methods, for example soil and foliar analysis, in the **nutritional diagnosis** of peatland forests. These methods are especially valuable when determining the need for fertilization at the stage of **refertilization**. Because drainage and first fertilization have changed the composition of ground vegetation (see e.g. Päivänen and Seppälä 1968; Päivänen 1970; Reinikainen et al. 1984; Vasander et al. 1993), site classification based on ground vegetation alone is inadequate.

As already mentioned in Section 3.3, there is a rather clear correlation between peatland site type and the concentrations and amounts of certain nutrients, such as nitrogen and calcium, in peat. Numerous fertilization experiments have indicated which nutrients and amounts are needed on different site types to increase the growth of peatland forests.

In the following discussion, most of the presented results are from Finnish fertilization experiments with Scots pine (*Pinus sylvestris*). These results do not differ markedly from those received in the experiments with Scots pine and other tree species in Sweden (Malmström 1952; Tamm 1965; Holmen 1969, 1985; Möller 1978; Carlsson and Möller 1985; Almqvist 1990), Norway (Haveraaen 1967; Meshechok 1968, 1971; Braekke 1977a,b, 1979, 1983b), the British Isles (Edwards 1959; Jack 1965; Dickson 1971; Adams and Dickson 1973; Dickson and Savill 1974), Russia (Ipatiev et al. 1984) and Canada (Wells 1991; Sundström 1992b,c; Sundström and Jeglum 1992).

Huikari (1952) has developed a classification system for Finnish conditions which is based on ground vegetation and tree stand. According to this classification (see also Sect. 5.2.1), there is usually no need for fertilization on peatlands dominated by eutrophic mosses, grasses and herb vegetation (Table 5.1). On less fertile peatlands phosphorus and potassium deficiency is very common, and PK fertilization is therefore recommended. On nutrient-poor, ombrotrophic sites

Table 5.1. Peatland site type classification (adapted from Huikari 1952) and recommendations (see e.g. Paavilainen 1979b) for practical forest fertilization

Peatland site type	Need for fertilization treatment
Old drainage areas	
Eutrophic peatlands	No need for fertilization
Herb-rich and *Vaccinium myrtillus* types	(N) PK
Vaccinium vitis-idaea and *Ledum – Empetrum* types	NPK
Cladonia type	Not suitable for fertilization
Hardwood-spruce and pine mires	
Eutrophic and herb-rich hardwood-spruce mires	No need for fertilization
Vaccinium myrtillus spruce mires, eutrophic, herb-rich and tall sedge pine mires	PK
Vaccinium vitis-idaea and cottongrass spruce mires, *Carex globularis* pine mires	[N] PK
Cottongrass and dwarf shrub pine bogs	NPK
Sphagnum fuscum pine bogs	Not suitable for fertilization
Treeless mires (planted usually with Scots pine)	
Ombrotrophic bogs	To be fertilized (NPK) only exceptionally
Minerotrophic mires	PK

Other characteristics: The need for nitrogen is inreased in peatlands with a shallow peat layer or *fuscum*-rich hummocks, and for PK in flarky areas; ( ) at refertilization; [ ] at refertilization, recommendable also at primary fertilization.

dominated by *Eriophorum vaginatum*, dwarf shrubs and often *Sphagnum fuscum*, nitrogen is also needed to increase tree growth. The need for nitrogen often also increases with increasing drainage age and decreasing peat depth.

5.1.2 Symptoms of Nutrient Deficiency

The symptoms of **nutrient deficiencies** are fairly easily recognizable in trees growing on peat soils. They are, however, not always specific and can indicate a lack of more than one nutrient (Reinikainen 1967, 1968; Kolari 1983; Veijalainen et al. 1984; see also Binns et al. 1980).

A deficiency of nitrogen and phosphorus is generally indicated by poor tree vitality, stunted needle growth and a yellowish needle colour.

The specific symptoms of phosphorus deficiency in conifers are a slenderness and bending of the stems and twigs and a small needle biomass. Severe phosphorus deficiency leads to the **necrosis** of the previous year's needles, beginning from the tip, which can, in the worst case, lead to the death of the shoot leader

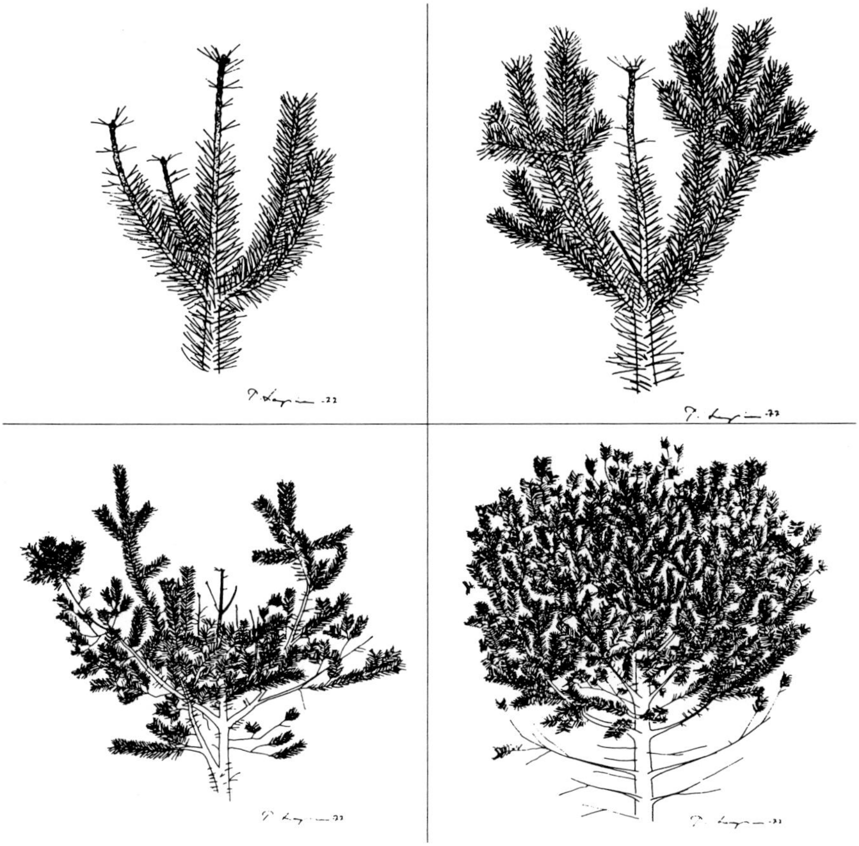

Fig. 5.1. Growth disorder caused by boron deficiency in a Scots pine (*Pinus sylvestris*) stand. (Courtesy of Päivi Lempinen)

and the new buds. Leaf necrosis is a phosphorus deficiency symptom shown by birch and it may lead to the death and shedding of the leaves.

Potassium deficiency in conifers is indicated by the yellow colour of the needle tips or whole needles (**chlorosis**). The symptom is the same in birch, where the chlorosis begins from the tips or edges of the leaves. The chlorosis may be followed by gradual necrosis of needles or leaves.

Chlorosis is also a symptom of magnesium deficiency but it differs from that caused by potassium deficiency in that the needle colour is somewhat reddish and the border between the yellow tip and the healthy part of the needle is sharp.

Under Finnish conditions, where there is very little boron in the rainfall, it is also useful to give boron. The deficiency of this micronutrient easily causes **growth disturbances** (Huikari 1977a; Braekke 1979, 1983a,b; Kolari 1983; Veijalainen et al. 1984). These growth disturbances include: bud damage, abnormal development of needles, loss of apical dominance and bushy growth of originally well-growing trees (Fig. 5.1). See Section 6.4.1 for more details.

5.1.3 Foliar Analysis

The chemical analysis of needles and leaves is a widely used method in the nutritional diagnosis of tree stands. The concentration of different nutrients in the foliage is mainly dependent on the nutrient status of the soil, and to a minor extent on other soil factors. In peatland forests, foliar analysis has been used with good results, especially concerning the nutrient requirements of Scots pine (*Pinus sylvestris*) (Paarlahti et al. 1971; Veijalainen 1977, 1984; Reinikainen 1988b; Vasander and Lindholm 1991).

The diagnostic value of foliar analysis is highest when the nutrient deficiency is modest. When growth increases after fertilization, the concentration of one or more nutrients may decrease (dilution effect, see Fig. 5.2). For example, phosphorus fertilization decreases the nitrogen concentration of needles on peatlands which are poor in nitrogen and phosphorus.

The foliar nutrient contents indicating a need for fertilization in peatland forests vary with tree species, developmental stage of the stand, and geographic location. These factors have so far not been studied in detail and the recommendations for practical forestry are based on mean values from a number of investigations (e.g. Puustjärvi 1962a,b,c; Paarlahti et al. 1971; Veijalainen 1977; Reinikainen 1988). The values presented in Table 5.2 are used when estimating the nitrogen, phosphorus, potassium and boron status of Scots pine (*Pinus sylvestris*) stands growing on peatland (Metsänterveys . . . 1988, p. 156–158, Vasander and Lindholm 1991).

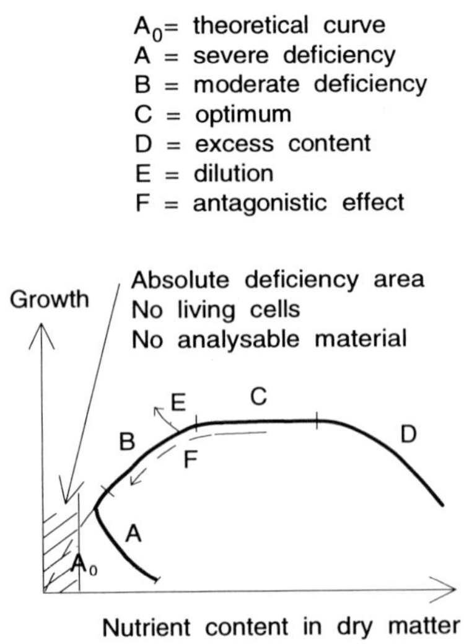

Fig. 5.2. Relationship between the nutrient concentration of the needles and the growth of trees. (Redrawn from Veijalainen 1977)

Table 5.2. Nutrient status of Scots pine (*Pinus sylvestris*) on peatlands estimated by foliar analysis

Nutrient status	Nitrogen ($g\ kg^{-1}$)	Phosphorus ($g\ kg^{-1}$)	Potassium ($g\ kg^{-1}$)	Boron ($mg\ kg^{-1}$)
Low	< 11.9	< 1.39	< 3.49	< 4.9
Fair	12.0–12.9	1.40–1.69	3.50–4.49	5.0–7.4
Good	> 13.0	> 1.70	> 4.50	> 7.5

Source: Metsänterveys . . . (1988), see also Vasander and Lindholm (1991).

For other nutrients, it has been possible to present only critical values with regard to the nutrient status of the trees. Critically low values are: magnesium $< 1.0\ g\ kg^{-1}$, copper $< 2.5\ mg\ kg^{-1}$, zinc $< 40\ mg\ kg^{-1}$. The needle manganese concentration is critically high (indicating poor aeration) when it exceeds 600 $mg\ kg^{-1}$.

Nitrogen deficiency occurs in birch (*Betula pendula*) when foliar concentrations are less than 18–24 $g\ kg^{-1}$ (Tamm 1964). These values are considerably higher than those for Scots pine. The foliar analysis values indicating the need for fertilization are almost the same for Norway spruce (*Picea abies*) as for Scots pine, namely: nitrogen 12.0 $g\ kg^{-1}$, phosphorus 1.5 $g\ kg^{-1}$, and potassium 4.0 $g\ kg^{-1}$ (Paavilainen 1975a).

The ratio between the different nutrient concentrations is also important when estimating the need for fertilization. For example, the needle N/P concentration ratio best explained the disorders and mortality of Scots pine transplants after nitrogen fertilization on sites poor in phosphorus (Paavilainen 1976; Kaunisto and Paavilainen 1977). Disorders occurred when the N/P ratio was higher than 11–13. The optimal ratios for nitrogen, phosphorus and potassium presented by Puustjärvi (1962a,b,c) are: N/K 3.5, N/P 11.

For foliar analysis it is worthwhile to determine not only the nutrient concentrations but also the needle dry mass and nutrient content (e.g. amount of nutrients in 100 needles). By using these parameters with the **vector analysis approach** (Timmer and Stone 1978; Timmer and Morrow 1984; Timmer and Ray 1988; Wells 1991; Mugasha et al. 1993), the diagnosis of nutrient deficiency, luxury consumption etc. is less constrained by variability in genetic, climatic, and stand conditions, and sampling and analysis protocols than when comparing the critical nutrient levels or nutrient ratios alone. The vector analysis technique has additional advantages: (1) it allows simultaneous comparison of multiple nutrients in a single nomogram; (2) it identifies nutrient interactions; and (3) it is effective where simultaneous excess and deficiency of several nutrients may be involved (Timmer 1985).

In addition to the nutrient concentrations, the concentrations of amino acids and enzymes of needles can also be used to estimate the nutrient status and fertilization needs of peatland forests. For example, enzyme activity in the needles indicates the nitrogen status (Pietiläinen and Lähdesmäki 1986, 1988; Pietilä et al. 1989; Lähdesmäki and Pietiläinen 1990) and the putrescine concentration is correlated to potassium deficiency in Scots pine (Sarjala and Kaunisto 1993).

Because of the large variability in foliar nutrient concentrations due to the age of the trees, location of the stand and the time of the year, foliar sampling procedures should be standardized. There is also a great variability between trees of the same species and age in one single stand. To obtain reliable results, samples should be taken from 10–15 dominant trees in the stand from the upper part of the crown. The best time for sampling for conifers is during the dormant period, and for decidious trees it is 2–3 weeks before the leaves turn yellow.

The use of foliar analysis alone is not recommended in practice. The quality of the site and the tree stand should also be studied carefully and the possible nutrient deficiency symptoms determined.

5.1.4 Soil Analysis

Using soil analysis to determine the need for fertilization involves difficulties in sampling, choosing the methods of analysis and interpreting the results.

As regards sampling depth, the surface peat layer where the nutrient-uptaking roots are located (usually 0–20 cm) is most important. The peat has to be carefully separated from the living mosses and litter layer, and in the case of drained peatlands also from raw humus that has accumulated on the original peat surface (Kaunisto and Paavilainen 1988). Each preferably volumetric peat sample should consist of at least five to ten subsamples to reduce the effects of local variation in the bulk densities and nutrient concentrations of peat soil. The best sampling time is the dormant period of the year, when the uptake and microbial activity in the peat is low.

The total amounts of nutrients in peat indicate the potential capacity of the site for forest production, and also give some guidance about the need for fertilizer, particularly nitrogen (Kaunisto 1984a; Kaunisto and Paavilainen 1988). According to Holmen (1969), the nitrogen concentration in the 0–20 cm peat layer needed for satisfactory tree growth is 1.3–1.5%, and the corresponding amounts of phosphorus and potassium are 150–200 kg ha^{-1} and 100–125 kg ha^{-1}, respectively (see also Kaunisto 1982).

The extractable amounts of nutrients are also useful in the nutritional diagnosis of peatland sites, although interpretating the results may be difficult. For example, Holmen (1964) and Kaunisto and Paavilainen (1988) observed that there was a negative correlation between the amount (kg ha^{-1}) of soluble phosphorus in peat (0–20 cm) and tree growth, whereas the correlation between the total amount of phosphorus and tree growth was just the opposite.

To increase the diagnostic value of soil analysis, the methods for the handling and analysis of peat samples need further development. For example, little is known about how the living roots, which are included in the soil sample, affect the results, or whether some other extractant could give a better correlation with tree growth than those used at present.

5.1.5 Other Types of Analysis

The use of chemical analysis of plant tissues other than foliage, for example buds, bark or roots, for the determination of fertilizer requirements has been studied. According to one study on peatlands, the concentrations of phosphorus and potassium in the surface soil and the phosphorus concentration and N/P ratio of Scots pine (*Pinus sylvestris*) roots correlated positively with tree growth (Paavilainen 1968). It seems, however, that the chemical analysis of roots as well as other tree parts is useful only to a limited extent, perhaps for the sampling of big trees where needle sampling is difficult.

It may be possible to estimate the nutrient status of peatland forests using remote sensing in association with field measurements (Pietiläinen and Kuivamäki 1982). The vitality of forests and the need for fertilizer improving the vitality of trees, especially on peatlands, could be determined simultaneously. One of the advantages of remote sensing is that it may indicate nutritional disturbances before they are visible.

5.2 Factors Affecting the Tree Growth Response to Fertilization

5.2.1 Peatland Site Type

The growth response of trees to fertilization is dependent on the nutrient status of the site. On nutrient-poor oligotrophic peatlands in Finland, nitrogen, phosphorus and potassium are needed. The effect of NPK fertilizer at normal doses (100 kg N ha^{-1}, 40–50 kg P ha^{-1} and 80 kg K ha^{-1}) on the growth of Scots pine usually lasts for less than 10 years (e.g. Huikari 1961, 1973; Paavilainen 1972, 1974, 1977b, 1984b; Karsisto 1974b; Ipatiev and Paavilainen 1975; Kaunisto and Paavilainen 1977; Kaunisto 1977, 1982).

Refertilization of Scots pine on nutrient-poor peatland sites usually involves NPK. In young Scots pine stands studied by Kaunisto (1982, 1987a), the application of NPK compared with PK increased the growth of trees when the total peat nitrogen content was less than 1.2–1.3%. Refertilization with nitrogen alone, on the other hand, caused needle damage and increased the mortality of seedlings when the N/P ratio in needles was higher than 11–13 (Paavilainen 1976; Kaunisto and Paavilainen 1977). At some sites, however, the use of nitrogen alone may lead to a good growth response (Moilanen 1993).

On peatlands rich in nitrogen but low in phosphorus and potassium (mainly *Vaccinium myrtillus* spruce mires and eutrophic, herb-rich and tall sedge pine mires), PK fertilization generally has a strong positive effect on tree growth (Fig. 5.3), lasting 15–20 years (Huikari 1973; Paavilainen 1978, 1979a, 1984b; Moilanen 1984, 1993; Penttilä 1984).

At refertilization both phosphorus and potassium are usually applied to nitrogen-rich mires, even though the effect of potassium often ends earlier than that of phosphorus (Paavilainen 1979a; Moilanen 1984; Kaunisto and Tukeva 1984;

Fig. 5.3. Unfertilized *(left)* and with PK-fertilized *(right)* 20-year-old Scots pine (*Pinus sylvestris*) stand on drained peatland. (Photo: Courtesy of Kemira Ltd.)

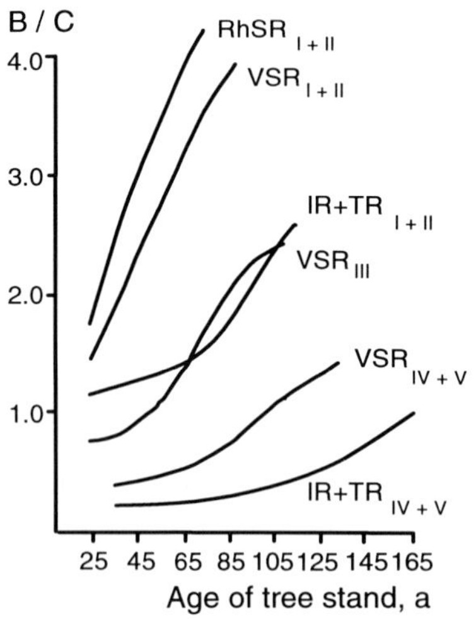

Fig. 5.4. Benefit/cost ratio (B/C) of PK fertilization in different peatland sites (for site types, see Table 3.17) by climatic regions (I-V) and stand age. Interest rate = 3%. (Redrawn from Keltikangas and Seppälä 1973)

Kaunisto 1989; Zalitis 1990a). A shortage of potassium is most likely to occur on sedge and/or herb-rich sites of fenlike origin with a thick peat layer. In some cases, potassium application alone is all that is needed (Paavilainen 1979a; Kaunisto 1989).

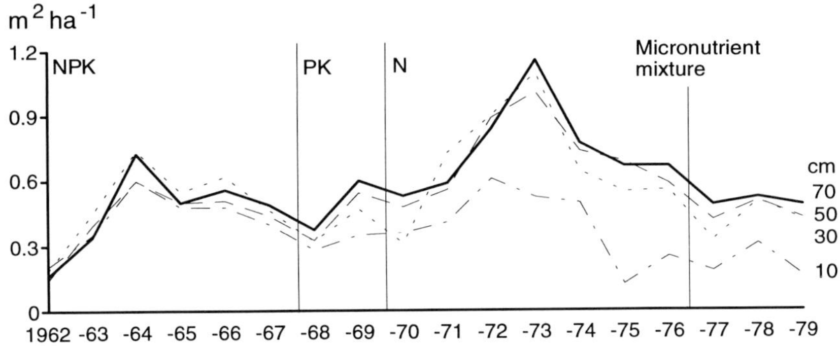

Fig. 5.5. Annual basal area growth of a Scots pine (*Pinus sylvestris*) stand on sample plots with different fertilizations and water regulation depths (10-70 cm). (Redrawn from Silfverberg 1984)

It is important to give boron together with phosphorus and potassium on nitrogen-rich mires in order to maintain a balance between nutrients (Veijalainen 1975, 1977; Paavilainen 1979a; Veijalainen et al. 1984).

It is recommended to fertilize only those sites and stands that can be profitably drained. The returns of PK fertilization distinctly decrease with decreasing fertility of the site (Fig. 5.4).

5.2.2 Drainage

In general, the fertilization of peatland forests results in a good response only on properly drained areas. On undrained or poorly ditched sites, the roots suffer from excess water and are unable to effectively take up fertilizer nutrients. For example, the increase in growth after repeated applications of fertilizer was much greater on sample plots where the ditch water was regulated down to a depth of 30–70 cm from the ground surface than on plots where the ditch water was regulated at 10 cm (Fig. 5.5). The best growth response is usually achieved when fertilizer is given some years after drainage, at least in nutrient-poor sites (Karsisto 1974b; Paavilainen 1986).

5.2.3 Climate

In the northern part of the boreal zone, the low temperatures limit tree growth directly and indirectly by slowing down the mineralization, and hence availability, of the originally bound nutrients. Thus, Starr and Westman (1978) found that the amount of available nutrients, especially nitrogen and phosphorus, in northern Finland was smaller than in a corresponding peatland site type in southern Finland.

According to some studies on Scots pine (*Pinus sylvestris*), the slow mineralization of nutrients, particularly nitrogen, can be compensated for by adding NPK fertilizer (Seppälä and Westman 1976; Kaunisto 1985a; Moilanen 1993). In

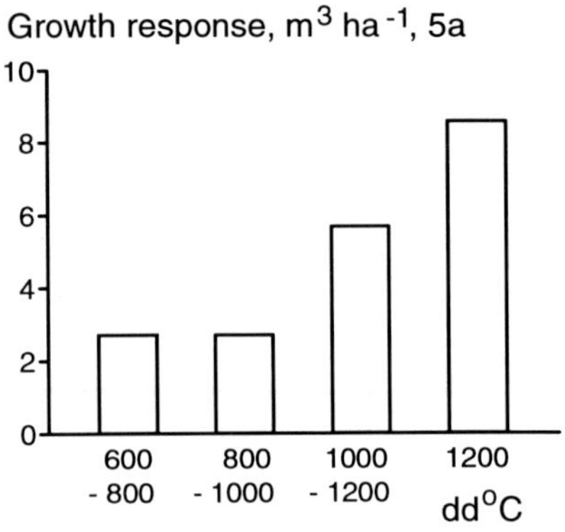

Fig. 5.6. Dependence of growth response (in 5 years) to NPK fertilization and temperature sum for drained Scots pine (*Pinus sylvestris*) stands. (Redrawn from Möller 1978)

contrast, other studies have found the effect of NPK fertilizer to be small (Heikurainen and Laine 1976; Heikurainen et al. 1983), or to be no greater than the effect of PK alone (Paavilainen and Simpanen 1976; Paavilainen 1978; Penttilä 1984). The profitability of fertilization was more favourable in southern Finland where growth conditions are more favourable than in northern Finland (Keltikangas and Seppälä 1973).

In general, the effect of NPK fertilization on the growth of Scots pine is better, the higher the effective temperature sum (Fig. 5.6). This dependence is, however, caused not only by the differences in temperature but also by differences in the initial stand volume and increment at the time of fertilization.

5.2.4 Tree Stand

5.2.4.1 Tree Species

The growth responses of different tree species to fertilizer is primarily dependent on their nutrient requirements and the effectiveness of nutrient cycling. As already mentioned in Section 3.3, birch (*Betula pubescens*), for example, takes and uses considerably more nutrients than Scots pine (*Pinus sylvestris*) to produce an equal amount of biomass. Norway spruce (*Picea abies*) does not require as much nutrients for biomass production as birch, but more than Scots pine.

Since birch leaves are shed every year, the internal cycling of nutrients in birch is weaker than in pine or spruce. This fact, and the high nutrient requirements, are the main reasons for the usually poor response of birch to fertilization, compared to that of conifers on peatlands (Oikarinen and Pyykkönen 1981; Moilanen 1985; Paarlahti and Paavilainen 1985). However, a positive response to fertilization has also been noted in some experiments (Valk 1980; Kaunisto 1987b; Ferm and Hytönen 1988; Lumme 1988; Paavilainen 1990a).

Scots pine (*Pinus sylvestris*) stands have been fertilized the most in Finland. The few results of the reactions of Norway spruce (*Picea abies*) to fertilizer so far indicate that it is of the same order or somewhat weaker than that in Scots pine stands (Paavilainen 1975a; Paarlahti and Paavilainen 1985).

In the British Isles, the biomass of young lodgepole pines (Pinus contorta) treated with phosphate fertilizer was about 30% more than that of Sitka spruce (*Picea sitchensis*), most of the difference being accounted for by needles and branches. Application of nitrogen fertilizer was essential to sustaining the growth of Sitka spruce (Carey et al. 1984).

5.2.4.2 Stand Age

The response to fertilizer is affected by the age (developmental stage) of the trees. After planting, the seedlings are able to take nutrients only from a very small soil volume and the competition with ground vegetation for nutrients is strong. At this stage, spot fertilization can be used on peatlands, but the mortality of the small seedlings is easily increased by the application of high levels of soluble nutrients, especially nitrogen (Heikurainen et al. 1966; Paavilainen 1970; Kaunisto 1971, 1982; Mannerkoski 1971; Seppälä 1971).

The nutritional requirements of a stand increase with age and are highest in young and advanced thinning stands and in mature stands. The relationship between the use of fertilizers and the developmental stage of closed stands is, however, not well known. Some results show that young Scots pine seedling stands probably do not need as much PK fertilizer as older stands (Finér 1986) and that nitrogen in addition to PK is needed in fast-growing thinning stands of Norway spruce but not in mature stands near regeneration (Paavilainen 1975a).

It is most important to ensure a sufficient amount of nutrients for wood production. For example, a lack of potassium may cause severe damage and even kill fully grown trees (Kaunisto and Tukeva 1984; Kaunisto 1989). On the other hand, the application of high nitrogen doses increases damage and mortality of mature tree stands on fertile sites and should therefore be avoided (Paavilainen and Pietiläinen 1983; Paavilainen 1984b; Veijalainen et al. 1984; Moilanen 1993).

The developmental stage (stand age) also affects profitability of fertilization (see Fig. 5.3). Mature stands which are growing well are economically most suitable for fertilization (Keltikangas and Seppälä 1973; Hämäläinen and Laakkonen 1983; Hämäläinen 1984; Salminen 1984; Hämäläinen et al. 1985).

Afforestation involving the use of fertilizer may be profitable under favourable climatic conditions. For instance, in Ireland peatland afforestation with Sitka spruce (*Picea sitchensis*) and lodgepole pine (*Pinus contorta*) yielded 4 to 8.4% interest depending on site, peat type and crop management (Gallagher and Gillespie 1984).

5.2.4.3 Volume and Increment

The growth response to fertilization is usually correlated with the initial stand volume at the time of fertilization. Therefore, stand volume or mean basal area

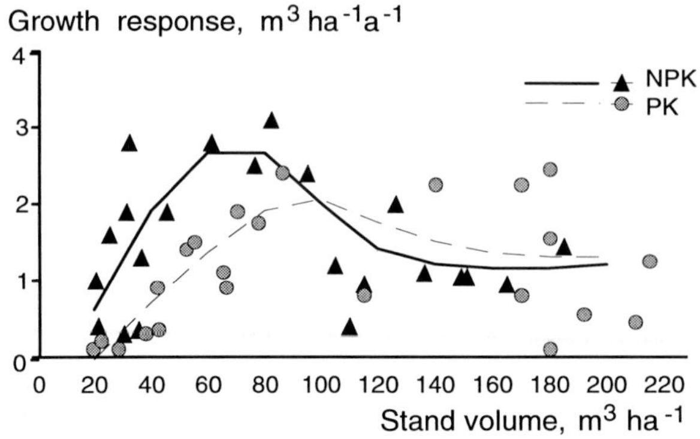

Fig. 5.7. The relationship between increase in growth and stand volume of Scots pine (*Pinus sylvestris*) stands after fertilization. (Data from the Finnish Forest Research Institute, redrawn from Paavilainen 1979b)

has been used to predict the response to fertilization (Keltikangas and Seppälä 1973; Almqvist 1990). The relationship between stand volume and increase in growth of Scots pine on peatlands after fertilization is presented in Fig. 5.7.

The absolute response to fertilizer is, to a certain extent, the higher, the better the trees were growing before fertilizer application. Only on the most fertile sites, where the growth of trees is not limited by nutrient deficiency, is the reaction to fertilization poor.

5.3 Fertilizers and Their Use

5.3.1 Types and Doses of Fertilizers

The following presentation and recommendations are based mainly on the results of Finnish experiments with Scots pine (*Pinus sylvestris*).

5.3.1.1 Nitrogen

The nitrogen in fertilizers is usually in a water-soluble form as nitrate (NO_3^-) or ammonium (NH_4^+) ions. **Nitrate-nitrogen** is more susceptible to leaching than is **ammonium-nitrogen. Urea-nitrogen** is an organic compound which is transformed under suitable moisture and temperature conditions to ammonium-nitrogen in the soil. In this process, ammonia gas (NH_3) is formed. This gas may escape to the atmosphere when the soil temperature is relatively high and the peat is drying up (Paavilainen 1973, 1975b). Other fertilizers which could be useful on peatlands are ammonium sulphate and very slow-reacting fertilizers such as ureaform and nitroform, and nitrogen compound fertilizers containing phosphorus, potassium and other macro- and micronutrients. Slowly soluble nitroform; together with slowly soluble phosphorus and potassium has a longer-lasting effect on tree growth on ombrotrophic mires than readily soluble urea with fast-dissolving P and K fertilizers (Lindholm and Vasander 1988).

In practice, ammonium nitrate containing lime (27.0% N – 50% NO_3–N and 50% NH_4^+-N – and 6.5% Ca and 1.0% Mg) and urea (46.3% N) are used. These nitrogen fertilizers are almost always given together with phosphorus and potassium. It is also recommended to use a compound fertilizer containing 13.0% N, 5.0% P, 11.0% K, 12.0% Ca, 1.3% Mg, 0.2% B, 0.2% Zn and 1.3% S.

Urea has been found to give a somewhat weaker growth response than ammonium nitrate with lime (Paavilainen 1972). This result is the same as on mineral soils (Malm and Möller 1974; Gustavsen and Lipas 1975; Lipas and Levula 1980). Nitrogen may have been volatilized from the urea but the difference may also be due to the strong biological binding of nitrogen after the application of urea (Schalin 1967; Paavilainen and Norlamo 1975). The use of urea is, however, recommended because both the price and risk of leaching are lower.

Growth increases with the dose of nitrogen applied at least to a level of 125 kg ha^{-1} (Huikari 1973, 1977b; Paavilainen 1971, 1972, 1974). Doses greater than 200 kg ha^{-1} have still been found to increase tree growth on nutrient-poor bogs (Paavilainen 1974), but are not recommended for minerotrophic mires because of the risk of damage and increased mortality of the trees (Paavilainen 1984b; Moilanen 1993).

5.3.1.2 Phosphorus

Rock phosphate has a low solubility and is suitable for peatland forests (Edwards 1959; Paarlahti and Karsisto 1968; Karsisto 1968, 1976a, 1977; Dickson 1971; Carlsson and Möller 1985), especially when finely ground. Superphosphate, with its high amount of water-soluble phosphorus, affects tree growth more rapidly, but the long-term effect is usually of the same order as that of rock phosphate.

There is little difference in the growth reaction between phosphorus sources. However, low-soluble P fertilizer is preferred because both fertilizing cost and risk of leaching are lower (see Sect. 7.2). The most widely used fertilizer in Finland for peatland forests is PK fertilizer containing 9.7% P, 16.0% K, 24.5% Ca, 0.8% Mg, 0.2% B, and 2.0% S. The phosphorus in this fertilizer is in a water-insoluble form as apatite.

The recommended dose of phosphorus is 40–50 kg P ha^{-1}, as broadcast fertilization. The application of higher doses may increase not only the duration of the fertilization effect (Paavilainen 1978), but also the risk for damage on nutrient-rich sites (Paavilainen 1979a).

At planting or sowing, **spot fertilization** is used. An area of 0.25 m^2 around the seedling is fertilized with phosphorus or PK fertilizer. To save the seedlings from root damage water-soluble fertilizer should be avoided. The effect of spot fertilizer usually lasts 5 to 10 years, after which broadcast or strip fertilization is needed.

5.3.1.3 *Potassium*

Potassium in currently available fertilizers is fully soluble in water and thus readily available to the trees, but also susceptible to leaching. When the effects of different potassium sources (potassium chloride, potassium sulphate, double salt of potassium and magnesium sulphates and so-called cement potassium) were studied, there were no significant differences in the peat or needle potassium concentrations 20–30 years after fertilization, nor in growth response (Kaunisto 1992).In practical forestry, potassium is generally applied using PK fertilizer or a fertilizer containing 30.0% K, 1.5% Ca, 7.0% Mg, 0.4% B, 0.4% Zn, and 6.0% S. When potassium alone is given, to correct P/K imbalance for example, potassium chloride is used as it is the cheapest K fertilizer.

The potassium dose recommended for practice is about 80 kg K ha^{-1}. When using this amount, the fertilization effect lasts for about 10 to 15 years (Kaunisto and Tukeva 1984; Kaunisto 1989, 1992). Shortening the interval between applications or increasing the amount of potassium does not greatly affect fertilization response, since much of the applied potassium is leached away. For example, Kaunisto (1992) estimated that, 20-30 years after fertilization, only 11–25% of the maximal dose of potassium given (664 kg K ha^{-1}) remained fixed in the tree stand and peat (0–30 cm).

Perhaps the best way to provide potassium would be to give smaller amounts more frequently or to use a slowly soluble form of potassium such as phlogopite. Experiments with phlogopite show promising results (Moilanen et al. 1987; Kaunisto et al. 1993).

5.3.1.4 *Micronutrients*

The deficiency of boron is so common in northern European peatlands that this micronutrient is added to some fertilizers. If boron is to be given alone, fast-affecting borate (14% B) fertilizer is recommended. A suitable dose of boron is about 1 kg ha^{-1}.

Only in exceptional cases is there a need for other micronutrients (Veijalainen 1977, 1980, 1981; Kolari 1983; Reinikainen et al. 1984). To correct nutrient imbalance the following doses are recommended: 5–10 kg Cu ha^{-1}; 20 kg Mn ha^{-1} and 20 kg Zn ha^{-1} (Veijalainen 1988). These micronutrients are usually given as water-soluble sulphates.

5.3.1.5 *Wood Ash*

Except for nitrogen, wood ash contains all the important nutrients, including microelements, which are needed for tree growth. In experiments, wood ash has had a great and long-lasting effect on tree growth (Silfverberg and Huikari 1985; Silfverberg 1991). Wood ash also has a strong acid-neutralizing effect; it increases the microbial activity in surface peat and changes the composition of the vegetation to include more easily decomposing herbs and grasses (Lukkala 1951;

Huikari 1953; Karsisto 1979b; Silfverberg and Huikari 1985). Therefore, wood ash is a suitable fertilizer on nitrogen-rich peatlands, particularly on tall sedge and more fertile mires.

Samples of ash from different origins show considerable differences in nutrient content. According to Paarlahti (1980), the variation in nutrient contents and dry mass of ash of wood bark of different origins was as follows:

P	K	Ca	Mg	Cu	B	Dry wt.
			g kg^{-1}			kg m^{-3}
10.0–12.3	16.0–65.8	266–321	24.4–30.8	0.16–0.26	0.20–0.27	390–923

Thus, if wood ash is used as a fertilizer, its nutrient content should be analyzed to determine the right dose. The dosage is usually calculated on the basis of the amount of phosphorus needed (40–50 P kg ha^{-1}). Based on the nutrient contents presented by Paarlahti (1980), 3000–5000 kg ha^{-1} of wood ash would be required. In most cases, the nutrient concentrations of the ash are lower, and the amount of ash fertilizer needed is more than 5000 kg ha^{-1}.

5.3.2 Fertilizers and Atmospheric Deposition

The increased **nitrogen** and **sulphur load** resulting from atmospheric deposition affects soil properties by speeding up the **soil acidification** process and by changing the ratios between different nutrients available to the trees (see Sect. 3.2).

Peatlands with high acidity and a high cation exchange capacity are rather tolerant against acidification. For surface peat (0–20 cm) of drained oligotrophic pine mires in Finland, Pätilä and Nieminen (1990) presented **exchangeable base cation contents** of 17.65–24.95 keq ha^{-1} (Table 5.3). The maximum **proton deposition load** is 0.20–0.40 keq H^+ ha^{-1}a^{-1} (Kulmala et al. 1990). The amount of base cations in the 0–20 cm layer would thus be sufficient to neutralize current levels of hydrogen deposition for the next 50 years at least.

The amount of base cations in deposition reach 0.5 keq ha^{-1} (Järvinen 1986). This load is higher than the proton load and is approximately equal to the amount of base cations annually bound to the tree stand and lost through runoff on ombrotrophic mires (Pätilä and Nieminen 1990).

Table 5.3. Exchangeable calcium, magnesium and potassium contents (keq ha^{-1}) in the 0–20 cm peat layer in different pine mires

Site type	Ca	Mg	K	Total
Cottongrass pine bog	11.45	4.81	1.39	17.65
Dwarf-shrub pine bog	12.34	5.23	1.53	19.10
Low-sedge pine mire	17.51	5.68	1.76	24.95

Source: Pätilä and Nieminen (1990).

Wood ash, with its long-lasting neutralizing capacity, is a very efficient fertilizer for compensating for proton deposition on peatlands. Liming may have a slightly negative effect on tree growth, at least in the short term, and is therefore less suitable (Meshechok 1968, 1971; Gardiner 1975; Kaunisto and Norlamo 1976; Kaunisto 1982; Dickson 1984).

There is no evidence so far of significant changes in the nutrient ratios of peat soils caused by the atmospheric nitrogen load. In spite of this, the use of potassium and micronutrients has been studied where a deficiency of these nutrients has caused poor health and lowered resistance to the adverse effects of air pollutants on the tree stands. The application of fertilizer containing potassium, boron and zinc in the most susceptible sites has already been recommended.

5.3.3 Application Date

In deciding when to apply fertilizer, both the expected growth response and environmental effects should be taken into consideration. If the fertilizer is applied before the onset of the growing season, the trees can already utilize the fertilizer nutrients in the same year (Karsisto 1967; Paarlahti 1967). NPK fertilizer given in autumn has resulted in a slightly better growth response than when given in the summer (Paavilainen 1969b, 1977a). The use of urea is to some extent dependent on weather conditions (Paavilainen 1972, 1973, 1975b; Moilanen and Issakainen 1985), but the risk of volatilization is lower on peatlands than on mineral soils.

The spreading of fertilizers onto snow cover during winter is generally not recommended because of the greater risk for leaching of nutrients, especially nitrogen and potassium (Ahti 1983; Ahti and Paarlahti 1988; Nieminen and Ahti 1993). The tree growth response to fast dissolving PK or NPK fertilizers applied in winter is also weaker compared to applications in the snow-free period (Paavilainen 1977a). Only fertilizers which contain nutrients in water-insoluble forms (e.g. apatite) may be applied at any time throughout the year except on melting snow.

5.3.4 Spreading Methods

The spreading of fertilizers from the air by mechanical methods is usually not as even as spreading by hand. Uneven spreading may cause a loss in annual tree growth of up to 0.5 m^3 ha^{-1} (Paavilainen and Virtanen 1977, 1979). Careful planning and control of spreading evenness is therefore important when mechanical spreading methods are used.

Manual work is limited to small woodlots. The amount that can be spread daily is about 2000–3000 kg. On medium-sized peatland areas, snowmobiles can be used for spreading in winter and tractors during summer. The spreading equipment is usually fitted with centrifugal blowers which fertilize a strip of 10–25 m width.

Fig. 5.8. Helicopter leaving the loading place where the fertilizer spreader has been changed. (Photo: Courtesy of Kemira Ltd.)

In difficult terrain and for large areas, aerial spreading is the most suitable method. When using aircraft, the amount to be spread from the same starting point is recommended to be at least 20 000 kg and the main part of the area to be fertilized at a distance of less than 7 km. When spreading by helicopter (Fig. 5.8), a flying distance of less than 1.5 km and a minimum amount of 10 000 kg fertilizer for a working area is recommended.

For the protection of surface waters, a sufficiently broad zone (at least a 50-m-broad strip) should be left along streams and around lakes. The spreading of fertilizers into the ditches should be avoided as much as possible by leaving an unfertilized 5-m-broad strip on both sides of the ditches.

6 Forest Management on Peatlands

6.1 Virgin Mires

6.1.1 Tree Stands

In this chapter, two different cases of tree stands on virgin peatlands and their management are discussed. The northern European approach, based on the studies and observations made in Finland, is explained, and the description of the Canadian situation is mainly based on research and observations made in Ontario.

Very few studies in **Finland** have dealt with the growth and yield of tree stands growing on virgin mires (Heikurainen 1971b; Gustavsen and Päivänen 1986; Finér et al. 1988). The following review is mainly based on data collected in the period 1951–1953 in connection with the 3rd National Forest Inventory (NFI) (Gustavsen and Päivänen 1986). At that time, most of Finland's mires were still in a natural state.

Some of the genuine, forested mire site types (see Sect. 3.4.1) support commercial-size tree stands. Spruce mires are usually dominated by Norway spruce (*Picea abies*), although the proportion of hardwoods, mainly pubescent birch (*Betula pubescens*), in the total stand volume may be considerable (24–61%). With higher trophic status the share of hardwoods increases. In pine mires, Scots pine (*Pinus sylvestris*) predominates and the proportion of hardwoods in the stand volume is usually under 15%.

The trophic status of virgin mires is reflected in the number of stems per hectare. In sites rich in nutrients, the stem numbers are considerably higher than in poor sites. Typically for virgin mire tree stands, the number of stems is very high in small diameter classes and decreases abruptly with increasing diameter (Figs. 6.1 and 6.2). Thus, only those trees growing on the very best micro-sites are able to continue their growth. Due to the high water table and overgrowth of *Sphagnum* mosses, most of the trees die sooner or later. The result is that on virgin mires the tree stands have an uneven age structure. The volume growth of the stands growing on virgin mires is better explained by the moisture conditions of the peat than by its nutrient or trophic status (Heikurainen 1971b; Gustavsen and Päivänen 1986).

Canada has a total area of 9.98 million km^2, from which about 4.40 million km^2 is considered forest land (Bonner 1982). However, only less than 61% of

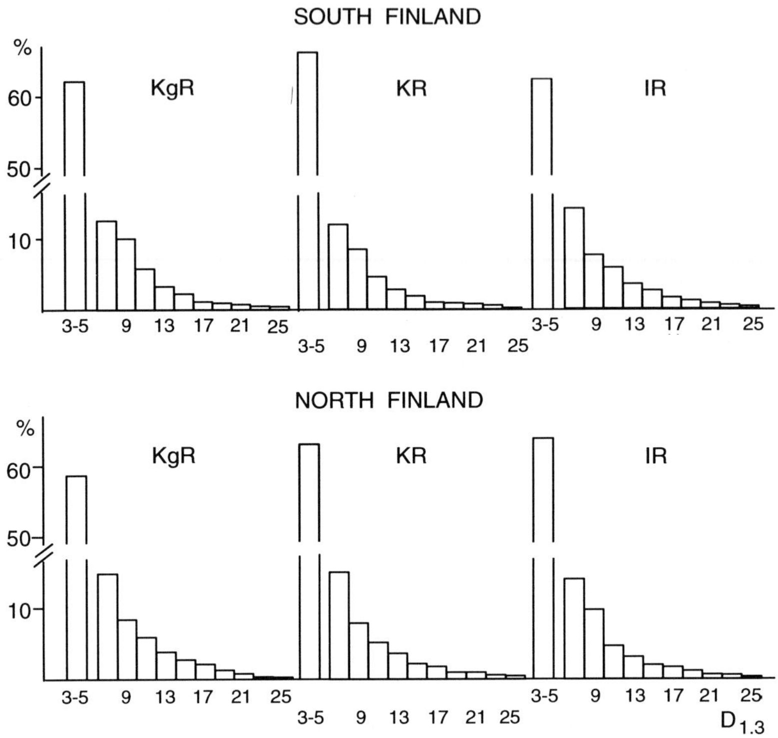

Fig. 6.1. Diameter *($D_{1.3}$)* distribution (percent of stem number) in tree stands on virgin forested pine mires on productive forest land in Finland; for site types, see Table 3.17. (Redrawn from Gustavsen and Päivänen 1986)

the forest land is considered productive, i.e. capable of producing a merchantable crop of trees in a reasonable length of time (Haavisto and Jeglum 1991).

Because of the climate, Canadian mires and wetlands occupy a vast area (170 million ha) of the country, especially in the boreal zone (Rowe 1972; Zoltai and Pollett 1983). Black spruce (*Picea mariana)* is the predominant tree species on mires and occurs alone or mixed with tamarack (*Larix laricina*). In western parts of Canada, lodgepole pine (*Pinus contorta*) and several other species of economic importance also grow on mires (Haavisto and Jeglum 1991).

Besides growing on mineral soil, black spruce (*Picea mariana*) is also found in areas ranging from nutrient-rich mires to nutrient-poor *Sphagnum* peats (Vincent 1965; Jeglum et al. 1974; Jeglum et al. 1982). On mires, black spruce often grows in pure stands, which are considered to be the result of fire (Vincent 1965).

The preliminary studies conducted in black spruce (*Picea mariana*) and tamarack (*Larix laricina*) stands on virgin mires showed that the differences in tree growth were attributed rather to the variable water table levels than to variation

Fig. 6.2. A characteristic feature of tree stands on virgin mires is their highly uneven age structure. Cottongrass pine bog in southern Finland. (Photo: J. Päivänen)

in peat nutrient contents (Jeglum 1974; Härkönen 1985; Mäkitalo 1985). Also, according to Payandeh (1986), soil factors, ground vegetation and nutrient concentration are not very good predictors of site quality as measured by site index for black spruce growing on virgin mires. This is supported by study results on the growth of tree stands on virgin peatlands in Finland (Heikurainen 1971b).

The timber resources on Canadian virgin mires is unknown as there has never been a systematic inventory. However, it is obvious that mires are producing wood at less than the optimum rate because of excessive water in the rooting layer (Haavisto and Jeglum 1991).

6.1.2 Forest Management

A characteristic feature of tree stands on virgin mires is their highly ***uneven age structure***. The tree stands preserve the same external appearance over decades, and increment is eliminated by natural removal, mainly through the death of older trees. Thus, tree stands on virgin mires, both Norway spruce (*Picea abies*) on richer sites and Scots pine (*Pinus sylvestris*) on poorer sites, can be considered climax forests (Heikurainen 1971b; Gustavsen and Päivänen 1986).

Very little tree harvesting is carried out on virgin, forested mires in **Finland**. If it is done, the cuttings should be "light", otherwise there is the danger that the site will become wetter due to the rise in groundwater level, and the growing conditions for trees will become worse (Päivänen 1982a). Only trees which would die in the near future anyway should be harvested. Indeed, virgin, forested mires may be one of the few cases where silviculture to promote an uneven-aged stand structure might be recommended although Lähde (1992) is expanding the method also for fertile sites of mineral soils. The idea in this kind of "**single tree selection** (**continuous cover**) is to retain all-sized (uneven-aged) mixed stand structure and develop it further in this direction" (Lähde et al. 1992). However, the harvesting of trees on virgin mires is not usually economic (Päivänen 1990).

In **Canada**, extensive forestry is practiced on both uplands and peatlands. The harvesting and regeneration of black spruce (*Picea mariana*) are significant, especially in Ontario and Quebec. Black spruce is a major source of fibre for the pulp and paper industry and a high proportion of the volume of this species comes from forested mires (Haavisto and Jeglum 1991). Guidelines for the management of black spruce stands on mires are under development in Ontario (Haavisto et al. 1988b; Jeglum and Kennington 1993). The first step has been the development of a classification system (Jones et al. 1983), in which the mire sites supporting merchantable or submerchantable black spruce are identified as operational groups.

On mire sites it is important to minimize soil damage that can result from the use of heavy harvesting equipment. This is even more important because forest drainage is not practiced on an operational scale, and in damaged soil the natural water movement is impaired causing a rise in the groundwater level (Groot 1987). Inventories of the vegetation changes on post-logged black spruce wetlands show that mechanical logging places fast-growing species, mainly broadleaved seedlings, at a competitive advantage over black spruce. For this reason, the maintenance of an intact organic layer and the preservation of **advance growth** have been considered to be essential features of black spruce regeneration in wetland habitats (Brumelis and Carleton 1988). Management practices have, indeed, been developed to minimize damage to soil, and to promote natural regeneration (Harvesting and . . . 1987).

In general, the possibilities for successful natural regeneration in black spruce stands are good – vegetative reproduction by **layering** (Stanek 1968b) and seed store in its semi-serotinous cones (Skeates and Haavisto 1987; Haavisto et al. 1988a) – only if the harvesting methods conserve the advance growth, maintain the exposed receptive seedbed with less logging slash on the sites, etc. Living *Sphagnum* moss has been shown to be an excellent **seedbed** for black spruce (Fleming and Groot 1984; Wood and Jeglum 1984; Jeglum 1987). On sites without sufficient advance growth good results have been achieved with both seed-tree groups and clear-cut strips (Haavisto 1979; Haavisto et al. 1988b; see also Sect. 6.2.3).

Typical for peatland forestry in Ontario is that intermediate cuttings and tending of young stands are hardly practiced. However, it has been suggested that more intensive management techniques such as forest drainage, fertilization, precommercial thinning and tree harvesting applying thinning cuttings may come more common in the future (Honer and Bickerstaff 1985; Jeglum 1991a). It seems that Canadian forestry is changing from exploitation towards a planned utilization of forest resources on a renewable basis.

6.2 Silviculture in Connection with Ditching

6.2.1 Growing or Regeneration

6.2.1.1 Capability for Revival

In draining genuine forested mires or treed fens or treed bogs, it has to be decided whether the present tree stand is capable of responding, and whether the stand density is high enough. It may be advisable to change the species if it is not suited to the site. The technical quality of the timber should also be taken into consideration (Päivänen 1990).

Studies concerning the post-drainage **revival of tree growth** outside of Fennoscandia are scanty. Black spruce (*Picea mariana*) has shown a response to drainage in the forest drainage experiments in the **USA** and **Canada** (e.g. Averall and McGrew 1929; Stanek 1968a; Payandeh 1973; Dang and Lieffers 1989; Hillman 1991; Trottier 1991a; Sundström 1992b). Observations concerning the response of tamarack (*Larix laricina*) to drainage are still more sporadic,usually having been made along drainage canals constructed for highways or railways (Päivänen and Wells 1977; Lieffers and Macdonald 1990). Therefore, if not otherwise expressed the following discussion is based on experience in **Finland**.

The term **revival** refers to the difference between post-drainage and pre-drainage tree growth. The term capability for revival expresses the tree's capacity to increase growth after drainage (Heikurainen and Kuusela 1962). The revival of radial and height growth of Scots pine (*Pinus sylvestris*) and Norway spruce (*Picea abies*) caused by drainage, and the dependence of this revival on the height, diameter and age of the trees has been studied on the basis of material from 20-year-old forest drainage areas (Heikurainen and Kuusela 1962).

The results of this extensive study showed that: (1) the height growth revival of small trees and trees with a small diameter was better than that of tall trees;

(2) the radial growth revival of small trees and trees with a small diameter was better than that of tall trees; (3) the radial growth revival was not dependent on the age of the tree to the same extent as on its size; and (4) the height growth revival was partly dependent on the age of the tree.

Thus, the volume growth revival is more dependent on tree size than age, and the height of a tree at the time of draining can be used as an indicator of its capability for revival. Heikurainen and Kuusela (1962) suggested that a maximum height of 10–12 m for *Pinus sylvestris* and *Picea abies* could be used as a standard.

However, the study of single growth characteristics (height growth, radial growth) does not reveal the changes that have taken place in a tree's volume growth. Seppälä (1976) showed that the post-drainage growth rhythm of trees varies with diameter. Basal areal growth, which reflects tree's volume growth, is greater the larger the diameter of the tree (in a certain age class) at the time of draining.

In conclusion, tree size cannot simply be used as the indicator of a tree's capability of revival. The existence of a clearly discernible top shoot, sufficient crown size as well as a green needle colour may be better indicators of a tree's physiological condition and its capability for revival.

6.2.1.2 Stand Density

It is usually only genuine forested site types that have stands with an acceptable stand density. However, gaps in the uneven-aged stands on sparsely forested composite site types (treed fens and bogs) are filled up after drainage when small and young trees strongly increase their growth. The in-filling rate of tree stands is more dependent on the fertility of the site than on the original volume of the growing stock at the time of drainage.

6.2.1.3 Tree Species

Normally, the tree species forming a stand on virgin mires is also suited for the site after drainage, at least in Finland (Heikurainen 1959b). Exceptions are weakly minerotrophic sites stocked by Norway spruce (*Picea abies*), the revival of which is very slow.

6.2.1.4 Technical Quality of the Timber

Trees, especially *Pinus sylvestris*, growing on virgin mires are typically very branchy. Various site factors such as excess water and a poor nutrient regime in the peat may cause the stands to be sparse and branchy. Flat-topped crowns are common and indicate that growth has ceased.

6.2.2 Improvement Cutting

The tree stands on virgin mires are rarely growable as such. When opening the ditch lines, there is often reason to carry out **improvement cutting.** This resembles intermediate cutting as it aims at changing the tree species composition and stand structure between the ditches to be dug.

If we compare the dominant heights and basal areas of stands in genuine forest-covered mires in a virgin state with those of thinning models that have been worked out for naturally originated and at least satisfactorily tended stands on mineral soils, mire forests on the average are sparse. The need for thinning is consequently negligible or non-existent. However, if the need for thinning is determined on the basis of the stem numbers, these mire forests are, in most cases, in need of thinning at the time of drainage (Gustavsen and Päivänen 1986). Improvement cutting may resemble a thinning from above because large trees that are incapable of revival are also removed. The technical quality of the growing crop is improved by removing defective and technically poor specimens. Simultaneous thinning from below regulates the stem number of the stand to be left.

Stands on virgin mires are usually uneven with openings. There may be a need to perform only light thinnings in the dense places. It is not necessary to cut in order to obtain a stand with an even tree height, because differences in height among the trees will be evened out with time after drainage.

Particularly in the case of the more nutrient-rich mire types in the category of genuine hardwood-spruce mires and composite site types, improvement cutting is necessary in order to direct the development of the species composition of the stand. The species composition of minerotrophic mires should be shifted towards Norway spruce (*Picea abies*) dominance, while weakly minerotrophic and ombrotrophic mires should be shifted towards Scots pine (*Pinus sylvestris*) dominance. The more nutrient-rich the site, the greater may be the admixture of birch (*Betula pubescens*). However, birch should not be allowed to replace the dominance of the coniferous species.

If the virgin mire tree stand is considered incapable of revival after drainage, it must be regenerated.

How well the recommendations given in this chapter are applicable to conditions outside of Fennoscandia is not known. From the ecological point of view, black spruce (*Picea mariana*) is more similar to Scots pine (*Pinus sylvestris*) than to Norway spruce (*Picea abies*). However, most of the drainage experiments in North America are still too recent to allow recommendations to be made. Little use has been made of improvement cuttings in Canada.

6.2.3 Natural Regeneration

6.2.3.1 Receptivity for Regeneration

Conditions for natural regeneration immediately after drainage are, in general, very good. The *Sphagnum* mosses in the bottom layer form a good moist medium (**seedbed**) for the germination of Scots pine (*Pinus sylvestris*) seeds (Sarasto and Seppälä 1964). Thus, the **receptivity for regeneration** has to be considered high. Competition from ground vegetation is generally smaller on mires to be drained than in regeneration areas on mineral soils. However, in the very richest hardwood-spruce mires, flourishing ground vegetation may lead to problems.

Even the seed production of old Scots pine (*Pinus sylvestris*) on Finnish virgin pine mires is rather good, although smaller than in stands of the same development class on mineral sites (Lukkala 1936). On genuine forested mire site types, natural regeneration of both *Pinus sylvestris* and *Picea abies* takes place all the time. Thus, there is changing seedling material (unstable seedlings less than 10 cm in height) on the site. Large average quantities of Scots pine seedlings have been counted on virgin mires both in southern and northern Finland (Heikurainen 1954):

	Seedling density, number per ha	
Site type[a)]	Southern Finland	Northern Finland
VLR	..	43 000
RaLR	..	38 000
IR	34 000	12 000
ITR	28 000	..
RaR	11 000	..

[a)] For the site types, see Table 3.17.

The age distribution of this changing seedling material is very large. Most are young seedlings (1–2 years) and the number of seedlings in each age class decreases with increasing age. Also in the case of Norway spruce (*Picea abies*) dominated stands, as many as 64% of the spruce seedlings have already appeared on the site prior to the strip felling aimed at regeneration (Lukkala 1946).

Characteristic of certain mires is their high number of pubescent birch (*Betula pubescens*) seedlings. Hundreds of thousands of seedlings per hectare may grow up during the course of a few years. This is because of the large seed production and effective distribution of the seeds. In fact, birch is a typical pioneer species which can easily colonize drainage areas rich in nutrients (Keltikangas and Seppälä 1977).

In **North America**, black spruce (*Picea mariana*) is the most important tree species growing on mires. It reproduces sexually as well as vegetatively. Black spruce cones persist on trees, and the viability of seeds in unopened cones is preserved for several years (Haavisto 1975, 1979). Regeneration from seeds takes place quite easily, especially following wildfire, leading to even-aged stands. Uneven-aged stands develop in the absence of fire, especially on poor, wet sites, where reproduction by **layering** (i.e. an attached branch that has rooted and is capable of independent growth) is common. The term "layering" designates both a tree of layer origin and the process of layer formation (Stanek 1968b, 1975). In overmature black spruce stands on peat, there is often **advance growth**, which may include both layerings and seedlings that pre-date harvesting (Haavisto

1979). When released, layer origin black spruce grow as well or even better than planted or seeded trees (Doucet and Boily 1988; Archibald and Hayhurst 1993). Natural regeneration from overmature black spruce stands could therefore be feasible after drainage.

6.2.3.2 Mire Tree Form – Genetic or Environmental?

It is uncertain if the poor external appearance of the trees (stem form and branchiness) on virgin mires is due to genetic properties or environmental factors related to site properties (e.g. Pihelgas 1963; Valk 1967; Roosaluste 1982). Lukkala (1952), in a far-sighted experiment established in the 1930s, followed the development of mire-originated Scots pine (*Pinus sylvestris*) on a mineral site. The tree stands have been measured several times since then, most recently when the stands were about 50 years old (Päivänen 1988).

The volume and total production development of the mire-originated pine stands were found to be somewhat better than that of average pine cultures on mineral soil. The dominant height of the mire-originated pine plantations covered by the study were at least within the limits of the dominant height variation in southern Finnish pine cultures on *Vaccinium* forest type (H_{100} = 24 m) at the biological age of 50 years. On the other hand, the development of the dominant height did not seem to reflect the original mire site type any more than the region from which the seeds originated (Fig. 6.3). The stem form of mire-originated pines was normal and even superior to that of average pine stands of mineral soil origin at the same stage of development.

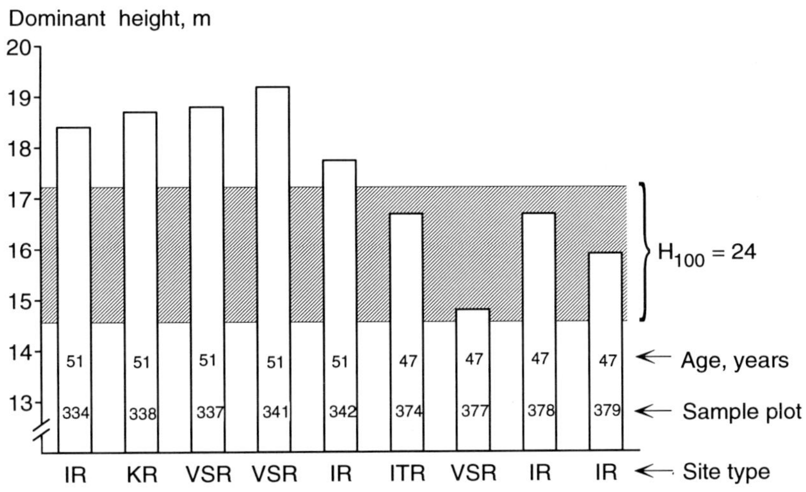

Fig. 6.3. Dominant heights of mire originated Scots pine (*Pinus sylvestris*) stands compared with the dominant heights of artificially regenerated Scots pine stands at the same age on *Vaccinium* forest type. Site index of H_{100} = 24 m (Vuokila and Väliaho 1980). For the original mire site types marked with symbols, see Table 3.17. (Redrawn from Päivänen 1988)

It may therefore be concluded that the poor growth and stem form of pines growing on virgin mires are not inherited properties. Estonian results also lead to this conclusion (Valk 1967; Roosaluste 1982). Such trees can be used for seeding in natural regeneration. Natural regeneration can also be based on the existence of the changing seedling material on the site at the time of drainage. Finally, it is necessary to point out that the inherited properties of mire trees were dealt with in the foregoing only indirectly, without a genetic approach.

In **North America**, black spruce (*Picea mariana*) grows under a wide range of climatic and edaphic conditions. Within a single locality it may occur in adjacent but very different habitats: peatlands and uplands. It has been speculated that these different environmental conditions may have led to ecotypic differentiation in black spruce (Parker et al. 1983). However, study results show that there is little ecotypic differentiation between peatland and upland black spruce origins (Fowler and Mullin 1977; Wang and Macdonald 1993a,b).

Both Scots pine (*Pinus sylvestris*) and black spruce (*Picea mariana*) are primarily outcrossing species. Thus, extensive gene exchange between adjacent populations may overcome the effects of selection so that little or no ecotypic differentiation is evident.

6.2.3.3 Methods of Natural Regeneration

Natural regeneration utilizes the already existing seedlings and those born after regeneration cutting to produce a new tree generation. In connection with drainage, the following regeneration cutting methods have been used in Finland (Kaunisto and Päivänen 1985): release of seedlings, seed-tree cutting, shelterwood cutting and strip cutting.

Mire stands dominated by *Pinus sylvestris* often have so-called **changing seedling material** (see the beginning of this chapter). In this case, regeneration is best carried out by releasing this seedling stock in connection with the opening of ditch lines. Formerly, the regeneration of mire pine stands was said to take place by clear-cutting (e.g., Heikurainen 1954). However, where the number of growable seedlings is sufficient, it would be more correct to talk about the ***release of the seedlings***.

If the seedling stock is insufficient, mire pine stands are regenerated by **seed-tree cutting**. In seed-tree cutting, 30–50 seed trees per hectare are left standing. The proportion of self-pollination is higher in a sparse stand (such as after seed-tree cutting) than in a fully stocked stand (Rudin et al. 1977). Seeds resulting from self-pollination do not germinate as well as those resulting from cross-pollination, and self-pollinated seedlings are inferior in growth to the cross-pollinated ones (Koski 1973). The proportion of cross-pollination is dependent on the size of the regeneration area, the number of seed trees and the tree stands growing around the area concerned.

The timing of regeneration is also an important issue. Regeneration is a sensible alternative in the case of old Scots pine (*Pinus sylvestris*) stands on peat that have almost ceased growing. In such cases, the regeneration should already be carried out in connection with drainage to avoid growth loss (Paarlahti and Takamaa 1984).

Shelterwood cutting is preferably used in the regeneration of Norway spruce-dominated stands on minerotrophic mires. About 150 shelter trees per hectare should be left standing. Apart from Norway spruce (*Picea abies*), Scots pine (*Pinus sylvestris*) and pubescent birch (*Betula pubescens*) should also be left as shelter trees when possible. These improve the light conditions of the stand and make the future harvesting of the shelter trees easier by reducing the amount of logging slash. If the shelterwood stand consists entirely of spruce, the future cutting of standards (shelter trees) will produce such an amount of slash that it will destroy a considerable part of the seedlings by burying them. The presence of birch in regeneration areas at the shelterwood stage ensures the rapid development of a nurse crop. Because of the risk of windfalls, a prerequisite for the use of shelterwood cutting is that the stand has been thinned previously (see also Hånell and Ottosson-Löfvenius 1993; Sect. 6.4.1). However, small depressions with hardwood-spruce stands on peat soil are often naturally sheltered from the wind.

Clear-cutting can be used to regenerate spruce-dominated stands on mires in narrow (< 50 m) depressions provided that the bordering mineral sites have enough seeding spruce. However, it is also presupposed that the hardwood thicket promoted with the clear-cutting will form a sufficient nurse crop on the site to protect the spruce seedlings from spring frosts.

In **strip-cutting**, strips of varying width are cut into the stand to be regenerated. The width of the clear-cut strips is on average 50 m, but depends on the size of the area to be treated and its geographical location. Strip-cutting is the preferred method of regeneration for large areas of spruce-dominated stands on mires, provided the strips will be able to develop a sufficient hardwood thicket as a nurse crop (Lukkala 1946). Due to the risk of windfalls, strip-cutting can be applied only in areas where the tree stand has undergone repeated thinnings.

In many cases, a substantial part of the spruce seedlings of a younger generation is already there prior to strip-cutting. In such cases, strip-cutting is used only for completing the seedling stand. The successful outcome of the regeneration depends on the seeding capability of the remaining stand strips. The number of seeds reaching the ground depends on the distance to the seeding trees. According to Lukkala (1946), the number of seeds falling to the ground at the edge of the strip is fourfold, and within the intermediate stand tenfold, of that in the middle of the strip. Moreover, spruce seedlings are less exposed to spring frost at the edge of the strip than in the middle. Wide strips are more exposed to wind than narrower ones. Therefore, the use of relatively narrow strips (40–50 m) has been recommended.

Strip width recommendations for black spruce (*Picea mariana*) stands on virgin mires in **North America** strip cuttings vary from 60–80 m (Jeglum 1987; Jeglum and Kennington 1993) to 120 m at the most (Verry and Elling 1978).

The strips should not be laid out in an east-west direction because snow and soil frost will remain along the southern edge for a longer time than on the other parts of the strip. The southern edge of a strip having an east-west direction will also be more shaded than the other parts of the strip. Factors affecting tree harvesting, however, must also be taken into consideration.

Soil preparation is not usually carried out in conjunction with natural regeneration. It is not generally necessary, but in areas where it would be desirable the methods are also lacking. Heavy scarification in treating very moist sites should be avoided in order to prevent seedbed degradation (Kaunisto and Päivänen 1985; Kolabinski 1991). Post-cut burning in aid of natural seeding on moist to wet lowland peat sites has been recommended for black spruce (*Picea mariana*) (Chrosciewicz 1990).

The need for **clearing the cutting area** should be considered from case to case. Particularly in connection with the natural regeneration of spruce, the spruce undergrowth that does not belong to the growable seedling stand should be removed. Hardwoods and thickets should be left to protect against spring frosts.

6.2.4 Artificial Regeneration

The need to resort to artificial regeneration in connection with drainage of forested mires seldom arises. However, in the case of sparsely stocked pine and hardwood-spruce mires, natural regeneration is not always successful. There may also be a need to change the tree species to reach optimum production after drainage.

The method of regeneration to be used is primarily dependent on the fertility of the site. If there is a choice between sowing and planting, the former is superior. In practice, this means that on sites poor in nutrients Scots pine (*Pinus sylvestris*) should be regenerated by sowing. Sites rich in nutrients should be planted. If artificial regeneration is chosen, Norway spruce (*Picea abies*) is usually only regenerated by planting. A further consideration is that the area to be regenerated must either already have or be able to support a hardwood thicket as shelter against frost.

Where planting is the form of regeneration on peat soils, bare-rooted seedlings should preferably be used (Kaunisto and Päivänen 1985; Hånell 1992). A problem connected with containerized seedlings is their susceptibility to soil frost heaving, especially in well-humified peat (Virgo 1975). If, however, paperpot-type seedlings are used, the paper container should be removed before planting to enable better root development and greater stability (Sundström 1992a).

In herb-rich sedge hardwood-spruce fens, natural spruce seedlings usually occur only on hummock surfaces. Thus, on these areas, complementary planting has to be done. The lawn-level surfaces are planted during the spring following drainage using tall spruce transplants.

6.2.5 Afforestation of Treeless Mires

The drainage and afforestation of treeless mires in **Finland** were quite extensive in the 1960s and 1970s. According to the information supplied by various categories of forest owners, the total area of formerly treeless afforested mires was about 235 000 ha in 1979 (Kaunisto and Tukeva 1984). Studies concerning the success of afforestation, ditching techniques and early responses of transplants to fertilization on treeless mires were also given strong support. Most of the afforestation experiments in treeless mires were carried out using Scots pine (*Pinus sylvestris*)(Kaunisto and Päivänen 1985).

On treeless mires of average to poor fertility sowing has been found to be a good afforestation method. **Sowing** is mainly carried out by hand or using hand tools. However, a prototype machine for simultaneous fertilization, soil preparation and sowing has also been developed (Kaunisto 1974).

Planting is preferred on treeless mires rich in nutrients where competition from the ground vegetation may hamper the development of the seedlings. Usually, planting is carried out manually using bare-root plants, and in this case the most widely used tool is a semicircular planting hoe. Planting carried out using planting pliers has also been successful (Seppälä 1968). Machine planting is easier on treeless mires than on mineral soils. This is why the earliest Finnish efforts to develop mechanization of artificial forestation were done explicitly on peat soil (Appelroth 1971).

Planting can begin in the spring as soon as there is no further soil frost. Planting in soil that is still frozen endangers the water uptake of the plants. Planting on drained mires can be carried out later than on mineral soils because, if it is done at the lawn level, water supply will not be a problem (Päivänen 1975).

On treeless mires fertilization is needed in connection with drainage and afforestation. Phosphorus, the most important main nutrient for the initial development of young tree plants, can be applied in the form of rock or fine-ground rock phosphate very near the plants – even into the planting hole. Water-soluble fertilizers, on the other hand, increase plant mortality if they come into contact with the root systems (Kaunisto and Paavilainen 1977). On drained tall-sedge fens and better site types, the development of young pine stands that have been artificially established and fertilized has been similar to that of plantations in *Vaccinium* forest type on mineral soil during the first 15 years of their existence (Kaunisto and Päivänen 1985).

In Finland, experimental afforestation of treeless mires with exotic tree species has also been carried out. In one experiment, the initial development during the first 9 years of growth of contorta pine (*Pinus contorta* var. *latifolia*) was clearly faster than that of our domestic Scots pine (Laine 1979). Later on, however, the contorta pine suffered more from dieback and canker fungus (*Ascocalyx abietina*, also called *Gremmeniella abietina*) than the Scots pine. Another exotic species which would be possibly suited for afforestation of treeless mires in Finland is black spruce (*Picea mariana*). Black spruce has showed little susceptibility to spring frosts. Moreover, it seems that moose (*Alces alces*) do not touch black spruce when searching for food in winter. However, a comparative experiment showed that the height development of Scots pine is somewhat superior to that of black spruce (Päivänen 1983). The slow initial development of black spruce may be partly due to growth stagnation after planting, which is typical of spruce species. Both black spruce and white spruce (*Picea glauca*) have thrived rather well, even in experiments carried out on treeless peatlands rich in nutrients in northern Finland (Numminen 1983).

It is important to consider provenance when planting exotic species. In some of the experiments, the provenances used have not necessarily been the best ones with regard to the climate of origin. Nevertheless, the importance of exotic tree species in the afforestation of mires in Finland is negligible.

In conclusion, the afforestation of treeless mires is technically rather easy. The problems that occur in this field are related to the later development of the established tree stands. One of the main problems is an adequate nutrient status. The progressive development of the stands requires repeated fertilizer application in most site types.

In certain cases, small areas of treeless mire may take a forest cover naturally through seeding from border stands. A sufficient pine seedling stand can be obtained on drained sites in this way up to a distance of 50–80 m from the seeding mother stand. Still larger numbers of birch (*Betula pubescens* and *B. pendula*) seedlings appear on drained and fertilized treeless mires.

Most of the treeless mires in Finland are located in northern Finland, where the climate restricts forestry. Furthermore, considerations relating to nature conservation, landscape protection and multiple use of mires speak in favour of leaving treeless mires outside of forestry use. In accordance with current recommendations, open peatlands are no longer drained and afforested in Finland.

In **Sweden**, most peatland drainage has taken place on originally forested mires. There have therefore been very few studies on the afforestation of treeless mires. Only preliminary results of the afforestation of abandoned peat harvesting areas and former agricultural land on peat have been published (Nilsson et al. 1987; Sundström 1991; Nilsson 1992). There is only one report giving the results of an afforestation experiment established on drained, treeless mires in five different areas of Sweden from the west coast to the Artic Circle (Sundström 1991). According to the results, Scots pine (*Pinus sylvestris*) up to the age of 18 years grew better the further south the site was located. Soil preparation in which the peat was turned over yielded better results than no preparation, and PK and NPK fertilization increased growth compared with non-fertilized controls (Sundström 1991).

In **Norway**, the bulk of the forest drainage that took place in the 1960s was directed towards nutrient-poor treeless sites which required fertilizers to become productive. Today, drainage is concentrated on paludified forest and forested mires (Braekke 1990b). In afforestation experiments on treeless drained mires, *Picea sitchensis* has been the most productive tree species in coastal areas. In central regions of southern Norway, *Pinus sylvestris* and *P. contorta* var. *latifo-*

lia are the most promising. *Picea abies* should only be planted on minerotrophic peats. Effective drainage (ditch spacing from 8 to 30 m) and, on ombrotrophic peat, also fertilization, is essential for successful afforestation results (Meshechok 1968, 1969; Braekke 1972, 1978).

In the **British Isles**, afforestation generally involves forest drainage (Fig. 6.4). Drainage and soil preparation are combined and since the time of the early afforestation experiments **turf planting** has been practiced on deep peat. The system involves opening a network of drains in the peat and using the spoil to make mounds on which to plant. The local draining of the turf by raising it above the level of the peat surface contributes to the improved conditions for growth (Zehetmayr 1954).

The recommendation is to plow furrows 60 cm deep in order to induce movement of water from the upper layers of the soil. Because shallow rooting is expected, wide platforms between the furrows are preferred to encourage stability. The type of plow used on deep peat is usually a double mouldboard (Thompson 1979). On upland heaths, mounted single plows with tines have been used to disrupt compacted layers (Neustein 1977) and to control *Calluna* (Thomson and Neustein 1973), which is known to have significant deleterious effects on young conifers (Malcolm 1975). It is sometimes also necessary to control the *Calluna* with herbicide (Mackenzie et al. 1976).

Fig. 6.4. Afforestation with *Pinus contorta* in Scotland. The furrows were made with a double moulboard plow. (Photo: J. Päivänen)

Exposure is a complex of temperature, wind, and rainfall. In addition to restrictions placed on the land used for forest purposes in Great Britain, wind is the major factor limiting growth. One method of measuring the relative exposure is to use "tatter flags", which are linen flags set up in the field. The relative exposure is estimated according to tearing of the flag (Booth 1974). For forestry purposes, a special windthrow hazard classification has been introduced (Booth 1977; Quine and White 1993).

Sitka spruce (*Picea sitchensis*) and lodgepole pine (*Pinus contorta)* are the main species used for planting. These species are considered to possess "pioneer" characteristics and are thus suitable for establishing plantations on land devoid of tree cover (Malcolm 1979). Lodgepole pine has been shown to be more tolerant of anaerobic conditions than Sitka spruce (Pyatt and Craven 1979). Sitka spruce is more productive, but on the poorest peats it has to be replaced or mixed with lodgepole pine due to its greater demand for nitrogen (Malcolm 1991). Almost all new planting takes place on infertile sites in the uplands and thus the use of broad-leaved trees is impossible.

It was discovered early that phosphate fertilizers are needed at the time of planting on almost all sites (Zehetmayr 1954; Everard 1974; Malcolm 1991).

In the southeast **USA**, afforestation of wetlands has been practiced by draining, bedding, fertilizing and planting, mainly with fast-growing pines (*Pinus taeda* and *P. elliottii*) (see Sect. 2.4.2).

In **Newfoundland, Canada** peatland afforestation began at the end of the 1960s. About 50 ha of treeless mires were afforested using a variety of species. Many later experiments proved unsuccessful because of the planting stock, unsuitable tree species, poor site selection in certain cases, and very often, inadequate drainage design (Wells 1991). One afforestation experiment on an ombrotrophic bog has been followed for 20 years. The tree species used was Japanese larch (*Larix leptolepis*). It was found that low levels of N, P and K in the soil all appeared to limit growth (Wells 1991). In other parts of Canada, very little afforestation has been done in connection with drainage.

6.2.6 Afforestation of Peat Cutover Areas and Peatland Fields

Peat **cutover areas** (**cut-away peatlands**) are former peatlands where peat harvesting has ceased. The term **peatland field**, in this connection, refers to **abandoned farmland** on peat soil. On both types of terrains the ditch network has usually deteriorated since cessation of the previous land use. The drainage has to be considered inadequate for tree growth. Long-term drainage may have caused peat subsidence and subsequently the hydraulic conductivity may be low (Eggelsmann and Mäkelä 1964; Eggelsmann 1973; Njøs 1978; Sorteberg 1978). Thus, the first phase of afforestation is ditch cleaning and, if necessary, complementary ditching.

Afforestation has been considered a feasible use for peat cutover areas in several countries (e.g. Mikola and Mikola 1958; Ferda 1972, 1975, 1986; Kaunisto 1986; Nilsson et al. 1987; Farrel 1990; Nilsson 1992).

In Finland, the afforested area of cut-away peatlands is still quite small (about 2500 ha), but the area available for forestry is increasing and research in this field has been intensified since the 1980s. The most important results are reviewed in this section.

Scots pine (*Pinus sylvestris*), silver birch (*Betula pendula*) and pubescent birch (*Betula pubescens*) can be grown on peat cutover areas if the drainage conditions and nutrient status are taken care of (Kaunisto 1986; Saarinen 1988). All the species can be sown or planted. The birch species can be grown for both timber and bioenergy. In some naturally born, unfertilized birch thickets, until an average age of 16 years, the annual above-ground leafless biomass increment ranged from 3.3 to 6.1 t ha^{-1} dry mass (Ferm and Kaunisto 1983).

However, an unbalanced nutrient status is typical of peat cutover areas. Peat from the bottom layers contains high amounts of nitrogen, but only neligible amounts of mineral nutrients. Peat thickness may vary from nought to more than 1 m, causing mosaic nutrient status in the surface peat (Kaunisto and Viinamäki 1991). Tree nutrition can be improved either by fertilization or by arranging mineral soil – spoil from the drainage ditches – into the root layer of trees. Spot fertilization with PKB fertilizer (P as apatite) is recommended at planting and sowing if there is no mineral soil in the vicinity (Kaunisto 1979, 1986, 1987b).

Because of agricultural over production, peatland farming is diminishing in several countries. The abandoned fields become colonized by bushes (e.g. *Salix* species) if not properly managed.

In Finland, it has been estimated that about 0.6 million ha of arable land should be withdrawn from farming and afforested. Peatland fields will play a substantial role here (Kaunisto 1991). However, very often afforestation has yielded unsatisfactory results due to ground vegetation competition, moose and vole damage (see Sect. 6.4.2), drainage problems and an unbalanced nutrient regime (e.g. Raitio and Rantala 1977; Veijalainen et al. 1984; Kaunisto 1991). Usually, there is plenty of nitrogen in the peat, which creates an imbalance between nitrogen and the other nutrients. Liming done when the area was farmed may have caused boron fixation. Potassium, phosphorus and boron deficiency can be taken care of by fertilization. Good results have also been achieved by using wood ash (Ferm et al. 1992). However, an excess of nitrogen causes problems in wood quality. For instance, Scots pine (*Pinus sylvestris*) develops thick branches (Kaunisto and Tukeva 1986). Norway spruce (*Picea abies*) cannot be recommended because of the risk of spring frost. Black spruce (*Picea mariana*) could be an alternative, but there are too few studies to make any recommendations.

6.3 Silviculture on Drained Peatlands

6.3.1 Tree Stands

6.3.1.1 Condition of Forests on Drained Peatland

As the area of peatlands drained for forestry is greatest in Fennoscandia, the following text deals mainly with drained peatland forests in this region. The description of the tree stands growing on drained peatlands is based particularly on studies done in Finland.

In peatland, forestry activities have been concentrated so far on draining the peatlands and, to some extent, fertilization and treatment of the young stand. At this stage thinnings and regeneration cuttings have only been carried out to a small extent. In estimating the silvicultural condition and future need for treatment of the forests, the development class distribution and the average volumes in each development class can be used as a basis (Table 6.1; Keltikangas et al. 1986).

On drained peatlands the combined proportion of shelterwood stands and stands which are mature for regeneration is slightly less than 4% of the total area (Fig. 6.5). Even that of advanced thinning stands is less than 10% of the total area. The large proportion of seedling and sapling stands as well as stands in the young thinning stage (more than 75%) means that extensive cuttings resembling first commercial thinnings have to be performed in the near future. The proportion of more mature development classes is largest in the oldest drainage areas and in the southern parts of the country (for more detailed information, see Keltikangas et al. 1986).

Table 6.1. Development class distribution and average volume of stands in various development classes in drained peatlands. (Data from Keltikangas et al. 1986)

Development class	Area (%)	Average volume, (m^3 ha^{-1})
Treeless areas and sparse seed-tree stands	7.2	1
Seedling and pole-stage stands	46.2	11
Young thinning stands (the next cutting producing mainly pulpwood)	29.3	59
Advanced thinning stands (the next cutting producing saw-timber, too)	9.5	122
Stands mature for regeneration	2.9	149
Shelterwood stands	0.9	52
Low-yielding stands	4.0	15
Total and average	100.0	42

Fig. 6.5. Minerotrophic mire site types will develop mixed stands after drainage. *V. myrtillus* spruce swamp 50 years after drainage supporting a mixed stand of *Betula pubescens*, *Pinus sylvestris* and *Picea abies*, southern Finland. (Photo: J. Päivänen)

Low-yielding stands (estimation based on tree species composition, stand density and occurrence of damage) occupy only 4% of the total area of drained peatlands (Keltikangas et al. 1986). A similar figure was derived from the 7th National Forest Inventory (Paavilainen and Tiihonen 1988). The proportion of tree stands on mires classified as low-yielding has decreased considerably compared to the earlier inventories. However, the decrease is mostly due to a change in the definition of "low-yielding" between the inventories.

In forest stands in the pole and thinning stage, the low-yielding category is usually a result of low stand density or unsuitability of the tree species for the site. Except for transformed drained peatlands, pubescent birch (*Betula pubescens*) is considered to be an ecologically proper tree species for drained peatlands. However, taking economic aspects into consideration, the second tree generation after drainage should be dominated by a coniferous species.

Sites described as **unfit for maintenance** (Keltikangas et al. 1986) are drained sites that do not meet the criteria of suitability for forestry use, neither in terms of site type nor climatic region. In southern Finland, the proportion of the drained area currently considered unfit for maintenance is small (< 5%), but in the northern part of the country it is noteworthy (> 20%) (Keltikangas et al. 1986; Päivänen 1990). Data from a botanical survey concentrated between the latitudes 60–66°N indicated that the proportion of drained area unprofitable for forestry was 17% on average (Eurola et al. 1991).

6.3.1.2 Structure of Forests Growing on Drained Peatland

The typical features of peatland forests at the time of drainage, i.e. their great variation in age and tree size, remain or are even emphasized during the first decades after draining. This is due to several factors (Hökkä and Laine 1988). Apart from genuine forested hardwood-spruce mires, the tree cover on natural

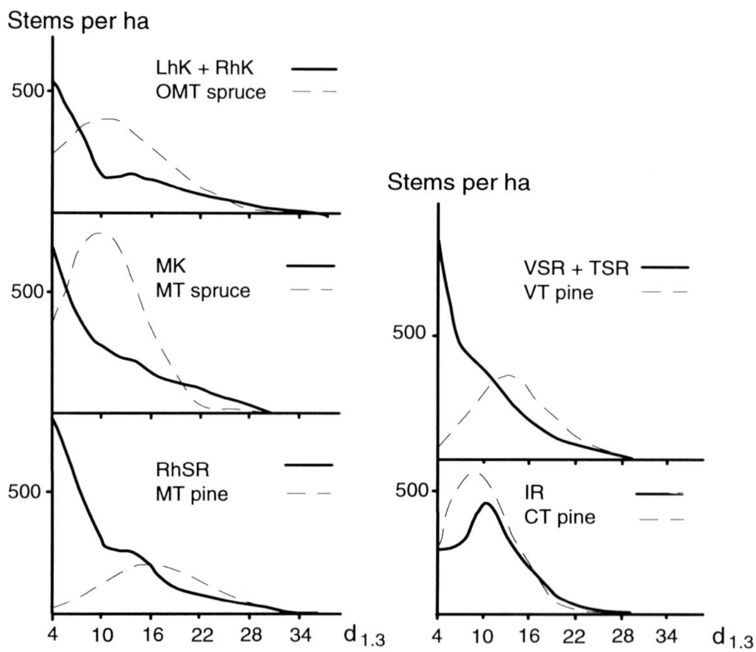

Fig. 6.6. Diameter *($D_{1.3}$)* distributions from peatland sites drained 31–50 years previously as compared with the diameter distributions from 60–70 year old upland site stands of similar fertility in southern Finland. For the original mire site types, see Table 3.17. Upland site types: *OMT spruce* Norway spruce on *Oxalis-Majanthemum* type; *MT spruce* Norway spruce on *Vaccinium myrtillus* type; *MT pine* Scots pine on *Vaccinium myrtillus* type; *VT pine* Scots pine on *Vaccinium vitis-idae* type; *CT pine* Scots pine on *Calluna* type. (Redrawn from Hökkä and Laine 1988)

Table 6.2. The increase in the number of sawlog stems during a 50-year period after drainage on various site types in southern Finland. (Data from Hökkä and Laine 1988)

Site type	Sawlog stems per ha		
	Virgin sites	50 years after drainage	Increase
LhK, RhK[a)]	150	290	140
MK	110	280	170
KgK	110	250	140
KgR	55	150	95
IR	30	100	70
RhSR	25	240	215
VSR, TSR	15	150	135

[a)] For the site types, see Table 3.17.

mires is sparse and unevenly scattered over the area. The latter situation is particularly true for sparsely forested composite pine mires and hardwood-spruce mires. As growth conditions improve after ditching, openings fill up with seedlings and the density of the stands increase. In the richest composite types in particular, an immediate reaction to drainage is observed – a strong increase in the quantity of trees with a small diameter, particularly birch. Diameter distributions show that even 31–50 years after drainage most of the trees forming the stand are relatively small compared to trees on upland sites (Fig. 6.6). The increase in stem number slows down when the stands have reached full density with respect to their development class.The number of stems reaching saw-timber dimensions in the first stand after draining depends on both the geographical location and the fertility of the site type (Table 6.2). In both southern and northern Finland the number of trees that have reached saw-timber dimensions by the time of draining has been greater in hardwood-spruce mires than in pine mires. In older drainage areas (50 years after draining on average), the number of trees of saw-timber dimensions is highest on the best hardwood-spruce mires in southern Finland. However, the increase in the number of stems reaching saw-timber dimensions has been largest in herb-rich sedge, birch-pine fens. The high potential capacity of composite mires to produce saw-timber after drainage is obvious (Hökkä and Laine 1988).

Also typical of tree stands on drained areas is that the trees grow in groups. This is particularly true for the composite types of hardwood-spruce and pine mires. The gaps are gradually filled in after drainage. However, drainage may also lead to unevenness in the growth of the trees within the stand. Tree growth is usually better close to the ditches (Seppälä 1972). Thus, the trees growing in the middle parts of the strips between ditches will be outgrown by those growing near the ditches. The difference in growth between trees growing near to and further from the ditches becomes clearer the greater the ditch spacing or the more defective the drainage for some other reason.

6.3.1.3 The Pubescent Birch on Peat

Pubescent birch (*Betula pubescens*) is more common than silver birch (*B. pendula*) in Finland. By volume, pubescent birch exceeds silver birch threefold. The frequency of pubescent birch is relatively highest in the region of Ostrobothnia, Finland (Fig. 6.7), where there is also plenty of drained peatland (Ferm 1989). These birch species differ from each other ecologically; silver birch usually dominates on uplands and pubescent birch on mires and other wetlands. Birch trees growing in peatlands are almost without exception pubescent birch, although silver birch may be found to some extent mixed in stands on the very best transformed drained peatlands.

In drained peatlands, as on mineral soils, pubescent birch is a typical pioneer tree species, which in a short time is able to occupy sites which for some reason or other are free. However, as it is a rather short-lived, light-demanding species, it gives way to spruce in time. Ditching or clear-cutting may create free-growing space. Pubescent birch is able to colonize the clear-cuts in hardwood-spruce mires at a very fast rate. The fertility of the best pine mires and treeless mires is sufficient for birch, and consequently it tends to colonize such sites after drainage (Keltikangas and Seppälä 1977).

Pubescent birch possesses several properties which mean that this species can be used to improve the soil. Birch transpires large quantities of water. The biological drainage produced in this way is particularly siginificant and important in

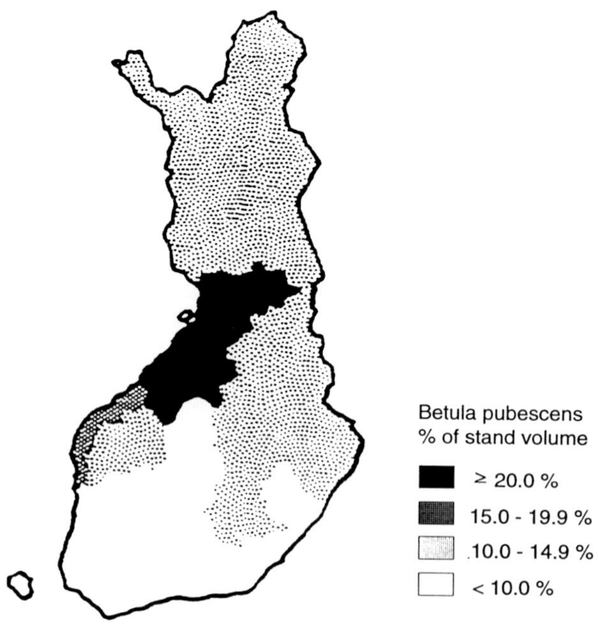

Fig. 6.7. Share of *Betula pubescens* of the total standing volume of the stands in different parts of Finland. (Redrawn from Ferm 1989)

the summer. When the soil is warm and the microbic activity lively, drainage should be effective to allow sufficient soil aeration.

In drained peatlands, the root systems of pubescent birch penetrate deeper into the soil than do the roots of conifers (Huikari 1959c; Paavilainen 1966b). It has been established that pubescent birch tolerates anaerobic conditions better than other tree species. In pure pine stands growing on peat, the density of mycorrhizae decreases at a considerable rate with increasing peat depth. On the other hand, for mixed stands with birch and conifers, it has been established that the numbers of mycorrhizae in the peat profile of both pine and spruce remain unchanged down to a depth of as much as 20 cm (Heikurainen 1955, 1959a). Thus, birch seems to promote the mycorrhizal infection of the root systems of conifers.

Birch litter physically hinders the growth of mosses. In birch stands, litter yields are greater and nutrient cycling more effective than in coniferous forests (see Sect. 3.3; Mikola 1973; Paavilainen 1984a). The production of pubescent birch in drained peatlands will be dealt with in Section 6.3.3.

6.3.2 Treatment of Young Stands

In broad terms, the treatment of young stands refers to a series of measures that ensure the success of natural or artificial regeneration. After reforestation the first young stand treatment measures are the prevention of excess competition from ground vegetation, the recording of successful reforestation and – if necessary – stand supplementation or, in the worst case, replanting. More specifically, however, the **treatment of young stands** refers to measures aimed at controlling the development of tree species and the growth density of stands regenerated naturally or through artificial regeneration, which do not yet produce commercial wood (pre-commercial thinning). **Young stand cleaning** refers to the removal of harmful trees which differ from the species under regeneration. **Young stand thinning** refers to the removal of excessive density of the regenerable tree species (Päivänen 1980c).

On meso-oligotrophic or poorer peat soils there is usually no urgent need for cleaning, but on better site types *Betula pubescens* has to be controlled if the aim is to grow Scots pine (*Pinus sylvestris*). From the wood production point of view, birch-pine mixed stands are not recommended on drained peatlands even though evapotranspiration during the growing season is higher in mixed stands than in pure pine stands (Zalitis 1989). Light-demanding Scots pine is too easily suppressed by birch.

As mentioned earlier, it would also be feasible to retain the birch, particularly on rich sites, as a nurse stand initially, and then to thin it successively to allow the naturally regenerated Norway spruce (*Picea abies*) understorey to develop (Seppälä and Keltikangas 1978). If aiming at a predominantly coniferous stand as soon as possible, the Norway spruce should be released at a height of 4 m (Heikurainen 1985). However, it is also possible to grow two-storey stands where a part of the birch will later reach timber size (Päivänen 1990).

Considerations relating to water regime (Päivänen 1982a), to improving the quality of trees (Kaunisto and Tukeva 1986) and perhaps also to anticipating possible vole (*Microtidae*) (Korhonen 1987) and moose (*Alces alces*)(e.g. Edenius 1991) damage speak in favour of dense pine or pine-dominated young stands on drained peatlands. There seems to be no ecological need to treat pure pine stands before the first commercial thinning, but diameter growth slows down as the number of live seedlings increases (Kaunisto and Tukeva 1986). Mechanization of the first thinning may, however, presuppose pre-commercial thinning.

In Finland, most of the young stand thinnings and cleanings are done manually using a brush saw. The effective working time may be greater than on mineral sites as the felled trees cannot be left in the ditches, but, on the other hand, there are no obstacles such as stones on mire sites (Kaunisto and Päivänen 1985). In the past, hardwoods were controlled by spraying herbicides, but the use of this treatment has been decreasing recently because of legislation concerning aerial spraying (Päivänen 1980c).

When pre-commercial thinning is carried out, the target is to leave the following numbers of young trees per hectare:

Pinus sylvestris	2000–3000 (depending on the site)
Picea abies	2000
Betula pubescens	2000–2500

The general idea is that, after pre-commercial thinning, the growth loss in the young stand should not be too high and the harvesting yield in the coming first commercial thinning should be sufficient from an economical point of view (e.g. Niemistö 1991).

Although the biological benefits of thinning in overly dense young conifer stands have been fairly well demonstrated in Canada (Riley 1973), the practice has not been widely applied there.

6.3.3 Intermediate Cuttings

Most of the **intermediate cuttings** on drained areas have the character of **first thinnings**. They may involve both the felling of standards (shelter trees) and normal thinnings. The felling of standards is necessary particularly in the case of old, tall trees which remain in the stand after draining and are incapable of revival. The need for cutting in tree stands growing on drained peatland areas should be based on the development class and density of the whole stand even if the sparsest parts of the stand do not yet require thinning.

Our knowledge about the effects of cuttings of varying intensity on the timber production of the stands in forest drainage areas is not complete. Thinning models that have been worked out for even-aged upland forests on the basis of dominant height and basal area are not suited for the first thinning in drained areas. This is because the stands on drained peatlands are very uneven: they usually consist of several tree species, the stems are unevenly distributed and the variation in age is large. Therefore, the first thinnings of tree stands on drained areas should be performed using stem number as the basis for thinning intensity (Table 6.3). Later, thinnings in stands dominated by coniferous tree species may be based on thinning models worked out on the basis of the dominant height and basal area.

Usually, the amount of removal in intermediate cuttings on drained peatlands is relatively low, although the opening of the **strip roads** for the first harvesting will increase it somewhat. After intermediate cutting there are both harvesting strip roads and **ditch openings** in the tree stand. In a case study their share of the area was found to be 14 and 6%, respectively (Rantonen and Päivänen 1989). The distance between the strip roads is normally 20 to 30 m (Eeronheimo 1991). Strip roads do cause some production loss in the stand, although the trees at the edge of the strip roads react positively to the asymmetrically increased growing space (Isomäki and Niemistö 1990). **Harvesting damage** caused to the above-ground parts of the remaining trees on drained peatlands have not been found to be greater than on uplands. However, severe ground damage (deep ruts on the strip roads) may be more frequently found than on mineral soils (Rantonen and Päivänen 1989).

On drained peatlands of medium or better fertility the development of pubescent birch (*Betula pubescens*) does not seem to lag behind that of pine growing on similar sites. The growth of pubescent birch revives soon after draining and its productivity is similar to or even better than that of Scots pine (*Pinus sylvestris*) (Keltikangas and Seppälä 1977; Saramäki 1977). Radical thinning in a pubescents birch stand leads to production loss (Fig. 6.8). For northern Finland, it has been suggested that if the aim is to produce pulpwood in *Betula pubescens* stands on peat, the optimum density is 2000 stems ha^{-1} up to the first commercial thinning, at a dominant height of 15 m. The density after this first thinning is 900 stems ha^{-1}. A second thinning is usually not even needed (Niemistö 1991).

When considering the intensity of thinning – irrespective of the tree species to be grown – one must know how and when possible improvement ditching will be carried out (see Sect. 4.2). If thinning is not followed by improvement ditching,

Table 6.3. Recommendations for the number of stems per ha after the first thinning on drained peatland forest site types in southern Finland. (Data from Päivänen 1990)

Tree species	Site type[a)]	Stems per ha	
		Dominant height	
		< 11 m	> 12 m
Pinus sylvestris	*V. myrtillus* type	1400	1100
Pinus sylvestris	*V. vitis-idaea* type	1300	1000
Pinus sylvestris	Dwarf-shrub type	1100	800
Picea abies	Herb-rich type and *V. myrtillus* type	1400	1200
Betula pubescens	*V. myrtillus* type	1600	1200

[a)] For the drained peatland forest site types, see Sect. 3.4.2.

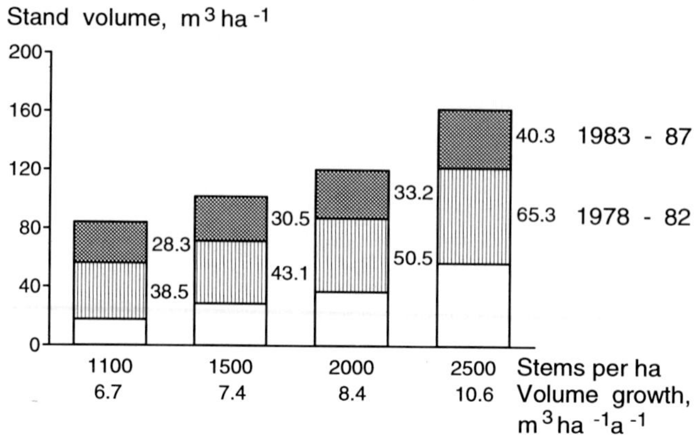

Fig. 6.8. The effect of thinning intensity on the stand volume development of *Betula pubescens* on drained peatland *Vaccinium myrtillus* type. At the time of the treatments, the stand age was 20 years and the dominant height 9–10 m. (Redrawn from Ferm 1989)

the treatment of the tree stand should be light in order to prevent a significant rise in the groundwater level (Päivänen 1982a). On the other hand, due to the fact that timber harvesting in peatland forests is expensive, thinnings should be of an intensity which allows as small a number of thinnings as possible.

Thinning experiments on drained peatlands under North American conditions are only in the beginning stages (Mugasha et al. 1991).

6.3.4 Regeneration

6.3.4.1 Maturity

Regeneration maturity usually refers to the age at which a tree stand can be regenerated. The average age of a tree stand on forest drainage areas is, however, obscure due to the uneven age structure at the time of drainage and the differing time elapsed since ditching. Thus, during the first tree generation, regeneration maturity on drained mires should be based on the average stem diameter and growing vigour of the trees rather than on biological age.

6.3.4.2 Natural Regeneration

Some of the tree stands on drained peatlands gradually regenerate naturally without regeneration cuttings. This is partly because the stands were sparse and the spatial distribution of trees uneven at the time of drainage. Although there is some loss of **receptivity for regeneration** after drainage when *Sphagnum* mosses are replaced by feather mosses, regeneration is still satisfactory, especially on the better sites. The receptivity for regeneration can be estimated on the basis of the number of seedlings – the so-called changing stock of seedlings or advance growth less than 10 cm in height (Kaunisto and Päivänen 1985).

The second tree generation on drained mires should be more or less even-aged. Exceptions are sites where *Betula pubescens* is a pioneer tree species with an understorey of Norway spruce (*Picea abies*) (Fig 6.9). Spruce, being shade-tolerant, can grow as the lower storey in such two-storey stands, with birch acting as a nurse stand for the regenerating spruce (Fig 6.10). The best quality birches may even be grown up to timber size. The standards have to be harvested in winter during thaw so that spruce forming the next generation on the site will not be damaged.

Stands predominated by Norway spruce (*Picea abies*) may be regenerated naturally according to the same principles as described in Section 6.2.3, see also Fig. 6.11. The results of experiments carried out in different parts of Sweden in-

Fig. 6.9. Spruce (*Picea abies*), being shade-tolerant, can grow as a lower storey under the pioneer tree species (*Betula pubescens*). Herb-rich sedge fen, originally open, 30 years after drainage, southern Finland. (Photo: J. Päivänen)

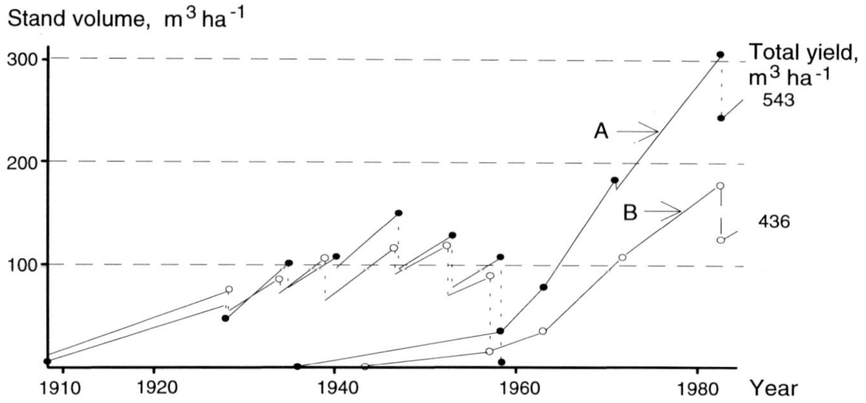

Fig. 6.10. The development of the stand volume on two sample plots on fertile sites in southern Finland since drainage. Norway spruce (*Picea abies*), naturally regenerated under the pioneer birch (*Betula pubescens*), is forming the second rotation. *A* herb-rich type; *B Vaccinium myrtillus* type II, for drained site types, see Sect. 3.4.2. (Redrawn from Paarlahti and Takamaa 1984)

Fig. 6.11. Especially on the better sites, natural regeneration is satisfactory even in old forest drainage areas. *Picea abies* seedlings on a herb-rich site, southern Finland. (Photo: J. Päivänen)

Fig. 6.12. Seed tree position with the aim of natural regeneration of *Pinus sylvestris* on a dwarf-shrub type, southern Finland. However, without scarification the stocking is insufficient. (Photo: J. Päivänen)

dicate that on fertile, drained mires (originally "low herb type", see Hånell 1991a) about 200 stems ha^{-1} should be left when applying shelterwood cutting (Hånell 1991b; Hånell 1993). This will leed to big enough stocking of Norway spruce regeneration and, at the same time, the extent of windthrow will still be acceptable. For the windthrow hazard, see Section 6.4.1.

In the naturally regenerated spruce mires, the old ditches should not be cleaned until after the harvest of the standards (shelter trees).

Natural regeneration of Scots pine (*Pinus sylvestris*) on poorer sites in old drainage areas is not as easy as that of spruce on better sites. The feather mosses, especially *Pleurozium schreberi*, are known to form a poor germinating substrate for coniferous seeds (Sarasto and Seppälä 1964; Wood and Jeglum 1984). In many cases, a raw humus layer of varying thickness develops on oligotrophic and meso-oligotrophic drained mires approaching the final stage of post-drainage succession. This seems to impair regeneration (Kaunisto and Päivänen 1985). Thus, a light scarification in connection with seed-tree cutting on old drainage areas where Scots pine is predominant is recommended (Kaunisto 1984b, see also Fig. 6.12). However, suitable equipment for this purpose is lacking.

A similar recommendation exists in Canada. When regenerating black spruce mires there, one philosophy in mechanical site preparation is not to cut too deeply, i.e. to "scarify delicately" (Jeglum et al. 1982).

6.3.4.3 Site Preparation for Artificial Regeneration

Soil preparation improves the shoot/root ratio of seedlings and stimulates growth by decreasing the competition from ground vegetation and improving the biological and physical properties of the substrate. The correct form of microsite for planting is highly site-dependent (e.g. Nelson and Quine 1990). Soil preparation on peatlands has positive and negative effects on the conditions for growth at the regeneration spot (Kaunisto and Päivänen 1985):

Growth conditions	Change: good (+)/poor (–)	
Temperature conditions	Improved	+
Aeration	Improved	+
Competition from ground vegetation	Decreased	+
Insect damages	Decreased	+
Nutrient cycling	Accelerated	+
Local drainage	Improved	+
Availability of water	Impaired	–
Erosion	Increased	–
Soil cracking	Increased	–
Peat adhering to seedlings	Increased	–
Soil frost heaving	Increased	–
	or decreased	+

However, the long-term beneficial effects are considered to dominate over the short-term negative ones.

Mechanical soil preparation on peatlands is hampered by the poor bearing capacity of the peat. Brush, stumps and roots impede deep cultivation such as plowing (Nelson and Quine 1990). At the moment, the only soil preparation method applicable on an operational scale to forested peatlands after clear-cutting is mounding by tractor digger or excavator (Fig. 6.13). However, this is an expensive method. In connection with clear-cutting and planting, the mounding should be done in such a way that the excavator digs shallow ditches at 10–12 m spacings at the same time. Despite the obvious benefits of mounding for the success of planting and the growth of seedlings (e.g. Mannerkoski 1975; Hånell 1992), much more research is required in order to find the biologically and economically best soil preparation techniques.

Prescribed burning could be an effective measure for eliminating the raw humus layer that covers the poorer sites on old drainage areas to be regenerated

Fig. 6.13. Before planting the site was prepared with mounding. Scots pine (*Pinus sylvestris*) transplant starting its first growing season in the field, northeastern Finland. (Photo: J. Päivänen)

with Scots pine (*Pinus sylvestris*)(Yli-Vakkuri 1958). In Canada as well, the best results in regenerating black spruce (*Picea mariana*) have been produced by post-cut burning in order to aid natural seeding on moist to wet lowland peat sites (Chrosciewicz 1990). There is, however, the danger of causing a wild peat fire.

6.3.4.4 Artificial Regeneration

Artificial regeneration in connection with the final cuttings on peatlands has only been practiced on a small scale up to now. This is because most of the old drainage areas where the tree stands have reached maturity are fertile sites where receptivity for regeneration remains fairly high and natural regeneration is possible.

A long-term project to evaluate alternative methods of regenerating old drainage areas is being carried out (Saarinen 1989). At the moment, **planting** is recommended for nutrient-rich sites where the competition from ground vegetation after clear-cutting may hamper the development of seedlings. For Norway spruce (*Picea abies*), it is essential to have a nurse stand to shelter the young trees from spring frost damage. Frost-tolerant Scots pine (*Pinus sylvestris*) cannot be planted on sites which are too fertile because the technical quality of the trees would not be acceptable.

An artificial regeneration experiment has been established in different locations in Sweden. The experiment includes site preparation carried out as mounding on both clear-cut areas and under shelter trees. Planting without site preparation on the clear-cuts was the least successful, while the best survival was observed under the shelter trees with mounding (Hånell 1992). In some of the experimental sites spruces planted on the clear-cuts suffered heavily from spring frost. There were also dramatic changes in the field layer vegetation caused by the different cutting measures. Clear-cutting resulted in an increase in the ground cover of the field layer but species diversity was decreased (Hannerz and Hånell 1993). The effect of the field layer competition on the development of the planted spruce was not studied, however.

Bare-root seedlings are preferred on peatlands. *Betula pendula* can be planted only on shallow peat soils where the roots are able to penetrate down to the mineral soil.

6.4 Damage in Peatland Forests and Its Control

6.4.1 Abiotic Damage

6.4.1.1 Growth Disturbances

Growth disturbances induced by unbalanced nutritional conditions in tree stands on drained mires have been reported in several studies (e.g. Kolari 1983; Veijalainen et al. 1984; Kolari 1988). The growth disturbance symptoms in Scots pine (*Pinus sylvestris*) start with a loss of apical dominance or dieback and lead to bushy growth. Growth disturbances are rarer in Norway spruce (*Picea abies*) and in the birches. The symptoms have, however, also been described for these tree species (Veijalainen et al. 1984). Trees growing in growth disturbance areas are characterized by high levels of foliar nitrogen ($N > 1.7\%$) and low levels of boron ($B < 5–10$ mg kg^{-1})(Kolari 1988).

The total area affected by nutritional growth disturbances in Finland was estimated to be about 100 000 ha in 1983 (Veijalainen et al. 1984). Most of the growth disturbance stands consisted of 3-to 8-m-tall Scots pine trees planted on abandoned agricultural peats or on nutrient-rich drained mires. Wood ash and borate fertilizer have proved beneficial in controlling the growth disturbances (Kolari 1988; Ferm et al. 1992). In the severest cases, giving up the cultivation of Scots pine or any commercial tree species should be considered, particularly if the afforestation attempts have repeatedly failed (Kaunisto and Päivänen 1985). Nowadays all commercial fertilizers for peat soils contain boron in order to prevent growth disturbances.

6.4.1.2 Wind Damage

Tree root systems in peat are usually superficial (Heikurainen 1955, 1959a; Paavilainen 1966a) and peat strength properties may be poor for tree anchorage. Stands on peatlands are therefore potentially susceptible to windthrow (Goodwin 1968; Fraser 1969, see also Fig. 6.14).

Wind has both physiological and mechanical effects on tree growth and ultimately tree stability. Tree species also differ in their susceptibility to windthrow. For example, it has been found that the rooting zone of *Pinus contorta* is smaller than that of *P. sylvestris* (Sundin 1982) and therefore the former may be more susceptible to windthrow.

Wind is considered to be the most limiting single factor in British silviculture (Malcolm 1991). Thus, the windthrow hazard has to be taken into consideration when an area is evaluated for afforestation, and silvicultural treatments such as site preparation, drainage and thinning are planned (Booth 1974; Ford 1979). The British **windthrow hazard classification** is mainly based on elevation above sea level, general exposure and soil type (Booth 1977). Soils subjected to waterlogging often get the highest score values when estimating the hazard class.

In the boreal zone, peatlands are low-lying and therefore less exposed to frequent strong wind (Kaunisto and Päivänen 1985). However, when deciding upon which regeneration method to use, due attention should be paid to the wind resis-

Fig. 6.14. Shelter tree position presupposes intermediate cuttings, otherwise the stand may be susceptible to windthrow. *Picea abies* stand in southern Finland. (Photo: J. Päivänen)

tance of the remaining stand. If natural regeneration is carried out by means of shelterwood or strip-cuttings, the stand should have undergone earlier thinning cuttings (Lukkala 1946). The risk of windthrow is also smaller if the number of the remaining Norway spruce (*Picea abies*) standards is high enough (> 200 stems ha^{-1}, Hånell 1991b; Hånell and Ottosson-Löfvenius 1994). The relation of the borderline length to the area of the remaining stand in the strip-cutting of black spruce (*P. mariana*) also affects windthrow damage (Elling and Verry 1978). The borders of clear- and strip-cutting areas should follow the terrain in order to minimize the risk of windthrow.

6.4.1.3 Spring Frost

Because of their low-lying locations, mires are often susceptible to **spring frost**. Freezing temperatures, especially at the time of budding and early shoot development, are detrimental to some tree species. Water is moved from cells to intercellular gaps where it forms ice crystals and ruptures the bud. This "drying of cells" causes shoot dieback.

In Fennoscandia, early shoots of Norway spruce (*Picea abies*) are damaged when the temperature falls below –3°C (Multamäki 1942). Spring frost damage is more common in southern Finland than in the northern part of the country (Leikola 1975). In the north tree growth starts later in the spring and the nights are too short to allow the temperature to fall to the freezing point. Scots pine (*Pinus sylvestris*) is less susceptible to frost damage; only exceptionally low temperatures in June may cause damage (Raitio 1985).

It has been found that not only night frosts early in the summer, but also constant sunshine during the following days may increase frost damage in Norway spruce (*Picea abies*)(Lundmark 1987). Thus, the positive effect of a nurse crop is not restricted to protection against cold night-time temperatures, but the shadowing effect during the day may be as important. The crown cover of a birch nurse crop (*Betula pubescens*) above a Norway spruce (*Picea abies*) regeneration area should be about 70% in order to give enough protection (Multamäki

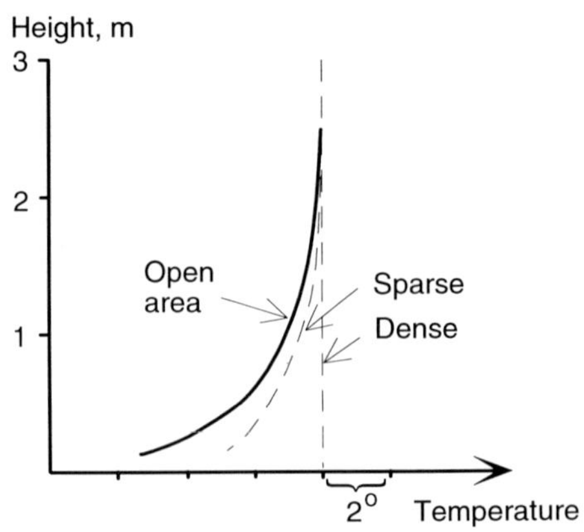

Fig. 6.15. The temperature gradient in air near the soil surface and its dependence on the density of the nurse stand during a typical spring frost night. (Redrawn from Odin et al. 1984)

1942; Leikola and Pylkkö 1969; Leikola 1975). The temperature under such a nurse crop may be 1–5°C higher than in the open area during nights, with heavy out-radiation (Multamäki 1942; Leikola and Pylkkö 1969; see also Fig. 6.15). The nurse crop can consist of pioneer *B. pubescens* forming the first generation on the site after drainage, or birch bushes originating from sprouts after clear-cutting.

The exotic tree species *Pinus contorta* and *Picea mariana* are tolerant to spring frosts and thus could be planted on drained treeless mires in Fennoscandia (Laine 1979; Päivänen 1983).

6.4.2 Biotic Damage

6.4.2.1 Mammal Damage

Damage caused by ***moose*** (*Alces alces*) in young Scots pine (*Pinus sylvestris*) stands has increased noticeably in the last few decades in Finland, on both mineral and peat soils (Fig. 6.16). This is partly due to the greater wintering moose population and partly due to the increase in the area of artificially regenerated young pine stands (Repo and Löyttyniemi 1985). A high proportion of peatland forests is found in areas of high moose density (Heikkilä and Härkönen 1993). Although quantitative aspects play an important role in the food consumption of the moose, it consumes more from the fertile-soil than from the poor-soil Scots pine (*Pinus sylvestris*) phenotype if both are available (Niemelä and Danell 1988). Forest fertilization appears to increase the risk of moose damage. On oligotrophic drained mires, nitrogen fertilization in particular has led to a greater risk of moose damage in Scots pine plantations (Laine and Mannerkoski 1980). However, nowadays sites needing nitrogen fertilization are not drained.

The moose selects between different tree species. When Scots pine (*Pinus sylvestris*) and black spruce (*Picea mariana*) were grown in the same drained mire site, only the former was damaged by moose (Päivänen 1983). The presence of birch (*Betula pubescens*) may increase moose damage in young Scots pine stands (Heikurainen 1982b). Removing the hardwoods in young Scots pine stands may reduce moose damage (Lääperi and Löyttyniemi 1988). However, contradictory results have also been presented (Edenius 1991). In response to the removal of the more palatable hardwood species, the moose may include lower ranked species like pine in its diet. Nevertheless, it is usually agreed that a dense stand is relatively resistant against moose damage as browsing is distributed over more trees than in a sparse stand (Edenius 1991; Heikkilä 1991; Heikkilä and Mikkonen 1992; Lyly and Saksa 1992). Young Scots pines do not usually die as result of moose damage, but stem defects caused by shoot breaking are still visible at the first commercial thinning (Heikkilä and Löyttyniemi 1992).

Some of the Fennoscandinavian **voles** (*Microtidae*) can be regarded as potential pests of seedling stands. Most of the damage is done during the winter when the voles gnaw and eat the bark, root collar, roots and even buds (Korhonen 1987). All commercial tree species are susceptible to vole damage, although Scots pine (*Pinus sylvestris*) is affected the worst.

Fig. 6.16. Scots pine (*Pinus sylvestris*) is not safe from moose (*Alces alces*) damage until it has passed the height of 3 m, southern Finland. (Photo: J. Päivänen)

In order to avoid vole damage, Norway spruce (*Picea abies*) should be planted whenever the site is fertile enough. On former agricultural peats in northern Finland, which seem to be the most vulnerable areas, the sowing of Norway spruce as a supplement to natural regeneration has been recommended (Kaunisto and Päivänen 1985). Vole damage could probably be prevented or at least reduced by planting in the spring following the decline in the vole population (Teivainen 1979).

6.4.2.2 Insect Damage

Young Scots pine (*Pinus sylvestris*) stands on drained peatlands in Finland suffer from the same insect damage as on mineral sites. However, stands established on open, drained mires may be less infected than those on mineral soil regeneration areas because of the absence of logging slash and stumps which form a suitable reproduction bed for certain insects. For example, damage caused by the pine weevil (*Hylobius abietis*) occurs less frequently in Scots pine seedlings on peat than on mineral sites (Heikkilä 1979). Recently, damage caused by the Aradus bug (*Aradus cinnamomeus*) have become more common in young Scots pine stands on both dry mineral soil sites and peat. The damage may be reduced by growing dense stands of Scots pine mixed with *B. pubescens* (Nuorteva 1982; Heliövaara and Väisänen 1983). Insect damage in young Scots pine stands usually results only in a loss of growth rather than increased mortality (Heliövaara 1982).

The fringe of Norway spruce (*Picea abies*) trees along the edge of strip cuts often become prone to infection by *Ips* species. Too heavy sunshine and attack by *Ips* species may even result in the death of trees (Nuorteva 1982). By directing the strip-cutting from south to north, the net radiation, and thus an outbreak of *Ips* infection, is reduced.

6.4.2.3 Fungus Damage

Snow blight fungus (*Phacidium infestans*), needle cast (*Lophodermella sulcigena*) and pine canker fungus (*Ascocalyx abietina* or *Gremmeniella abietina*) are the worst damage-causing fungi in young Scots pine (*Pinus sylvestris*) stands. Pine canker fungus also attacks lodgepole pine (*Pinus contorta*). In order to prevent snow blight fungus, it is important that the pine seedlings grow above the snow cover as fast as possible. Fungus damage on drained mires may be associated with growth disturbances caused by nutritional imbalance (Kurkela 1983). Therefore, taking care of balanced nutrient and water conditions helps to prevent the occurrence of fungus damage. *Chrysomyxa ledi* sporadically infects Norway spruce on mires, turning the recent-year needles yellow. Such epidemics are, however, usually harmless for tree growth.

Godronia multispora destroys young planted *Betula pendula* stands on peatlands by attacking the bark of the seedlings. This may be one reason why *B. pendula* does not thrive on drained mires (Kurkela 1973).

7 Effect of Forest Drainage and Fertilization on Wood Production Capacity

7.1 Increase in Productive Land Base

The most important aim of forest drainage is to increase the area of productive forest land and the growth of naturally tree-covered peatlands. Unproductive peatlands are mainly open mires, but there are also large areas of peatlands growing low-productive forest in Finland, Sweden, Russia and Canada, for instance. If climatic conditions are not limiting, the productivity of these peatlands can be increased through drainage.

The increase in the productive land base that can be brought about by forest drainage is well illustrated by the results of the National Forest Inventories (NFI) carried out several times since the 1920s in Finland (e.g. Paavilainen and Tiihonen 1988). In these inventories, peatlands are classified into three categories as follows:

1. **Forest land** (mean annual increment under optimal silvicultural conditions and with a normal rotation period at least 1 $m^3\,ha^{-1}$ over bark);
2. **Scrub land** (mean annual increment at least 0.1 $m^3\,ha^{-1}$, but less than 1 $m^3\,ha^{-1}$);
3. **Waste land** (mean annual increment below 0.1 $m^3\,ha^{-1}$).

To show the effect of climate, the results of the forest inventories from southern and northern Finland are dealt with separately in the following section. These regions are shown in Fig. 7.1.

As mentioned in Section 4.1, large-scale forest drainage activity in Finland began in the 1950s. According to the third inventory (1951–1953), the total area of drained peatlands in southern Finland at that time was 0.681 million ha, of which 0.546 million ha was forest land and 0.135 million ha scrub land or waste land (Fig. 7.2). In the eighth inventory (1986–1991), the total area of drained peatlands was 2.550 million ha, or an increase of 1.869 million ha compared to about 35 years earlier. The area of forest land on drained mires increased during this period by 1.746 million ha. Part of these drained mires is still scrub or waste land and too poor for wood production.

The foregoing results differ from those for northern Finland. In this climatically less favourable part of the country, the total area of drained peatlands at the time of the third inventory was 0.236 million ha, of which 0.146 million ha was

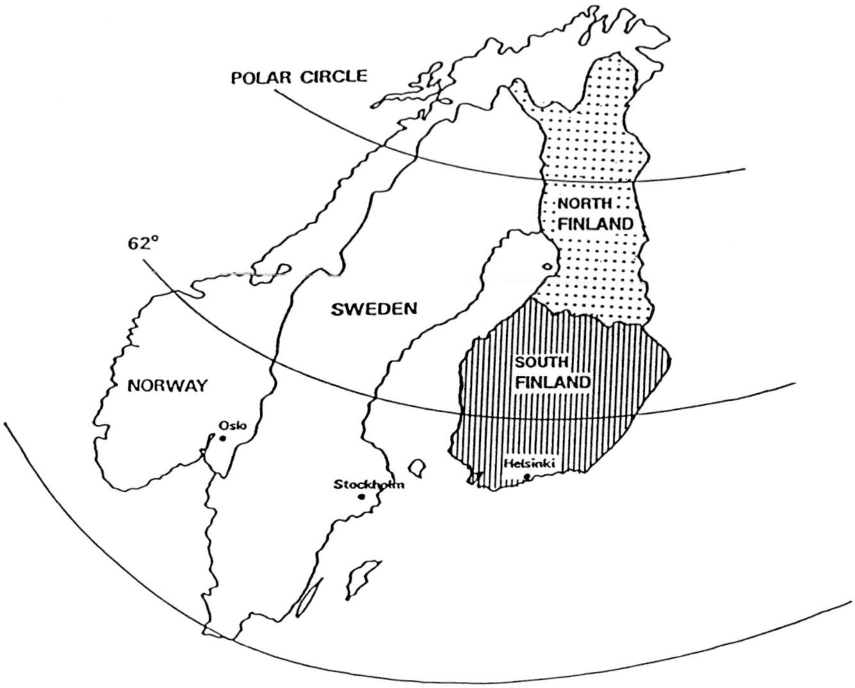

Fig. 7.1. The southern and northern regions of Finland as used in the National Forest Inventory (NFI)

forest land (Fig. 7.2). About 30 years later, at the time of the seventh NFI (1982–1984), the area of forest land on drained mires had increased to 1.532 million ha and the total area of drained mires was 2.034 million ha. The share of drained mires in the category of scrub land or waste land was much higher in northern Finland than in the southern part of country. This difference shows first of all, the effect of the shorter time period since drainage in northern Finland and the slow development of the stands under the harsher northern climatic conditions. It also shows that the productive potential of some northern site types, in particular, was overestimated at the time of drainage.

The guidelines for evaluating the profitability of forest drainage have changed in Finland because of changing drainage costs, timber prices and so on, and the need to increase wood production. According to 1987 guidelines, as much as 14.6% of ditching by the 1980s was unprofitable, but this figure dropped to only about 3% when the initial economical criteria for drainage from the 1960s were applied (Keltikangas et al. 1986; Keltikangas 1990).

The example from Finland shows that forest drainage is an effective measure for increasing the productive land base in the boreal zone. It also shows that the potential production capacity of the site (site type and climatic region as indicators) after drainage should be considered when deciding which areas will be ditched.

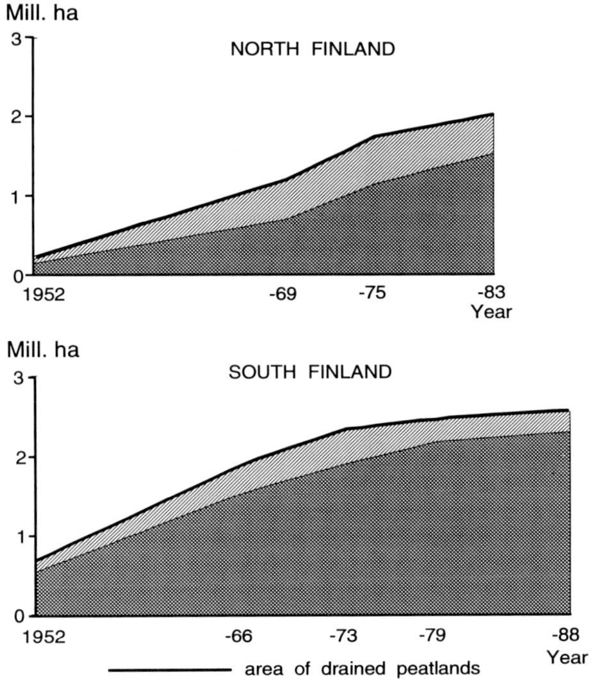

Fig. 7.2. The total area of drained peatlands and the share classified as forest land (mean annual increment > 1 m^3 ha^{-1}, *dark shading*) in Finland according to the National Forest Inventories. (Redrawn from Paavilainen and Tiihonen 1988 and preliminary data from 8th NFI)

7.2 Increase in Yield

7.2.1 Stand Growth

Post-drainage tree stand productivity is primarily dependent on the fertility of the site, the climatic conditions and the amount and quality of the existing growing stock. Other factors affecting tree growth include time since drainage, the spacing of ditches and the thickness of the peat. In northern European peatlands where Scots pine (*Pinus sylvestris*), Norway spruce (*Picea abies*) and birch (*Betula pubescens*) are the main tree species, the results of drainage are generally better, the more fertile the site is, the higher the temperature sum and the more timber there is on the site at the time of drainage (e.g. Heikurainen 1959b; Seppälä 1969; Heikurainen and Seppälä 1973; Laine and Starr 1979; Keltikangas et al. 1986; Hånell 1988a, 1990; Gustavsen et al. 1994).

Figure 7.3 shows the dependence of the post-drainage stand increment on the original site type and climatic conditions (indicated by temperature sum) based on the material of Hånell (1990) from Sweden. The annual increment for different site types, for example, at a temperature sum of 1300 dd°C is as follows:

Original site type, tree species	Annual increment ($m^3\,ha^{-1}\,a^{-1}$)
1. *Aconitum-Filipendula*/tall-herb type, spruce + hardwoods	10.2
2. *Vaccinium myrtillus*/bilberry-horsetail type, spruce + hardwoods	8.4
3. *Carex rostrata*/tall-sedge type, pine	3.4
4. *Eriophorum vaginatum*/low-sedge type, pine	2.2
5. *Calluna vulgaris*/marsh *Andromeda*-cranberry type, pine	0.4

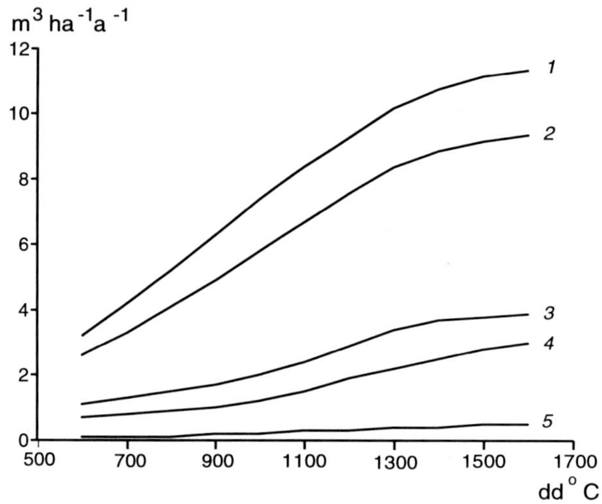

Fig. 7.3. The dependence of post-drainage stand increment on the original site type and temperature sum. (Data from Hånell 1991a). Site types: *1 Aconitum-Filipendula*/tall herb type, tree stand: spruce + hardwoods; *2 Vaccinium myrtillus*/bilberry-horsetail type, tree stand: spruce + hardwoods; *3 Carex rostrata*/tall-sedge type, tree stand: pine; *4 Eriophorum vaginatum*/low-sedge type, tree stand: pine; *5 Calluna vulgaris*/marsh *Andromeda*-cranberry type, tree stand: pine

The results of the 7th Finnish NFI (1977–1984) show a similar effect of site type and climatic conditions on the post-drainage stand increment (Fig. 7.4).

Generally, there is a strong correlation between the increment after drainage and stand volume (e.g. Heikurainen 1959b; Heikurainen and Seppälä 1965, 1973; Laine and Starr 1979; Gustavsen et al. 1994). On the basis of this relationship and some other easily measurable stand and environmental variables regression functions for predicting the volume and total yield of peatland stands on sites drained for forestry in southern Finland have been developed (Gustavsen et al. 1994). Independent variables used for modelling volume increment were stand volume and stem number, and for modelling total yield, stand volume and years since drainage was used. More accurate functions were developed using temperature sum, thickness of peat layer and ditch spacing as additional independent variables.

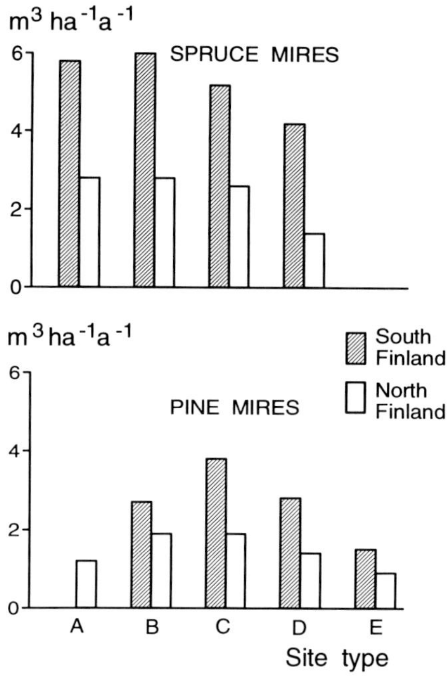

Fig. 7.4. Mean annual volume increment of different peatland stands in Finland. (Data from Paavilainen and Tiihonen 1988). Site type: *A* peatland with eutrophic vegetation; *B* herb-rich peatland; *C* tall-sedge and *Myrtillus* peatland; *D* small-sedge and *Vaccinium* peatland; *E* cottongrass and low-shrub peatland

In Fig. 7.5 the volume increment as a function of stand volume and age of drainage in four yield classes developed by Gustavsen et al. (1994) is presented. In this study, the four different yield classes included the following peatland site types (see Heikurainen and Pakarinen 1982; see also Table 3.17):

Yield class	
I	= eutrophic and mesoeutrophic types: LhK, VLR
II	= mesotrophic (mesooligotrophic) types: RhK, RhSK, MK, KgK, VSK, RhSN
III	= oligotrophic (mesotrophic) types: PK, PsK, RhSR, VSR, LkSR, PsR, KR, KgR, VSN
IV	= ombrotrophic types: IR, TR, RaR, LkN

The expected average annual increase in total yield brought about by forest drainage varies at an initial stand volume of 100 $m^3 ha^{-1}$ from 1.0 $m^3 ha^{-1}$ (yield class IV) to 2.6 $m^3 ha^{-1}$ (yield class I). The corresponding change at an initial stand volume of 80 $m^3 ha^{-1}$ is from 0.9 to 2.1 $m^3 ha^{-1}$ (Gustavsen et al. 1994).

The increase in growth is also very clear when the increment in different yield classes of drained peatlands is compared with the study results from undrained peatlands (Fig. 7.6).

Results from Russian studies also clearly show the dependence of post-drainage tree growth on the age of drainage and site fertility classes (Knize and

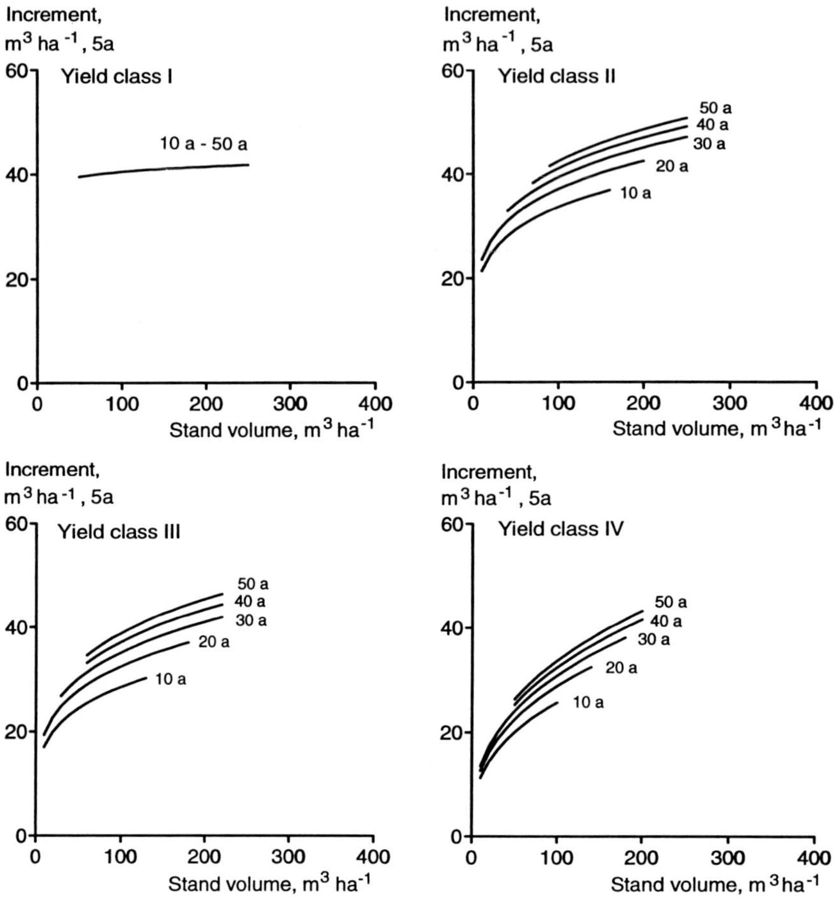

Fig. 7.5. Volume increment as a function of stand volume and age of drainage *(a)* in peatland *yield classes I–IV*. (Redrawn from Gustavsen et al. 1994)

Rubtsov 1974; Konstantinov 1979; see Fig. 7.7). Likewise there is a strict correlation between stand age and total yield after drainage (Fig. 7.8).

The effect of **fertilization** on the yield of peatland stands is also dependent on the quality of the site and tree stand (see Chap. 5). According to Moilanen (1993; see also Moilanen and Penttilä 1988), the average annual increase in yield after NPK fertilization of Scots pine (*Pinus sylvestris*) stands at thinning stage on Finnish ombrotrophic mires (not the poorest ones) was 0.75 m^3 ha^{-1} and the duration of the fertilization effect was 8–10 years. In nitrogen-rich sites, the corresponding figures after PK fertilization were 1.5 m^3 ha^{-1} and 20–25 years. In

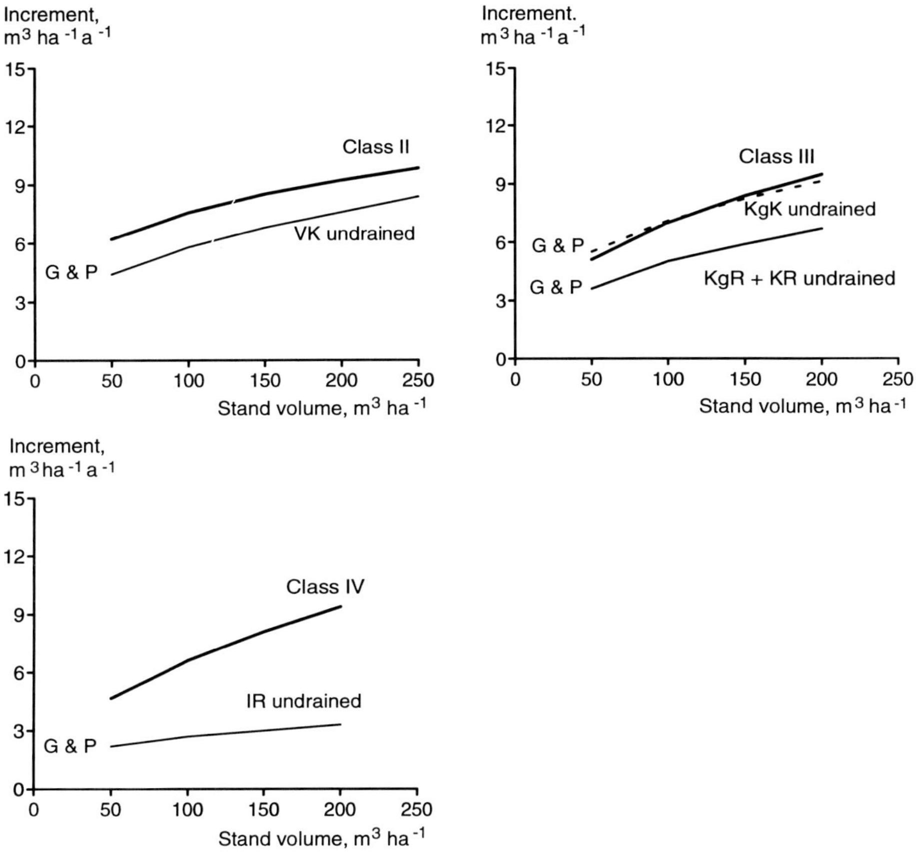

Fig. 7.6. Volume increment as a function of stand volume in peatland yield *classes II–IV* compared to the virgin mire types: *G & P* Gustavsen and Päivänen (1986); for site types, see Table 3.17. (Redrawn from Gustavsen et al. 1994)

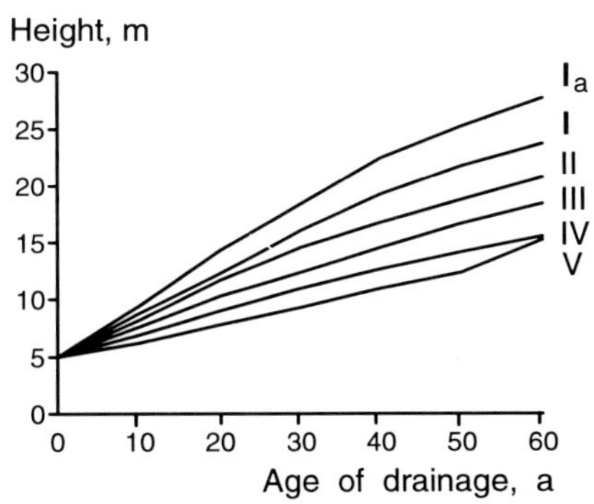

Fig. 7.7. The dependence of post-drainage height development of Scots pine (*Pinus sylvestris*) on the age of drainage in different site fertility classes in the former Soviet Union. (Data from Konstantinov 1979)

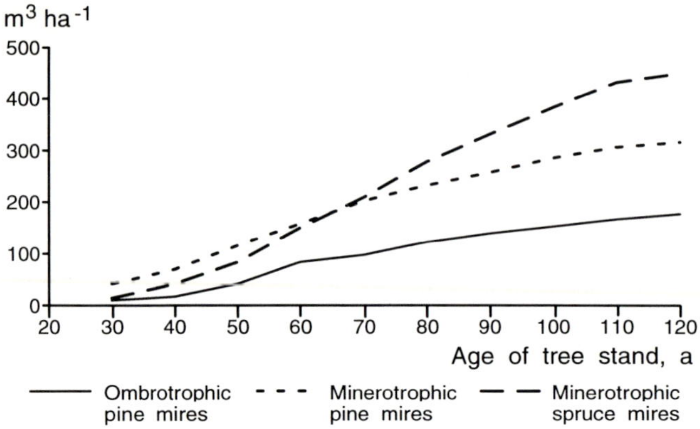

Fig. 7.8. The dependence of the total yield of Scots pine (*Pinus sylvestris*) after drainage on the age of the stands in the former Soviet Union. (Data from Sabo 1988)

mature spruce stands, the average increase in yield 5 years after fertilization was as follows (Paarlahti and Paavilainen 1985):

	Annual increase in yield (m^3 ha^{-1})
NPK	1.9
N	1.3
PK	0.5

The best responses are much higher. For example, in one Finnish experiment, PK fertilization increased the yield of a young Scots pine stand by 63 m^3 ha^{-1} in 18 years and the fertilization effect still continued (Paavilainen 1979a).

In Finnish experiments with wood ash, the annual growth increase has averaged 4–8 m^3 ha^{-1}. The highest total yield was 370 m^3 ha^{-1} after 44 years and the annual growth was 12–17 m^3 ha^{-1} 30–40 years after fertilization. The poorest results were from experiments with a low nitrogen level (< 1%) in the surface peat (Silfverberg and Huikari 1985).

In Sweden, models to estimate fertilization response on peatlands have been developed (Almqvist 1990). They are based on material collected from drained Scots pine stands (initial volume 13–124 m^3 ha^{-1}) of middle or low site fertility. The response with NPK is higher than with PK (Fig. 7.9). Correction factors, which take into account the basal area weighted mean diameter and latitude, are applied when using these prediction models.

Under more favourable climatic conditions, the post-drainage tree stand productivity is much higher than in Finland or in Sweden. For example, in Ireland, the maximum mean annual growth increment on drained blanket peatlands has been calculated at 13.3 m^3 ha^{-1} for *Picea sitchensis* and 10.6 m^3 ha^{-1} for *Pinus contorta* (Farrell 1990). Even production of over 20 m^3 ha^{-1} a^{-1} is possible in

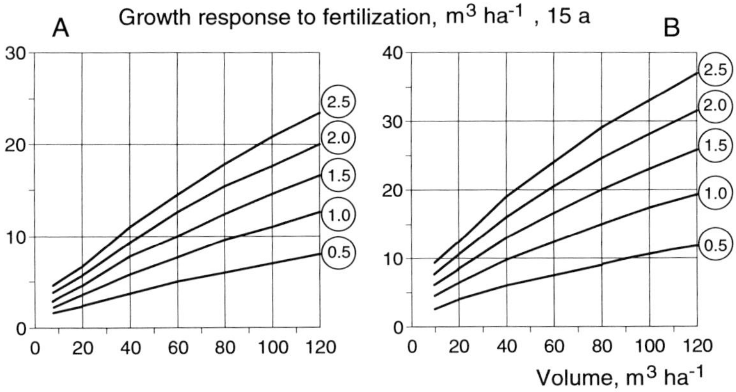

Fig. 7.9. Growth response (in 15 years) to PK (**A** 40 kg P, 80 kg K ha^{-1}) and NPK (**B** 60 kg N, 40 kg P, 70 kg N ha^{-1}) treatments as a function of pre-fertilization stand volume and basal area growth (0.5, 1.0, 1.5, 2.0, 2.5 m^2 ha^{-1} in 5 years before treatments). (Adapted from Almqvist 1990)

Sitka spruce stands (Carey and Griffin 1981; Carey et al. 1984). In the carbon balance studies, the average yield class of 12 m^3 ha^{-1} a^{-1} has been used for *Picea sitchensis* for Britain (Cannell et al. 1993).

7.2.2 Change in National Timber Balance

The large-scale amelioration of peatlands is one possible way to positively change the national timber balance. However, only in few countries has the potential capacity of peatlands for wood production been used.

In **Finland**, according to the 7th NFI (1977–84), the standing crop of peatland forests has increased by 39 million m^3 since the beginning of the 1950s. In southern Finland, the standing crop has increased from 161 million m^3 to 194 million m^3 and in northern Finland from 91 million m^3 to 97 million m^3 (Fig. 7.10). The increase in total annual increment during the same time for the whole country was 50%, from 6.49 million m^3 to 10.65 million m^3 in southern Finland and from 3.37 million m^3 to 4.21 million m^3 in northern Finland. The growth of peatland forests made up 21.7% of the total growth of all Finnish forests. Results from the eighth inventory in southern Finland indicate that the increase in growth continues (Fig. 7.10; see also Paavilainen 1991). The total annual increment of stands on peatlands and wet mineral soils in Finland may reach a level of 16–18 million m^3 at the end of this century, as was predicted by Heikurainen (1982c) at the beginning of the 1980s.

In **Norway**, where the current annual increment of all forests according to the recent estimates is 16.5 million m^3 (under bark) and that of peatlands and wet mineral soils 0.94 million m^3, the annual increment could be potentially raised on these sites to 6.4 million m^3 and theoretically even up to 8.9 million m^3 (Braekke 1990b).

The current annual increment on peatlands and wet mineral soils in **Sweden** is estimated at about 13 million m^3 which accounts for about 16% of the growth of all forests (Hånell 1986). The drainage of peatlands could increase the annual increment by 5.7–7.8 million m^3, depending on the ditching intensity.

In the **former Soviet Union**, the drainage of about 5.5 million ha has increased the wood resources of peatland forests by 110 million m^3 (Sabo 1988). Yet, compared with the large area of peatland, drainage is only locally important.

There are about 800 000 ha of man-made forests in **Great Britain** and in **Ireland**. The most important peatland forestry regions are in Scotland, the Republic of Ireland and Northern Ireland. In the Republic of Ireland, for instance, the utilization of peatlands for forestry has become important after the development of suitable plows and the use of phosphate fertilizer. In 1983, forests covered 170 000 ha, or 14.4% of the total peatland area, and peatland forests accounted for 51% of the state-owned forests (O'Carroll 1984).

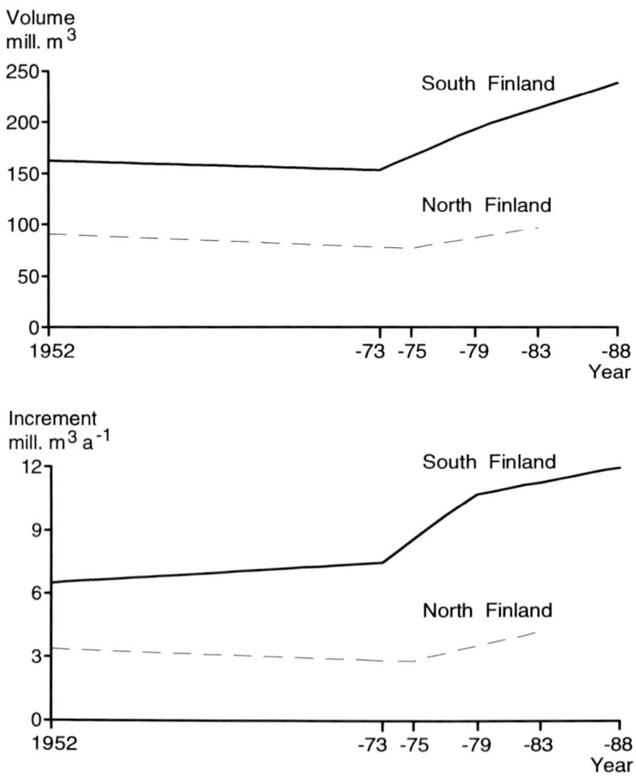

Fig. 7.10. Total volume and annual increment of peatland forests in Finland. (Redrawn from Paavilainen and Tiihonen 1988 and preliminary data from 8th NFI)

8 Environmental Effects of Peatland Forestry

8.1 Hydrological Effects

Forest drainage and other amelioration measures on peatlands not only improve the conditions for increased wood production but also cause many environmental changes in and outside the peatland ecosystem. The best possible methods should be developed and used to avoid the harmful effects of peatland forestry on the environment.

As already mentioned (Sect. 3.5.2), forest drainage increases the annual runoff, the runoff peaks caused by heavy summer rains and usually also the spring maximum runoff and the minimum runoff during dry periods. Cutting, particularly clear-cutting, noticeably increases the amount of runoff. The improved tree stand growth after drainage, on the other hand, increases the evapotranspiration. For this reason, the direct influence of forest drainage on runoff levels up and the annual runoff restores to its former condition in a few decades (e.g. Mustonen and Laikari 1961; Heikurainen and Päivänen 1970; Mustonen and Seuna 1971; Verry 1972, 1980; Päivänen 1974a, 1982a; Seuna 1982, 1988; Rosén 1984; Vompersky et al. 1991).

The **increase in flooding** is regarded as one of the most harmful of the hydrological changes to the environment. The occurrence of flooding is, however, affected by many factors which are not dependent on forest drainage. Such factors include the size of the catchment area, the proportion of lakes and amount of agricultural land in it, the volume and growth of initial tree stand, soil type and the water content of the snow cover (Kaitera 1939; Mustonen 1965). As the ditch network becomes denser, the risk of flooding grows (Ferda and Novak 1976; Ahti 1980; Ferda and Cermak 1981, 1982). The probability of harmful flooding increases when clear-cutting is carried out in connection with drainage. In this case, the interception of the tree stand is excluded, transpiration decreases and the soil water content increases (Heikurainen 1967a,b; Heikurainen and Päivänen 1970; Päivänen 1974a, 1982a; Verry 1976).

The damaging effect of short-term flooding which occurs before the onset of the growing season is usually of little importance as far as tree growth is concerned (Pelkonen 1975; Päivänen 1984b). Only protracted flooding in late summer or in autumn weakens the growth of the trees.

The increase in the amount of runoff during dry periods is, on the other hand, considered beneficial, e.g. for the production of hydroelectric power and the domestic water supply.

8.2 Quality of Water

8.2.1 Suspended Solids

The most detrimental environmental effect of forest drainage is usually the increase in suspended solids. **Suspended solids** refer to organic or inorganic matter that occurs in water in suspended or particle form. Solid **organic matter** silts up watercourses more than **inorganic matter**. Moreover, organic matter is biologically active, thus consuming the oxygen resources of watercourses as it is decomposed.

In some instances, drainage results in the washing out of a large amount of **solid matter** (Table 8.1). The ditching operation itself seems to be of considerable importance for the washing out of solid organic matter, whereas post-drainage high runoff periods are important in the washing out of inorganic matter (Sallantaus 1986). The increase in the humus load resulting from drainage seems to be primarily a consequence of an increase in runoff due to drainage in most cases. This is at its greatest immediately after ditching, particularly in very wet peatlands with a thick peat layer. Long-lasting effects of drainage on the humus load in watercourses are generally negligible (Heikurainen et al. 1978). On the basis of measurements carried out on watercourses over long periods of time, no substantial changes have been observed in the suspended solid load, irrespective of the considerable increase that has taken place in the total area of peatlands drained for forestry in Finland.

Ditch cleaning also increases the amount of suspended solids in runoff. In a study by Ahti et al. (1993), the concentration of solid material before ditch cleaning was 1.7 mg l^{-1} and 2 years after cleaning the value was 28.3 mg l^{-1}.

Finely dispersed soils, such as fine sand and silt, are the most sensitive to erosion. The **erosion** of peatlands is the greatest from very wet peatlands and from areas where there are layers of mud or highly decomposed peat (Sallantaus 1986).

Table 8.1. Increases in suspended solids after drainage. (Data compiled by Sallantaus 1986)

Publication	Increase in the suspended solids load (kg ha^{-1})	Observation period after drainage
Seuna (1982)	980	2 years
Seuna (1982)	2300	Spring
Hynninen and Sepponen (1983)	4500	3 years
Lundin (1984)	< 200	1 year
Berquist et al. (1984)	< 240	Spring

In a study by Ahtiainen et al. (1988), a combination of clear-cutting, ditching, and soil preparation increased the amount of suspended solids from 4 kg ha^{-1} a^{-1} to 1010 kg ha^{-1} a^{-1} at its highest. Five to 8 years later, the amount of suspended solids was still approx. 60 kg ha^{-1} a^{-1}. In an area similarly clear-cut and plowed, but where buffer strips without any cuttings were used, changes in water quality were small (Ahtiainen et al. 1993).

To prevent an increase in suspended solids in adjoining watercourses, protection methods other than buffer strips have been developed and applied. These include, for example, the use of **temporary dams** and **sedimentation pits** or **ponds** (Chap. 4). The **timing of the ditching work** also has an important effect on the suspended solid load. The draining ditches and the main ditches in the upper parts of the catchment area should be dug first and the main ditches discharging to the lower watercourses last (Sallantaus and Pätilä 1983; Ahtiainen 1988; Seuna 1988; Joensuu 1990, 1992; Ahti 1991; Berry and Jeglum 1991b).

8.2.2 Soluble and Colloidal Organic Material

Runoff, particularly from ombrotrophic bogs, contains **humic material** in **soluble** or **colloidal form**, giving rise to its typical brown colour. Of various forestry measures, drainage and clear-cutting have the strongest effect on the amount of organic material in runoff (Sallantaus 1986).

The results of studies into the effects of forest drainage on the load of organic material are contradictory. Factors leading to an increase in the amount of organic material are increased runoff, the discharge of subsurface waters which often contain much dissolved humus, and increased rates of decomposition of surface peat. Drainage, on the other hand, may bring soil water into contact with mineral soil and thus decrease the amount of humus. According to most studies, the amount of organic material increases during ditching work but then decreases rapidly (e.g. Heikurainen et al. 1978; Kenttämies 1980, 1981; Ferda and Cermak 1982).

In one study area in Sweden, the amount of organic material increased by 170% during 3 years after clear-cutting (Grip 1982). It was concluded that the sudden rise in the groundwater level resulted in the death of roots and mycorrhizae and thus a source of erodable organic material. Similar results have been reported in Finnish studies on peatland-dominated catchments where clear-cutting has been carried out. The **COD_{Mn} -values** (chemical oxygen demand, oxidation with permanganate) doubled after cutting (Ahtiainen 1988). The amount of organic material in water flowing from peatlands after cutting is higher than in the runoff from mineral soils (Haveraaen 1981; Meyer and Tate 1983; Micka et al. 1985).

8.2.3 Phosphorus

Phosphorus concentrations in the soil water of natural peatlands are low. Mean concentrations are typically about 20 µg l^{-1}. Assuming an annual runoff value of

300 mm, the annual loss of phosphorus is 0,06 kg ha^{-1} (Kenttämies 1980, 1981; Sallantaus 1986).

Forest drainage only increases runoff phosphorus concentrations if erosion increases at the same time (Rekolainen 1989). The amount of phosphorus bound to suspended solids may be considerable. For instance, Kenttämies (1977) observed an increase in total phosphorus concentrations in runoff in the year following drainage from 21 to 58 µg l^{-1}, and even values up to 280 µg l^{-1} have been measured (Saura et al. 1993). In old drainage areas which have not been fertilized, the phosphorus concentrations in runoff are of the same order of magnitude as in runoff from natural peatlands.

Fertilization increases the risk for the **leaching of phosphorus** from drained peatlands. When water-insoluble rock phosphate was used, the increase in the total annual phosphorus leaching reached 0.5-2.0% of the amount applied under Finnish conditions (Särkkä 1970; Karsisto and Ravela 1971; Kenttämies 1977, 1981; Heikurainen et al. 1978). This is equivalent to some 100 g ha^{-1} $year^{-1}$. The duration of the leaching of fertilizer phosphorus may exceed 10 years (Kenttämies 1980, 1981; Malcolm and Cuttle 1983; Ahti 1983, 1984).

The spreading of fertilizer containing about 20% water-soluble phosphorus onto the snow cover increased total phosphorus leaching from May to July of the same year by as much as 5 kg ha^{-1}. A similar dose applied in the summer resulted in an increase of only 0.8 kg ha^{-1} (Ahti and Paarlahti 1988). In another study, the amount of phosphorus leached in the first year after fertilization was 2.7 kg ha^{-1} a^{-1}, which was 7% of the amount applied (Saura et al. 1993). Similar results have been reported for Scottish peatlands (Malcolm and Cuttle 1983). In practice, fertilizers containing water-insoluble phosphorus (e.g. apatite) are recommended (see Chap. 5).

Knighton and Stiegler (1980) reported an increase in phosphorus leaching during the first year after clear-cutting of 0.5 kg ha^{-1} for an ombrotrophic bog and 2.5 kg ha^{-1} for a minerotrophic mire. On the ombrotrophic site, the load returned to normal in the third year after cutting, but on the minerotrophic site the values were still high in the sixth year after cutting. In a peatland-dominated catchment Ahtiainen (1988) found that the phosphorus load caused by cutting reached more than 1 kg ha^{-1}, most of the phosphorus being in a soluble form. Much of the phosphorus in runoff probably originates from cutting residues which contain considerable amounts of phosphorus (e.g. Paavilainen 1980; Finér 1989, 1991a).

The mean values for **phosphorus load** after silvicultural measures, as estimated by Ahti (1991) on the basis of Finnish studies, are presented in Table 8.2. The phosphorus in runoff causes **eutrophication** and increases the production of algae in forest brooks collecting drainage waters (Huttunen et al. 1988; Holopainen and Huttunen 1992; Ahtiainen et al. 1993).

Table 8.2. The influence of different silvicultural measures on phosphorus load on peatlands

Measure	Load ($kg\ ha^{-1}a^{-1}$)	Duration (years)
New ditching	0.5	5
Cleaning of ditches and additional ditching	0.3	3
PK-fertilization (incl. water-soluble P)	1.5	15
Regeneration cuttings	2.0	5
Thinnings	0.3	3

Source: Ahti (1991).

8.2.4 Potassium

Potassium, all of which is almost totally water soluble, is easily leached from drained peatlands (Haveraaen and Steenberg 1967; Haveraaen 1978, 1986; Ahti 1983; Malcolm and Cuttle 1983; Kaunisto and Tukeva 1984; Braekke 1987). Leaching losses after application of potassium fertilizers are also high. Ahti and Paarlahti (1988) reported the leaching of 2.7 kg K ha^{-1} between March and July when 83 kg K ha^{-1} had been applied in the summer, and 10.8 kg K ha^{-1} when applied in the winter on snow. Cuttings also increase the leaching of potassium. Although cutting-related losses of potassium are not very important from the standpoint of water quality, they affect the nutrient balance of peatlands, which often suffer from potassium shortage (see Sect. 3.3).

8.2.5 Nitrogen

Most of the **nitrogen** that is leached from natural peatlands is **in an organic form**. The concentration of the organic material nitrogen in runoff on minerotrophic *Carex*-dominated peatlands may be over 2% (Lundin 1984), whereas on an ombrotrophic bog the concentration is less than 1% (Sallantaus 1983). The concentration of ***inorganic nitrogen*** in runoff from natural peatlands is usually less than 0.1 mg l^{-1} (e.g. Hynninen and Sepponen 1983; Berquist et al. 1984; Lundin 1984; Sallantaus 1986).

Forest drainage increases the leaching of both organic and inorganic nitrogen. The increase in annual leaching of organic nitrogen caused by drainage is of the order of 10 kg ha^{-1} and that of ammonium and nitrate nitrogen under 1 kg ha^{-1} (Hynninen and Sepponen 1983; Kenttämies and Laine 1984; Lundin 1984). The risk of leaching of nitrate nitrogen is especially high after clear-cutting (Kubin 1992). However, most of the nitrogen that is released after drainage, which may be as high as 100 kg N ha^{-1} (Silvola 1988), is fixed by the vegetation.

After application of nitrogen fertilizer, leaching losses of up to 5% of the amount used have been measured (Saura et al. 1993). Fertilizers containing nitrate nitrogen, for example, ammonium nitrate and calcium ammonium saltpeter, are more susceptible to leaching than are urea-based fertilizers (e.g. Karsisto 1974a; Wiklander 1977; Nömmik and Möller 1981). Because of the lower risk of nitrogen leaching losses, the application of urea is therefore recommended in peatland forestry, even though the increase in tree growth may be somewhat less than with calcium ammonium saltpeter (Paavilainen 1973).

The use of unfertilized strips next to the ditches and other protective measures, as described in Chapter 5, should be used to reduce the risk of leaching of nitrogen fertilizer. The urea should also be applied during the snow-free period because of the risk of nitrogen losses after spreading onto snow cover (Ahti and Paarlahti 1988).

The nitrogen contained in logging residues after clear-cutting often exceeds 100 kg ha^{-1} (Paavilainen 1980; Finér 1989, 1991a), and is subject to leaching. The increased decomposition of the peat, which is speeded up by the rise in soil temperature, adds to the risk of nitrogen leaching after cutting. Thus, Ahtiainen (1988) found that the leaching of nitrogen from a peatland-dominated catchment was higher than from the control area on mineral soils after clear-felling. However, it is also possible that part of the released nitrogen is immobilized after cutting because of the large amount of organic material and the effect of anaerobic conditions on the biological activity in peat, caused by the rise in groundwater level.

8.2.6 Heavy Metals

The results of studies concerning the leaching of **iron** from peatlands after drainage vary. Some show an increase in concentrations (Kenttämies and Laine 1984), some no change (Ferda and Cermak 1982; Berquist et al. 1984), and others, a decrease in concentrations after drainage (Hynninen and Sepponen 1983). A similar range in response to drainage is shown for **aluminium** and other metals. This is understandable, since the risk for leaching of heavy metals is dependent on whether there is leaching of these elements from deep peat or from the underlying mineral soil, and on oxidation conditions, for example.

Airborne **mercury** has been found to be the most important source of an increase in the mercury load. It seems that forest drainage increases the washing out of mercury bound to humus into watercourses, but it becomes sedimented at a fast rate (Simola and Lodenius 1982; Verta and Rekolainen 1985; Rekolainen et al. 1986). Mercury concentrations in fish in lakes located in undisturbed catchments did not differ from the concentrations measured in fish from lakes around which extensive peatland forestry had been practiced (Verta and Rekolainen 1985; Verta 1990).

8.2.7 Acidity

The influence of forest drainage on the **acidity of water** discharging from peatlands has been classified by Sallantaus (1986) as follows:

1. Nutrient-poor peatlands with a thick peat cover – temporary decrease in pH values.

2. Peatlands with acid surface peat in which the ditches reach down to sedge peat or mineral soil - increase in pH, particularly during periods of low runoff.
3. Nutrient-rich peatlands, also rich in sulphur or coastal sulphate soils – significant decrease in pH during periods of high runoff following dry spells.

Forest drainage is generally assumed to lead to the acidification of watercourses downstream. However, the neutralizing effect of groundwater flow may lead to a rise in the pH of runoff (Ramberg 1981; Sallantaus 1986; Kenttämies and Laine 1984; Lundin 1984; Braekke 1987; Ahtiainen et al. 1993). The interception of more neutral groundwaters after drainage may explain the increase in the pH of runoff observed in several studies (e.g. Simonsson 1987).

8.3 Multiple Use of Peatlands

Peatlands play a significant role not only in the production of wood but also as sites for conservation and recreation. In addition, peatlands are sources of other useful products; berries, mushrooms and game, for example. The **multiple use of peatlands** should be taken into account when planning forest amelioration activities.

Forest amelioration does not prevent the **recreational use of peatlands**. On the contrary, drained mires with managed forests are often more suitable for recreation than undrained wetlands.

In many countries, programmes for conserving the original species composition and diversity of mires have been planned and carried out (Sect. 2.5, Peatland Conservation).

The **picking of berries** growing on certain undrained mires, e.g. cloudberry (*Rubus chamaemorus*, see Fig. 8.1) and cranberry (*Vaccinium oxycoccos*), is often economically important for the local population. Forest drainage has an adverse effect on cranberry but a positive effect initially on the yield of cloudberry. This positive effect on cloudberry often turns negative at a later stage of the vegetational succession (Veijalainen 1974a, 1979; Sepponen 1979; Salo 1982, 1988; Huttunen 1983). The ditching of peatlands renowned for their berries should therefore not be allowed. The effect of different silvicultural measures on the yield of these and other berries is summarized in Table 8.3.

Table 8.3. Effect of different silvicultural measures on the yield of berries on peatlands; +, positive; 0, not known; -, negative; --, very negative. 1, Short-time effects; 2, effects in the long run

Measure	*Rubus chamaemorus*		*Vaccinium oxycoccus*		*Vaccinium vitis-idaea*		*Vaccinium myrtillus*		*Rubus idaeus*	
	1	2	1	2	1	2	1	2	1	2
Drainage	+	±	--	--	+	–				
Fertilization	±	–	--	--	+	–	–	0	+	–
Clear-cutting	–	+			+	–	--	+	+	–

Source: Salo (1988).

Fig. 8.1. The moist *Sphagnum* moss carpet and the long day in the Arctic and boreal zones provide favourable conditions for the cloudberry (*Rubus chamaemorus*). (Photo: J. Päivänen)

Attempts have been made to increase the yields of mire berries by **cultivation** or **semicultivation**. Cranberry (*Vaccinium oxycoccos*) has been successfully cultivated on peatlands, for example in Estonia and Canada (Salo 1988). In the cultivation of cloudberry (*Rubus chamaemorus*), shallow drainage and application of fertilizer have given promising results (Kortesharju 1979; Salo 1988).

Forest amelioration usually increases the **yield of mushrooms** on peatlands (Huikari 1972; Veijalainen 1974a; Salo 1979, 1980, 1981, 1988). Drainage reduces the variety of mushrooms native to moist *Sphagnum* mosses but mycorrhizal mushrooms, commonly found on mineral soil sites, increase in number (Salo 1981, 1988).

Drained and fertilized peatlands are generally suitable **habitats for game** (Karsisto and Issakainen 1974; Veijalainen 1974b; Issakainen 1978; Rajala and Linden 1982). After drainage, most game species benefit from the vegetational changes, such as the increased amount of hardwoods. Indeed, the density of moose (*Alces alces*) may become so high on drained and fertilized peatlands that damage occurs to the new plantations (see Sect. 6.4.2).

8.4 Other Effects

The shallow ditches used in forest drainage are not known to affect the quality of the **regional groundwater** which is used for domestic water supply. Clear changes have only been noted in the quality of water taken from wells dug directly into the peat. Nutrient concentrations in water taken from wells are dependent on the nutrient status and floristic composition of the surrounding peat layer (e.g. Starr and Laine 1988) and are affected by forestry measures, especially fertilization (Ahti 1990). The quality of the water taken from the groundwater wells changed in much the same way as that of runoff water after fertilization (Ahti 1990).

The effect of peatland forestry on the **nutrient stores** and flows in mire ecosystems is dealt with elsewhere (Sect. 3.3.4). Attention is especially paid to the adverse influence of whole-tree harvesting on nutrient-poor peatlands. The possible effect of drainage on **carbon balance** and thus on **climatic change** is discussed in Section 3.1.

Species diversity in the peatland ecosystems is considerably affected by drainage and other silvicultural measures. After drainage, the species diversity of the field layer is decreased, but that of the ground layer is increased (Vasander 1984, 1990). In later stages of the vegetational succession, the ground layer also becomes poorer in species diversity (Sarasto 1952, 1957; Mannerkoski 1976; Reinikainen 1990; Vasander 1990). Some species, like *Eriophorum vaginatum* and *Rubus chamemorus*, respond positively to fertilization (Finér and Braekke 1991). The diversity of the faunal species of the peatland ecosystem is also decreased by drainage (Koponen 1980, 1985; Markkula 1986).

When extensive forest drainage is carried out, some natural peatland ecosystems become rarer and some are in danger of extinction (Eurola et al. 1991). It is therefore necessary to reserve representative sites as conservation areas and to prevent the useless drainage of small and floristically rich peatlands.

References

Aarne M (1993)(ed) Metsätilastollinen vuosikirja 1992 (Yearbook of forest statistics 1992). Finn For Res Inst, Helsinki, 317 pp

Adams SN, Dickson DA (1973) Some short-term effects of lime and fertilizers on a Sitka spruce plantation. I. Forestry 46 (1):31–37

Ahti E (1971) Maaveden jännityksen mittaamisesta tensiometrillä (Use of a tensiometer in measuring soil water tension). Folia For 112:1–10

Ahti E (1977) Runoff from open peatlands as influenced by ditching. I. Theoretical analysis. Commun Inst For Fenn 92(3):1–16

Ahti E (1978) Maaveden energiasuhteista ojitetulla suolla (Energy relationships of soil water on drained peat). Commun Inst For Fenn 94(3):1–56

Ahti E (1980) Ditch spacing experiments in estimating the effects of peatland drainage on summer runoff. Int Assoc Hydrol Sci, IAHS-AISH Publ 130:49–53

Ahti E (1983) Fertilizer-induced leaching of phosphorus and potassium from peatlands drained for forestry. Commun Inst For Fenn 111:1–20

Ahti E (1984) Fertilizer-induced leaching of phosphorus and potassium from peatlands drained for forestry. Proc 7th Int Peat Congr, Irish Nat Peat Comm, Dublin, Ireland, pp 153–163

Ahti E (1987) Water balance of drained peatlands on the basis of water table simulation during the snowless period. Commun Inst For Fenn 141:1–64

Ahti E (1988) Hydrologia ja turpeen vesitalous. Metsäntutkimusl tied 308:29–39

Ahti E (1990) Fertilizer-induced leaching of elements from oligotrophic peat. In: Hånell B (ed) Biomass production and element fluxes in forested peatland ecosystems. Swed Univ Agric Sci, Dept For Site Res, pp 139–156

Ahti E (1991) Metsätalouden ympäristövaikutukset. Rep Minist Agric For Finl, 55 pp

Ahti E, Paarlahti K (1988) Ravinteiden huuhtoutuminen talvella lannoitetulta metsäojitusalueelta (Leaching of nutrients from a peatland area after fertilizer application on snow). Suo 39:19–25

Ahti E, Päivänen J, Vuollekoski M (1988) Kunnostusojitus. Metsäntutkimusl tied 308:46–55

Ahti E, Alasaarela E, Ylitolonen A, Nieminen M (1993) Kunnostusojituksen vaikutus suon hydrologiaan. In: Saukkonen S, Kenttämies K (eds) Metsätalouden vesistöhaitat ja niiden torjunta. METVE-projektin väliraportit. Vesi- ja ympäristöhallituksen monistesarja 455:53–54

Ahtiainen M (1988) Effects of clear-cutting and forestry drainage on water quality in the Nurmes-study. Publ Acad Finl 4:206–219

Ahtiainen M, Holopainen AL, Huttunen P (1988) General description of the Nurmes-study. Publ Acad Finl 4:107–121

Ahtiainen M, Huttunen P, Holopainen AL (1993) Metsätaloustoimenpiteiden vaikutukset purovesien laatuun ja hydrobiologiaan. In: Saukkonen S, Kenttämies K, (eds) Metsätalouden vesistöhaitat ja niiden torjunta. METVE-projektin väliraportit. Vesi- ja ympäristöhallituksen monistesarja 455:35–46

Aitolahti M (1972) Usability of covered ditches in the drainage of peatlands. Proc 4th Int Peat Congr, Finland, 3:413–424

Allen HL, Campbell RG (1988) Wet site pine management in the southeastern United States. In: Hook DD, McKee WH, Smith J et al. (eds) The ecology and management of wetlands, vol 2. Croom Helm, Timber Press, Portland, Oregon, pp 173–184

Almqvist C (1990) Gödslingseffekter på dikade torvmarker – prognoskurvor för tall (Prediction equations for estimation of Scots pine (*Pinus sylvestris*) response to fertilizer on drained peatland in Sweden). Inform Växtnäring – skogsproduktion. Inst Skogsförbättring 1989/90:1–6

Anderson AR, Pyatt DG, Sayers JM, Blackhall SR, Robinson HD (1992) Volume and mass budgets of blanket peat in the north of Scotland. Suo 43:195–198

Andersson S, Wiklert P (1967) Markfysikaliska undersökningar i odlad jord. XVII. Om de vattenhållande egenskaperna hos rena system och blandsystem av sand, lera och torv (On the water-retaining properties in pure systems and mixed systems of sand, clay and peat). Grundförbättring 20:3–27

Andersson S, Wiklert P (1970) Markfysikaliska undersökningar i odlad jord. XX. Studier av några markprofiler i Norrland (Soil physical studies in cultivated soils. XX. Studies of some soil profiles in Norrland). Grundförbättring 23:3–76

Andersson S, Wiklert P (1972) Markfysikaliska undersökningar i odlad jord XXIII. Om de vattenhållande egenskaperna hos svenska jordarter (Soil physical studies in cultivated soils. XXIII) Grundförbättring 25 (2–3):53–143

Andrejko MJ, Fiene F, Cohen AD (1983) Comparison of ashing techniques for determination of the inorganic content of peats. In: Jarrett PM (ed) Testing of peats and organic soils. ASTM Spec Tech Publ 820:5–20

Appelroth S-E (1971) Tree planting machine for open peat land. Tree Planters' Notes 22(3):1–3

Archibald D, Hayhurst D (1993) Careful logging and forest level wood supply. Oshki, Northeast Sci, Tech Newsl 1(2):4–5

Armentano TV, Menges ES (1986) Patterns of change in the carbon balance of organic soil-wetlands of the temperate zone. J Ecol 74:755–774

Aslyng HC (1952) Characterization of soils. Årsskr Kong Vet o Landbohøjsk:20–56

Averell JL, McGrew PC (1929) The reaction of swamp forests to drainage in northern Minnesota. Minn Dep Drainage Waters Publ, pp 1–66

Baden W, Eggelsmann R (1963) Zur Durchlässigkeit der Moorböden (The hydraulic conductivity of the organic soils). Z Kulturtech Flurbereinig 4:226–254

Baden W, Eggelsmann R (1964) Der Wasserkreislauf eines nordwestdeutschen Hochmoores. Schriftenr Kuratoriums Kulturbauwesen 12:1–156

Baden W, Segeberg H (1949) Die Probenahme und Volumengewichtsbestimmung von Moorböden. Landwirtsch Forsch 1:147–162

Baize D (1993) Soil science analyses. A guide to current use. Wiley, Chichester, 192 pp

Bather DM, Miller FA (1991) Peatland utilisation in the British Isles. Cent Agric Strategy, Univ Reading 21:1–64

Bay RR (1967) Factors influencing soil-moisture relationships in undrained forested bogs. In: Sopper WE, Lull HW (eds) Forest hydrology. Pergamon Press, London, pp 335–342

Bay RR (1968) The hydrology of several peat deposits in northern Minnesota, USA. Proc 3rd Int Peat Congr, Nat Res Council Canada, pp 212–218

Bay RR (1969) Runoff from small peatland watersheds. J Hydrol 9:90–102

Bélair JL (1991) Low hydraulic conductivity in catotelm limits effectiveness of drainage. In: Jeglum JK, Overend RP (eds) Peat and peatlands – diversification and innovation. Can Soc Peat Peatlands 1:141–145

Benzler J-H, Finnern H, Müller W, Roeschmann G, Will KH, Wittmann O (1982) Bodenkundliche Kartieranleitung, 3. Aufl. Hannover, 331 pp

Berg B, Staaf H (1981) Leaching, accumulation and release of nitrogen from decomposing forest litter. In: Clark FE, Roswall T (eds) Terrestrial nitrogen cycles. Ecol Bull (Stockholm) 33:163–178

Berg B, Staaf H (1986) Release of nutrients from decomposing white birch leaves and Scots pine needle litter. Pedobiologia 30:55–63

Berquist B, Lundin L, Andersson A (1984) Hydrologiska och limnologiska konsekvenser av skogs- och myrdikning. Siksjöbäcksområdet (Effects of peatland drainage on hydrology. The basin Siksjöbäcken). Uppsala Univ, Limnol Inst, Forskningsrap 9, Rap LIU 1984 B 4:1–140

Berry GJ (1991) Hydrology of drained and undrained black spruce peatlands: streamflow and hydrologic balance. Can-Ontario For Resour Dev Agree COFRDA REP 3317:1–25

Berry GJ, Jeglum JK (1991a) Hydrology of drained and undrained black spruce peatlands: groundwater table profiles and fluctuations. Can-Ontario For Resour Dev Agree COFRDA Rep 3307:1–31

Berry GJ, Jeglum JK (1991b) Hydrology of drained and undrained black spruce peatlands: surface water and groundwater quality. Can-Ontario For Resour Dev Agree COFRDA Rep 3316:1–29

Binns WO, Mayhead GJ, MacKenzie JM (1980) Nutrient deficiences of conifers in British forests. An illustrated guide. For Comm Leaflet 76:1–23

Bliss LC (1975) Devon Island, Canada. Ecol Bull (Stockholm) 20:17–60

Boelter DH (1964a) Laboratory techniques for measuring water storage properties of organic soils. Soil Sci Soc Am Proc 28:823–824

Boelter DH (1964b) Water storage characteristics of several peats in situ. Soil Sci Soc Am Proc 28:433–435

Boelter DH (1965) Hydraulic conductivity of peats. Soil Sci 100:227–231

Boelter DH (1969) Physical properties of peats as related to degree of decomposition. Soil Sci Soc Am Proc 33:606–609

Boelter DH, Blake GR (1964) Importance of volumetric expression of water contents of organic soils. Soil Sci Soc Am Proc 28:176–178

Boelter DH, Verry ES (1977) Peatland and water in the northern Lake States. USDA For Serv Gen Tech Rep NC-31:1–22

Bogatyrev L, Berg B, Staaf H (1983) Leaching of nutrients from leaf and needle litter. Swedish Coniferous Forest Project. Swed Univ Agric Sci Uppsala Tech Rep 33:1–25

Bohlin E (1993) Botanical composition of peat – Implications on chemical, microbiological and physical properties. Swed Univ Agric Sci, Diss For Vegetat Ecol 3:1–30

Bohlin E, Hämäläinen M, Sundén T (1989) Botanical and chemical characterization of peat using multivariate methods. Soil Sci 147:252–263

Bohnstedt TL (1852) Förslag till förbättrad skogsvård. Inlemnadt till Nyköpings läns hushållningssällskap. Stockholm

Bonner GM (1982) Canada's forest inventory 1981. Can For Serv, 79 pp

Booth TC (1974) Topography and wind risk. Ir For 31:130–134

Booth TC (1977) Windthrow hazard classification. Res Inf Note For Comm 22:1–4

Botch MS, Masing VV (1983) Mire ecosystems in the U.S.S.R. In: Gore AJP (ed) Mires: swamp, bog, fen and moor. Ecosystems of the world, vol 4B. Elsevier, Amsterdam, pp 95–152

Brække FH (1970) Myrgrøfting for skogsproduksion. Innflytelse på vannhusholdning og flomfare. Tidsskr Skogbruk 78:227–238

Brække FH (1972) Varmehusholdning og mikroklima på ulike myrtyper (Energy balance and micro-climate on different peat land vegetation types). Medd Nor Skogforsoksves 119 (30):1–134

Brække FH (1977a) Fertilization for balanced mineral nutrition of forests on nutrient-poor peatland. Suo 28:53–61

Brække FH (1977b) Growth and chemical composition of Scots pine on nutrient-deficient peat after drainage and fertilization. Medd Nor Inst Skogforsk 33:1–8

Brække FH (1978) Afforestation on peatland in Norway. In: Landscaping of cut-over peatlands and soil conservation on cultivated peatlands. Int Peat Soc, Norw Nat Comm, pp 73–96

Brække FH (1979) Boron deficiency in forest plantations on peatland in Norway. Medd Nor Inst Skogforsk 35(3):213–236

Brække FH (1983a) Micronutrients prophylastic use and cure of forest growth disturbances. In: Kolari KK (ed) Growth disturbances of forest trees. Commun Inst For Fenn 116:159–169

Brække FH (1983b) Occurence of growth disturbance problems in Norwegian and Swedish forestry. In: Kolari KK (ed) Growth disturbances of forest trees. Commun Inst For Fenn 116:20–25

Brække FH (1983c) Water table levels at different drainage intensities on deep peat in northern Norway. For Ecol Manage 5:169–192

Brække FH (1987) Nutrient relationships in forest stands: Effects of drainage and fertilization on surface peat layers. For Ecol Manage 21:269–284

Brække FH (1988a) Biomass and nutrient distribution in middle-aged stands on afforested peatland-fields. Proc 8th Int Peat Congr, Leningrad, USSR, 3:103–113

Brække FH (1988b) Nutrient relationships in forest stands: field vegetation and bottom-litter layer on peatland. Medd Nor Inst Skogforsk 40 (7):1–20

Brække FH (1990a) Biomass and mineral elements in the root system of a low-shrub pine bog. In: Hånell B (ed) Biomass production and element fluxes in forested peatland ecosystems. Swed Univ Agric Sci, Umeå, Dep For Site Res, pp 109–121

Brække FH (1990b) Peatland and paludified forest on mineral soil in Norway – potentials for forest production. In: Hånell B (ed) Biomass production and element fluxes in forested peatland ecosystems. Swed Univ Agric Sci, Umeå, Dep For Site Res, pp 29–32

Brække FH, Finér L (1990) Decomposition of cellulose in litter layer and surface peat of low-shrub pine bogs. Scand J For Res 5:297–310

Brække FH, Finér L (1991) Fertilizations effects on surface peat of pine bogs. Scand J For Res 6:433–449

Brandesten C-O (1987) Runoff from mire catchments. A study of the mire complex Komosse southern central Sweden. R Inst Tech, Dep Land Improv Drainage, Medd Trita-Kut 1046:1–88

Brumelis G, Carleton TJ (1988) The vegetation of postlogged black spruce lowlands in central Canada. I. Trees and tall shrubs. Can J For Res 18:1470–1478

Bujsman E, Erisman JW (1988) Wet deposition of ammonium. J Atmos Chem 6:265–280

Burger D, Pierpoint G (1990) Trends in forest site and land classification in Canada. For Chron 66(2):91–96

Buringh P (1984) Organic carbon in soils of the world. In: Woodwell GM (ed) The role of terrestrial vegetation in the global carbon cycle. Scope 23:91–109

Burke W (1972) Aspects of the hydrology of blanket peat in Ireland. Int Symp Hydrol Marsh-Ridden Areas, Minsk, Byelorussian SSR, 16 pp

Burt TP, Heathwaite AL, Labadz JC (1990) Runoff production in peat-covered catchments. In: Anderson MG, Burt TP (eds) Process studies in hillslope hydrology. Wiley, Chichester, pp 463–500

Bush K (1961) Über die graphische Darstellung typologischer Schemata versumpfter und entwässerter Wälder. Latv SSR Zinat Akad Vestis 7(168):111–114 (in Russian with German Summary)

Bush K (1964) Puuston kasvukyvyn jatkuvuudesta ojitetuilla turvemailla. (Capacity of trees for continued growth on drained swamps). Suo 15:3–8

Cajander AK (1909) Über Waldtypen. Acta For Fenn 1(1):1–175

Cajander AK (1913) Studien über die Moore Finnlands. Acta For Fenn 2(3):1–208

Cajander AK (1926) The theory of forest types. Acta For Fenn 29(3):1–108

Cajander AK (1949) Forest types and their significance. Acta For Fenn 56:1–71

Canada Soil Survey Committee, Subcommittee on Soil Classification (1978) The Canadian system of soil classification. Can Dep Agric Ottawa, Ont Publ 1646:1–164

Cannell MGR, Dewar RC, Thornley JHM (1992) Carbon flux and storage in European forests. In: Teller A, Mathy P, Jeffers JNR (eds) Responses of forest ecosystems to environmental changes. Elsevier, Amsterdam, pp 256–271

Cannell MGR, Dewar RC, Pyatt DG (1993) Conifer plantations on drained peatlands in Britain: a net gain or loss of carbon. Forestry 66(4):353–369

Cantrell RL (1991) Peat. In: Grubich DN, Malterer TJ (eds) Proc Int Peat Symp, Peat and Peatlands: the resource and its utilization. Duluth, pp 167–192

Carey ML, Griffin E (1981) Sitka spruce or lodgepole pine? A financial appraisal. Ir For 38:61–77

Carey ML, McCarthy RG, Hendrick E (1984) Nutrient budgets in young Sitka spruce and lodgepole pine on a raised bog. Proc 7th Int Peat Congr, Irish Nat Peat Comm, Dublin, 3:207–218

Carlsson T, Möller G (1985) Effekter av svårlösliga fosforgödselmedel på torvmark (Growth response after treatment of peatland forests with phosphorus fertilizers of low solubility). Fören Skogsträdsförädl, Inst Skogsförbättr Årsbok, pp 81–109

Chapman RC, Hemond HF (1982) Dinitrogen fixation by surface peat and *Sphagnum* in an ombrotrophic bog. Can J Bot 60:538–543

Chason DB, Siegel DI (1986) Hydraulic conductivity and related physical properties of peat, Lost River Peatland, northern Minnesota. Soil Sci 142:91–99

Chrosciewicz Z (1990) Post-cut burning and black spruce regeneration. In: Titus BD, Lavigne MB, Newton, PF, Meades, WJ (eds) The silvics and ecology of boreal spruces. IUFRO Working Party S1.05-12 Symp Proc, Newfoundland, Aug 1989. For Can Inf Rep N-X-271:35–44

Clymo RS (1983) Peat. In: Gore AJP (ed) Mires: swamp, bog, fen and moor. Ecosystems of the world, vol 4A. Elsevier, Amsterdam, pp 159–224

Clymo RS (1984) The limits to peat bog growth. Philus Trans R Soc Lond B 303:605–654

Clymo RS (1992) Models of peat growth. Suo 43:127–136

Cooke J (1976) Ireland`s peatlands – extent and utilization. Proc 5th Int Peat Congr, Poznan, Poland, 1:315–327

Cross JR (1992) Man's impact on the peatlands in Ireland and its implications for conservation. In: Bragg OM, Hulme PD, Ingram HAP, Robertson RA (eds) Peatland ecosystems and man: an impact assessment. Univ Dundee, Dep Biol Sci, pp 345–350

Dai TS, Sparling JH (1973) Measurement of hydraulic conductivity of peats. Can J Soil Sci 53:21–26

Damman AWH (1979) Geographic patterns in peatland development in eastern North America. In: Classification of peat and peatlands. Int Peat Soc, Helsinki, pp 42–57

Damman AWH (1986) Hydrology, development, and biogeochemistry of ombrogenous peat bogs with special reference to nutrient relocation in a western Newfoundland bog. Can J Bot 64:384–394

Damman AWH (1990) Nutrient status of ombrotrophic peat bogs. Aquilo Ser Bot 28:5–14

Damman AWH, French TW (1987) The ecology of peat bogs of the glaciated northeastern United States: a community profile. US Fish Wildl Serv Biol Rep 85(7.16):1–100

Dang QL, Lieffers VJ (1989) Assessment of patterns of response of tree ring growth of black spruce following peatland drainage. Can J For Res 19:924–929

De Bakker H (1982) Soils and their geography. In: de Bakker H, van den Berg MW (ed) Proc Symp Peat lands below sea level, vol 30. ILRI, Wageningen, pp 85–97

Dickson DA (1971) The effect of form, rate and position of phosphatic fertilizers on growth and nutrient uptake of Sitka spruce on deep peat. Forestry 44:17–26

Dickson DA (1984) Effects of ground limestone and urea on growth of Sitka spruce planted on deep oligotrophic blanket peat in Northern Ireland. Proc 7th Int Peat Congr, Irish Nat Peat Comm, Dublin, 3:255–263

Dickson DA, Savill PS (1974) Early growth of *Picea sitchensis* (Bong) Carr. on deep oligotrophic peat in northern Ireland. Forestry 47:57–88

Doucet R, Boily J (1988) Développement des bouquets de marcottes d'épinette noire dans des coupes récentes au Québec. Naturaliste Can Rev Écol Syst 115:139–147

Douglass JE (1983) The potential for water yield augmentation from forest management in the eastern United States. Water Resour Bull 19(3):351–358

Duncan DV, Terry TA (1983) Water management. In: Stone EL (ed) The managed slashpine ecosystem. School For Resour Conserv, Univ Florida, Gainsville, pp 91–111

Edenius L (1991) The effect of resource depletion on the feeding behaviour of a browser: winter foraging by moose on Scots pine. J Appl Ecol 28:318–328

Edwards MV (1959) Effects of different forms and amounts of basic slag and mineral phosphate on the growth of Japanese larch planted on blanket bog. For Comm Rep For Res 3/1959

Eeronheimo O (1991) Suometsien puunkorjuu (Forest harvesting on peatlands). Folia For 779: 1–29

Eggelsmann R (1972) The water balance of lowlands areas in north-west coastal region of the FRG. Int Symp Hydrol Marsh-Ridden Areas, Minsk, Byelorussian, SSR, 15 pp

Eggelsmann R (1973) Dränanleitung für Landbau, Ingenieurbau und Landschaftsbau. Wasser Boden (Hamburg) 332 pp

Eggelsmann R (1976) Peat consumption under influence of climate, soil condition, and utilization. Proc 5th Int Peat Congr, Poznan, Poland, 1:233–247

Eggelsmann R, Mäkelä T (1964) Einfluss von Entwässerung und landwirtschaftlicher Nutzung auf die Durchlässigkeit des Moorbodens. J Sci Agric Soc Finl 36:77–84

Eggelsmann R, Heathwaite AL, Grosse-Brauckmann G, Küster E, Naucke W, Schuch M, Schweickle V (1993) Physical processes and properties of mires. In: Heathwaite AL (ed) Mires: process, exploitation and conservation. Wiley, Chichester, pp 171–262

Eichelsdörfer D (1990) Moor in der Kurortmedizin. In: Göttlich K (ed) Moor- und Torfkunde. Schweizerbart'sche Verlags, Stuttgart, pp 476–495

Elina GA, Kuznecov OL, Maksimov AI (1984) Strukturno-funktsional'naja organizatsija i dinamika bolotnyh ekosistem Karelii. Nauka, Leningrad, 128 pp

Elling AE, Verry ES (1978) Predicting wind-caused mortality in strip-cut stands on peatland black spruce. For Chron 54:249–252

Elonen P (1971) Particle-size analysis of soil. Acta Agric Fenn 122:1–122

Erviö R (1970) The importance of soil bulk density in soil testing. Ann Agric Fenn 9:278–286

Eurola S (1968) Über die Ökologie der nordfinnischen Moorvegetation im Herbst, Winter und Frühling. Ann Bot Fenn 5:83–97

Eurola S (1975) Snow and ground frost conditions of some Finnish mire types. Ann Bot Fenn 12:1–16

Eurola S, Hicks S, Kaakinen E (1984) Key to Finnish mire types. In: Moore PD (ed) European mires. Academic Press, London, pp 11–117

Eurola S, Aapala K, Kokko A, Nironen M (1991) Mire type statistics in the bog and southern aapa mire areas of Finland (60°–66°N). Ann Bot Fenn 28:15–36

Everard JE (1974) Fertilisers in the establishment of conifers in Wales and southern England. For Comm Booklet 41:1–49

Farnham RS, Finney HR (1965) Classification and properties of organic soils. Adv Agron 17:115–162

Farrell EP (1990) Peatland forestry in the Republic of Ireland. In: Hånell B (ed) Biomass production and element fluxes in forested peatland ecosystems. Swed Univ Agric Sci, Umeå, Dep For Site Res, pp 13–18

Faulkner JS, Weyl RS, Corbett PMCM, Standfield K (1992) Towards a conservation strategy for peatlands in Northern Ireland. In: Bragg OM et al (eds) Peatland ecosystems and man: an impact assessment. Univ Dundee, UK, Dep Biol Sci, pp 396–398

Ferda J (1972) Fertilizing and nutrition of forest woody plants in exploited fen deposits. Proc 4th Int Peat Congr 3:509–520

Ferda J (1975) Forestry reclamation of extracted peat deposits. Res Inst Land Reclam Improv Sci Monogr 6:1–192

Ferda J (1986) Reclamation and utilization of extracted peat deposits for forest planting. In: Socioeconomic impacts of the utilization of peatlands in industry and forestry. Int Peat Soc, Finn Nat Comm, Helsinki, pp 227–239

Ferda J, Cermak P (1981) Vliv advodneni lesnick pud na zmeny prostredi (The influence of forest soil drainage on environmental changes). Lesnictvi 27(3):271–284

Ferda J, Cermak P (1982) Obnova lesa na zamokrenyck lesnich pudach v imisnich oblastech Krusnych hor komplexni melioraci (Forest regeneration on water-logged forest soils in emission regions of the Krusnych hory Mts. by the method of complex reclaims). Lesnictvi 28(7):587–610

Ferda J, Novak M (1976) The effect of ameliorative measures on the changes of the quality of surface and ground waters in peat soils. Proc 5th Int Peat Congr, Poznan, Poland, 1:118–127

Ferm A (1989) Hieskoivun kasvatus soilla. Metsäntutkimusl tied 322:40–52

Ferm A, Hytönen J (1988) Effect of soil amelioration and fertilization on the growth of birch and willow on cut-over peat. Proc 8th Int Peat Congr, Leningrad, USSR, 3:268–279

Ferm A, Kaunisto S (1983) Luontaisesti syntyneiden koivumetsiköiden maanpäällinen lehdetön biomassatuotos entisellä turpeennostoalueella Kihniön Aitonevalla (Above-ground leafless biomass production of naturally generated birch stands in a peat cut-over area at Aitoneva, Kihniö). Folia For 558:1–32

Ferm A, Hokkanen T, Moilanen M, Issakainen J (1992) Effects of wood bark ash on the growth and nutrition of Scots pine afforestation in central Finland. Plant Soil 147:305–316

Finér L (1986) Tuloksia sararämeen fosforilannoitelajikokeesta (Results from a phosphorus fertilization experiment on a mesotrophic mire). Metsäntutkimusl tied 228:1–39

Finér L (1989) Biomass and nutrient cycle in fertilized and unfertilized pine, mixed birch and pine and spruce stands on a drained mire. Acta For Fenn 208:1–63

Finér L (1990) Nutrient content of a fertilized and unfertilized Scots pine stand on a low shrub pine bog. In: Hånell B (ed) Biomass production and element fluxes in forested peatland ecosystems. Swed Univ Agric Sci, Umeå, Dep For Site Res, pp 123–130

Finér L (1991a) Effect of fertilization on dry mass accumulation and nutrient cycling in Scots pine on an ombrotrophic bog. Acta For Fenn 223:1–42

Finér L (1991b) Root biomass on an ombrotrophic pine bog and the effects of PK and NPK fertilization. Silva Fenn 25(1):1–12

Finér L (1992) Nutrient concentrations in *Pinus sylvestris* growing on an ombrotrophic pine bog, and the effects of PK and NPK fertilization. Scand J For Res 7(2):205–218

Finér L, Braekke FH (1991) Understorey vegetation on three ombrotrophic pine bogs and the effects of NPK and PK fertilization. Scand J For Res 6:113–128

Finér L, Heimala-Raimas R, Päivänen J (1988) Tree stands and ground vegetation in two watersheds in the Nurmes-research area. Aqua Fenn 18:47–60

Finér L, Laine J, Halko L (1992) Fine root dynamics on two drained peatland sites. Suo 43:207–210

Fleming RL, Groot A (1984) Alternatives for regenerating black spruce clearcuts. Can For Serv For Newsl Summer 2–5

Fojt W (1992) British fens and their conservation problems. In: Bragg OM, Hulme PD, Ingram HAP, Robertson RA (eds) Peatland ecosystems and man: an impact assessment. Univ Dundee, Dep Biol Sci, pp 357–362

Ford ED (1979) An ecological basis for predicting the growth and stability of plantation forests. In: Ford ED, Malcolm DC, Atterson J (eds) The ecology of even-aged forest plantations. Inst Terr Ecol, Cambridge, pp 147–174

Forest 2000 Programme (1986) Guidelines for developing Finnish forestry and the forest industries. Silva Fenn 20(1):35–44

Foss PJ, O'Connell CA (1992) The conservation objectives of the Irish Peatland Conservation Council. In: Bragg OM, Hulme PD, Ingram HAP, Robertson RA (eds) Peatland ecosystems and man: an impact assessment. Univ Dundee, Dep Biol Sci, pp 402–405

Fowler DP, Mullin RE (1977) Upland-lowland ecotypes not well developed in black spruce in northern Ontario. Can J For Res 7:35–40

Francez A-J (1991) Production primaire et accumulation de matière organique dans les tourbières à Sphaignes des Monts du Forez (Puy-de-Dome). These de doctorat, Univ Paris 6

Francez A-J, Bignon J-J, Mollett A-M (1992) The peatlands in France: localization, characteristics, use and conservation. Suo 43:11–24

Fraser AI (1969) Tutkimuksia puiden latvuksen, rungon ja juuriston kehityksestä ohutturpeisilla soilla Britanniassa (Some investigations of the relative development of crown, stem and root of trees growing on shallow peat soils in Britain). Suo 20:1–8

Frenzel B (1983) Mires – repositories of climatic information or self-perpetuating ecosystems In: Gore AJP (ed) Mires: swamp, bog, fen and moor. Ecosystems of the world, vol 4A. Elsevier, Amsterdam, pp 35–65

Gadd PA (1773) Försök til en systematisk inledning i swenska landt-skötselen; lämpad efter rikets nordliga climate. Tryckt i Kongl Tryckeriet, Stockholm, Tom I:1–374

Gallagher GJ, Gillespie J (1984) The economics of peatland afforestation. Proc 7th Int Peat Congr, Irish Nat Peat Comm, Dublin, 3:271–285

Gardiner JJ (1975) The influence of fertilizers upon microbial activity in peat II. Calcium and nitrogen. Ir For 32:101–114

Glebov FZ, Melentyeva NV (1992) Plant dynamics, balance and destruction of biomass in the soil of drained bogged birch forest. Proc 9th Int Peat Congr, Int Peat Soc, Swed Nat Comm, Uppsala, 2:457–469

Glinski J, Stepniewski W (1985) Soil aeration and its role for plants. CRC Press, Boca Raton, Florida, 299 pp

Glob PV (1969) The bog people. Iron-Age man preserved. Faber and Faber, London, 200 pp

Glossary of Soil Science Terms (1979) Soil Sci Soc Am, Madison, Wisconsin, 37 pp

Goodall DW (1983) Conclusion – the future of mires. In: Gore AJP (ed) Mires: swamp, bog, fen and moor. Ecosystems of the world, vol 4B. Elsevier, Amsterdam, pp 395–396

Goodwin GE (1968) The influence of wind on forest management and planning. Forestry (Suppl):60–66

Gore AJP (1983) Introduction. In: Gore AJP (ed) Mires: swamp, bog, fen and moor. Ecosystems of the world, vol 4A. Elsevier, Amsterdam, pp 1–34

Gorham E (1991) Northern peatlands: role in the carbon cycle and probable responses to climatic warming. Ecol Appl 1(2):182–195

Göttlich K, Kuntze H (1990) Moorkultivierung für Land- und Forstwirtschaft. In: Göttlich K (ed) Moor- und Torfkunde. Schweizerbart'sche Verlags, Stuttgart, pp 385–410

Grip H (1982) Water chemistry and runoff in forest streams at Kloten. Uppsala Univ, Naturgeogr Inst UNGI Rap 58:1–144

Grip H (1987) Avrinningsförändringar efter kalavverkning. Sver Skogsvårdsförb Tidskr 2:43–49

Groot A (1987) Silvicultural consequences of forest harvesting on peatlands: site damage and slash conditions. Can For Serv, Great Lakes For Res Cent, Inf Rep O-X-384:1–20

Guide sur le drainage sylvicole (1989) Gouvernement du Québec, 53 pp

Günther J, Schmilewski G (1986) The present state of peatland utilization and use of peat in the Federal Republic of Germany and future perspectives of the German industry. In: Socio-economic impacts of the utilization of peatlands in industry and forestry. Int Peat Soc, Finn Nat Comm, Helsinki, pp 21–30

Gustafsson J (1980) Dikningsteknik (Drainage techniques). Skogsarbeten Rep 3:26–36

Gustavsen HG, Lipas E (1975) Lannoituksella saatavan kasvunlisäyksen riippuvuus annetusta typpimäärästä (Effect of nitrogen dosage on fertilizer response). Folia For 246:1–20

Gustavsen HG, Päivänen J (1986) Luonnontilaisten soiden puustot kasvullisella metsämaalla 1950–luvun alussa (Tree stands on virgin forested mires in the early 1950's in Finland). Folia For 673:1–27

Gustavsen HG, Heinonen R, Paavilainen E, Reinikainen A (1994) Growth and yield models for forest stands on drained peatland sites in southern Finland. Silva Fenn (submitted)

Haavisto VF (1975) Peatland black spruce seed production and dispersal in northeastern Ontario. In: Black spruce symposium. Can For Serv, Symp Proc O-P-4:250–264

Haavisto VF (1979) Some considerations for regenerating black spruce on peatlands in the northern clay forest section, Ontario. Can For Serv Rep O-X-295:1–32

Haavisto VF, Atkinson GT (1977) Capabilities of a modified "Timberjack" and a "Terrain-Master" for ditching on peatlands. Can For Serv, Great Lakes For Res Cent, Rep O-X-264:1–11

Haavisto VF, Jeglum JK (1991) Peatland potentially available for forestry in Canada. In: Jeglum JK, Overend RP (eds) Peat and peatlands – diversification and innovation. Can Soc Peat Peatlands 1:30–37

Haavisto VF, Päivänen J (1987) Planning operational drainage in boreal forest peatlands. Am Soc Agric Eng Pap 87-2517:1–13

Haavisto VF, Wearn VH (1987) Intensifying forestry use of peatlands with drainage in Ontario. In: Rubec CDA, Overend RP (eds) Proc Symp '87 Wetlands/Peatlands. Edmonton, Alberta, Canada, pp 249–256

Haavisto VF, Fleming RL, Skeates DA (1988a) Potential and actual yields of seed from black spruce cones. For Chron 64:32–34

Haavisto VF, Jeglum JK, Groot A (1988b) Management practices – black spruce ecosystem. In: Hook DD, McKee WH, Smith J et al. (eds) The ecology and management of wetlands, vol 2. Croom Helm, Timber Press, Portland, pp 195–202

Hakkila P (1989) Utilization of residual forest biomass. Springer, Berlin Heidelberg New York, 568 pp

Håland B, Braekke FH (1989) Distribution of root biomass in a low-shrub pine bog. Scand J For Res 4:307–316

Hämäläinen J (1984) Metsänlannoituksen kannattavuudesta erityisesti turvemailla (On the profitability of fertilization especially in forest stands on peatlands). Suo 35:132–136

Hämäläinen J, Laakkonen O (1983) Turvemaan varttuneiden männiköiden lannoituksen edullisuus (Profitability of fertilization in nature Scots pine stands on peatland). Folia For 570:1–32

Hämäläinen J, Paavilainen E, Salminen O, Heinonen R (1985) Tuloksia ojitettujen korpikuusikoiden lannoituksesta (Profitability of fertilization in drained spruce swamps). Folia For 623:1–26

Hammond RF (1975) The origin, formation and distribution of peatland resources. In: Robinson DW, Lamb JGD (eds) Peat in horticulture. Applied Botany Series 3. Academic Press, New York, pp 1–22

Hammond RF (1979) The peatlands of Ireland. An Foras Taluntais, Dublin. Soil Surv Bull 35:1–60

Hånell B (1984) Skogsdikningsboniteten hos Sveriges torvmarker (Post-drainage site index of peatlands in Sweden). Swed Univ Agric Sci, Rep For Ecol For Soils 50:1–121

Hånell B (1986) Skogliga våtmarker som resurs för ökad virkesproduktion. In: Skog på våtmarker. SJFRS Progr Rap 4:43–60

Hånell B (1988a) Postdrainage forest productivity of peatlands in Sweden. Can J For Res 18:1443–1456

Hånell B (1988b) Torvmark. In: Lundmark J-E (ed) Skogsmarkens ekologi – ståndortsanpassat skogsbruk, Del 2 – Tillämpning. Skogsstyrelsen, Jönköping, pp 287–305

Hånell B (1989) Skogliga våtmarker i Sverige (Peatlands in Sweden. A description of forest conditions on shallow and deep peatlands and their national and local distribution). Swed Univ Agric Sci, Umeå, Dep For Soils, Rep 60:1–85

Hånell B (1990) Present situation and future possibilities of peatland forestry in Sweden. In: Hånell B (ed) Biomass production and element fluxes in forested peatland ecosystems. Swed Univ Agric Sci, Umeå, Dep For Site Res, pp 45–48

Hånell B (1991a) Forest classification of peatlands in Sweden – a field quide. Swed Univ Agric Sci, Umeå, Dep Silvicult, 16 pp

Hånell B (1991b) Förnyelse av gransumpskog på bördiga torvmarker genom naturlig föryngring under högskärm (Shelterwood regeneration of spruce forests on productive peatlands). Swed Univ Agric Sci, Umeå, Dep Silvicult, Rep 32:1–35

Hånell B (1991c) Peatland forestry in Sweden. In: Jeglum JK, Overend RP (eds) Peat and peatlands – diversification and innovation. Can Soc Peat Peatlands 1:19–25

Hånell B (1992) Skogsförnyelse på högproduktiva torvmarker – plantering av gran på kalhygge och under skärmträd (Forest renewal on productive peatlands: planting of Norway spruce on clearcuts and in shelterwoods). Swed Univ Agric Sci, Umeå, Dep For, Rep 34:1–71

Hånell B (1993) Regeneration of *Picea abies* forests on highly productive peatlands – clearcutting or selective cutting? Scand J For Res 8:518–527

Hånell B, Ottosson-Löfvenius M (1994) Windthrow after shelterwood cutting in Norway spruce peatland forests. Scand J For Res 9:261–269

Hannelius S (1975) Ojitusalueiden kulkukelpoisuudesta puunkorjuussa (On the trafficability of drained peatlands in harvesting). Silva Fenn 9(3):181–211

Hannerz M, Hånell B (1993) Changes in the vascular plant vegetation after different cutting regimes on a productive peatland site in central Sweden. Scand J For Res 8:193–203

Härkönen K (1985) Soiden metsätaloudellista luokittelua Albertassa (Classifying peatlands for forest drainage and growth in Alberta). Suo 36:77–83

Hartge KH (1965) Die Bestimmung von Porenvolumen und Porengrössenverteilung (Determination of pore volume and pore size distribution). Z Kulturtech Flurbereinig 6:193–206

Harvesting and regeneration option (1987) Photo guide. Ecological Services for Planning Ltd, Minist Nat Resour, Ontario, 20 pp + App

Haveraaen O (1966) Skogdyrkning på myr – Produksjon og lønnsamhet. Tidsskr Skogbruk 74:142–163

Haveraaen O (1967) Vekst – og naeringsstudier i et gjødslingsforsøk med svartgran, *Picea mariana*, på myr (Growth and nutrient studies in a fertilizer experiment with *Picea mariana* on peatland). Medd Nor Skogforsøksves 85:137–175

Haveraaen O (1978) Nedvaskning av kalium i torvjord (Leaching of potassium in peat soil). Medd Norg Lantbrukhøgsk 57(43):1–12

Haveraaen O (1981) Virkning av hogst på rannmengde og vannkvalitet fra en østnorsk barskog (The effect of cutting on water quantity and water quality from an East-Norwegian coniferous forest). Medd Nor Inst Skogforsk 36(7):1–27

Haveraaen O (1986) Ash fertilizer and commercial fertilizers as nutrient sources for peatland. Medd Nor Inst Skogforsk 39(14):251–263

Haveraaen O, Stenberg K (1967) Nedvaskning av næringsstoffer i myrjord. Noen resultater ved bruk av radioactive isotopes (Leaching of nutrients from peat soi!. Some results by use of radioactive isotopes). Medd Norg Lantbrukhøgsk 46(21):1–25

Hayen H (1990) Moorarchäologie. In: Göttlich K (ed) Moor- und Torfkunde. Schweizerbart'sche Verlags, Stuttgart, pp 156–174

Heal OW, Jones HE, Whittaker JB (1975) Moor house, UK. Ecol Bull (Stockholm) 20:259–320

Heathwaite AL (1993) Mires: process, exploitation and conservation. Wiley, Chichester, 506 pp

Heathwaite AL, Eggelsmann R, Göttlich K–H, Kaule G (1993) Ecohydrology, mire drainage and mire conservation. In: Heathwaite AL (ed) Mires: process, exploitation and conservation. Wiley, Chichester, pp 171–262

Heikkilä H (1992) Eutrophic fens and their conservation in southern Finland. In: Bragg OM, Hulme PD, Ingram HAP, Robertson RA (eds) Peatland ecosystems and man: an impact assessment. Univ Dundee, Dep Biol Sci, pp 351–356

Heikkilä R (1979) Kituvien suotaimistojen tuhonaiheuttajista Pohjois-Suomessa (On the damages noted in pine seedling stands established on peatlands in northern Finland). Suo 30:7–11

Heikkilä R (1991) Moose browsing in a Scots pine plantation mixed with deciduous tree species. Acta For Fenn 224:1–13

Heikkilä R, Härkönen S (1993) Moose (*Alces alces* L.) browsing in young Scots pine stands in relation to the characteristics of their winter habitats. Silva Fenn 27(2):127–143

Heikkilä R, Löyttyniemi K (1992) Growth response of young Scots pines to artificial shoot breaking simulating moose damage. Silva Fenn 26(1):19–26

Heikkilä R, Mikkonen T (1992) Effects of density of young Scots pine (*Pinus sylvestris*) stand on moose (*Alces alces*) browsing. Acta For Fenn 231:1–14

Heikurainen L (1954) Rämemänniköiden uudistamisesta paljaaksihakkausta käyttäen (Über natürliche Verjüngung von Reisermoor-Kiefernbeständen unter Anwendung von Kahlschlag). Acta For Fenn 61(27):1–27

Heikurainen L (1955) Rämemännikön juuriston rakenne ja kuivatuksen vaikutus siihen (Der Wurzelaufbau der Kiefernbestände auf Reisermoorböden und seine Beeinflussung durch die Entwässerrung). Acta For Fenn 65(3):1–85

Heikurainen L (1957) Metsäojien syvyyden ja pintaleveyden muuttuminen sekä ojien kunnon säilyminen (Changes in depth and top width of forest ditches and the maintaining of their repair). Acta For Fenn 65(5):1–45

Heikurainen L (1959a) Sekametsiköiden juuristoista ojitetulla suolla (Der Wurzelaufbau in Mischwäldern auf entwässerten Moorböden). Acta For Fenn 67(2):1–32

Heikurainen L (1959b) Tutkimuksia metsäojitusalueiden tilasta ja puustosta (Über waldbaulich entwässerte Flächen und ihre Waldbestände in Finnland). Acta For Fenn 69(1):1–279

Heikurainen L (1963) On using ground water table fluctuations for measuring evapotranspiration. Acta For Fenn 76(5):1–16

Heikurainen L (1964) Improvement of forest growth on poorly drained peat soils. Int Rev For Res 1. Academic Press, New York, pp 39–113

Heikurainen L (1967a) Effect of cutting on the ground-water level on drained peatlands. In: Sopper WE, Lull HW (eds) Forest hydrology. Pergamon Press, London, pp 345–354

Heikurainen L (1967b) Hakkuun vaikutus ojitettujen soiden vesitalouteen (On the influence of cutting on the water economy of drained peatlands). Acta For Fenn 82(2):1–45

Heikurainen L (1967c) On the possibilities of optimum drainage in peat lands. 14th IUFRO-Congr 4:264–277

Heikurainen L (1968) Peatlands of Newfoundland and possibilities of utilizing them in forestry. For Res Lab, St John's, Newfoundland, Inf Rep N-X-16:1–57

Heikurainen L (1971a) Pohjavesipinta ja sen mittaaminen ojitetuilla soilla (Ground water table in drained peat soils and its measurement). Acta For Fenn 113:1–23

Heikurainen L (1971b) Virgin peatland forests in Finland. Acta Agric Fenn 123:11–26

Heikurainen L (1973a) Metsäojitustoiminta ja toimenpideraja (The profitability limit in forest drainage). Silva Fenn 7(4):310–339

Heikurainen L (1973b) Soiden metsänkasvatuskelpoisuuden laskentamenetelmä (A method for calculation of the suitability of peatlands for forest drainage). Acta For Fenn 131:1–35

Heikurainen L (1974) An attempt to survey the influence of forest drainage on the hydrology. Proc Int Symp For Drainage, Jyväskylä-Oulu, Finland, pp 365–371

Heikurainen L (1976) Comparison between runoff conditions on a virgin peatland and a forest drainage area. Proc 5th Int Peat Congr, Poznan, Poland, 1:76–86

Heikurainen L (1979) Peatland classification in Finland and its utilization for forestry. In: Classification of peat and peatlands. Int Peat Soc, Helsinki, pp 135–146

Heikurainen L (1980a) Effect of forest drainage on high discharge. Int Assoc Hydrol Sci, IAHS-AISH Publ 130:89–96

Heikurainen L (1980b) Kuivatuksen tila ja puusto 20 vuotta vanhoilla ojitusalueilla (Drainage condition and tree stand on peatlands drained 20 years ago). Acta For Fenn 167:1–39

Heikurainen L (1982a) Forest drainage activity in Europe in the 1980's. In: Sheppard JD, Musial J, Tibbetts TE (eds) A symposium on peat and peatlands. Shippagan, New Brunswick, Int Peat Soc, Can Nat Comm, pp 109–126

Heikurainen L (1982b) Ojitusalueiden taimistojen kehityksestä vuosina 1964–68 toimeenpannun suometsäkilpailun valossa (Development of seedling stands on drained peatlands). Silva Fenn 16(3):287–318

Heikurainen L (1982c) Peatland forestry. In: Laine J (ed) Peatlands and their utilization in Finland. Finn Peatland Soc, Int Peat Soc, Finn Nat Comm, Helsinki, pp 53–62

Heikurainen L (1985) Verhopuuston vaikutus kuusitaimikon kehitykseen (The influence of birch nurse crop (*Betula pubescens*) on the growth of spruce (*Picea abies*) seedling stands on drained peatlands). Silva Fenn 19(1):81–88

Heikurainen L (1986) Suo-opas. Kirjayhtymä, Helsinki, 51 pp

Heikurainen L, Huikari O (1952) Turvelajin mikroskooppinen määrittäminen (The microscopic determination of peat types). Commun Inst For Fenn 40(5):1–29

Heikurainen L, Kuusela K (1962) Revival of the tree growth after drainage and its dependence on the tree size and age. Commun Inst For Fenn 55(8):1–15

Heikurainen L, Laine J (1976) Lannoituksen, kuivatuksen ja lämpöolojen vaikutus istutus- ja luonnontaimien kehitykseen rämeillä (Effect of fertilization, drainage and temperature conditions in the development of planted and natural seedlings on pine swamps). Acta For Fenn 150:1–38

Heikurainen L, Päivänen J (1970) The effect of thinning, clear cutting and fertilization on the hydrology of peatland drained for forestry. Acta For Fenn 104:1–24

Heikurainen L, Pakarinen P (1982) Mire vegetation and site types. In: Laine J (ed) Peatlands and their utilization in Finland. Finn Peatland Soc, Int Peat Soc, Finn Nat Comm, Helsinki, pp 14–23

Heikurainen L, Seppälä K (1963) Kuivatuksen tehokkuus ja turpeen lämpötalous (The effect of drainage degree on temperature conditions of peat). Acta For Fenn 76(4):1–33

Heikurainen L, Seppälä K (1965) Regionality in stand increment and its depence on the temperature factor on drained swamps. Acta For Fenn 78(4):1–14

Heikurainen L, Seppälä K (1973) Ojitusalueiden puuston kasvun jatkumisesta ja alueellisuudesta (Regionality and continuity of stand growth in old forest drainage areas). Acta For Fenn 132:1–36

Heikurainen L, Päivänen J, Sarasto J (1964) Ground water table and water content in peat soil. Acta For Fenn 77(1):1–18

Heikurainen L, Päivänen J, Seppälä K (1966) Koetuloksia männyn kylvöstä ja istutuksesta ojitetuilla soilla (Some results of pine seeding and planting on drained peat soils). Silva Fenn 119(2):1–21

Heikurainen L, Kenttämies K, Laine J (1978) The environmental effects of forest drainage. Suo 29:49–58

Heikurainen L, Laine J, Lepola J (1983) Lannoitus- ja sarkaleveyskokeita karujen rämeiden uudistamisessa ja taimikoiden kasvatuksessa (Fertilization and ditch spacing concerned with regeneration and growing of young Scots pine stands on nutrient poor pine bogs). Silva Fenn 17(4):359–379

Heliövaara K (1982) The pine bark bug, *Aradus cinnamomeus* (Heteroptera, Aradidae) and the height growth rate of young Scots pines. Silva Fenn 16(4):357–361

Heliövaara K, Väisänen R (1983) Environmental changes and the flat bugs (Heteroptera, Aradidae and Aneuridae). Distribution and abundance in eastern Fennoscandia. Ann Entomol Fenn 49(4):103–109

Helmisaari H-S (1990) Nutrient retranslocation within *Pinus sylvestris*. PhD Thesis, Univ Joensuu, Joensuu, Finland 13 pp

Helmisaari H-S, Mälkönen E (1989) Acidity and nutrient content of throughfall and soil leachate in three *Pinus sylvestris* stands. Scand J For Res 4:13–28

Hendrickx JMH (1990) Determination of hydraulic soil properties. In: Anderson MG, Burt TP (eds) Process studies in hillslope hydrology. Wiley, Chichester, pp 43–92

Hibbert AD (1967) Forest treatment effects on water yield. In: Sopper WE, Lull HW (eds) Forest hydrology. Pergamon Press, London, pp 527–543

Hillebrand K (1993) The greenhouse effects of peat production and use compared with coal, oil, natural gas and wood. Tech Res Cent Finl, Res Notes 1494:1–50

Hillel D (1971) Soil and water. Physical principles and processes. Academic Press, London, 288 pp

Hillman GR (1987) Improving wetlands for forestry in Canada. Can For Serv, North For Cent, Inf Rep NOR-X-228:1–29

Hillman GR (1991) The Canada-Alberta wetlands drainage and improvement program for forestry: an update. In: Jeglum JK, Overend RP (eds) Peat and peatlands – diversification and innovation. Can Soc Peat Peatlands 1:54–61

Hillman GR (1992) Some hydrological effects of peatland drainage in Alberta's boreal forest. Can J For Res 22:1588–1596

Hirvas H (1991) Pleistocene stratigraphy of Finnish Lapland. Geol Surv Finl Bull 354:1–127

Hofstetter RH (1983) Wetlands in the United States. In: Gore AJP (ed) Mires: swamp, bog, fen and moor. Ecosystems of the world, vol 4B. Elsevier, Amsterdam, pp 201–244

Hökkä H, Laine J (1988) Suopuustojen rakenteen kehitys ojituksen jälkeen (Post-drainage development of structural characteristics in peatland forest stands). Silva Fenn 22(1):45–65

Holmen H (1964) Forest ecological studies on drained peat land in the province of Uppland, Sweden. Parts I–III. Stud For Suec 16:1–236

Holmen H (1968) Afforestation of peatlands in Sweden, classification and site research. Proc 3rd Int Peat Congr, pp 371–377

Holmen H (1969) Skogsproduktion på torvmark (Afforestation of peatlands). Kungl Skogs- o Lantbr-akad Tidskr 108:216–235

Holmen H (1985) De skogliga våtmarkernas växtnäringsförhållanden. In: Skog på våtmarker. Skogs- och jordbrukets forskningsråd Rapp 4:29–32

Holmes GD (1979) The ecology of even-aged plantations: an introduction to forestry in upland Britain. In: Ford ED, Malcolm DC, Atterson J (eds) The ecology of even-aged forest plantations. Inst Terrest Ecol, Cambridge, pp 7–19

Holopainen AL, Huttunen P (1992) Effects of forest clearcutting and soil disturbance on biology of small forest brooks. Hydrobiologia 243/244:457–464

Holstener-Jørgensen H (1958) Jordbundsfysiske undersøgelser i danske bogebevoksningar (Physical soil-investigations in Danish beech-stands). Forstl Forsogsv Danm 25:93–223

Honer TG, Bickerstaff A (1985) Canada's forest area and wood volume balance 1977–1981: an appraisal of change under present levels of management. Can For Serv, Pacific For Cent BC-X-272:1–84

Hotanen J-P, Nousiainen H (1990) Metsä- ja suokasvillisuuden numeerisen ryhmittelyn ja kasvupaikkatyyppien rinnastettavuus (The parity between numerical units and site types of forest and mire vegetation). Folia For 763:1–54

Hotanen J-P, Vasander H (1992) Eteläsuomalaisten metsäojitettujen turvemaiden kasvillisuuden numeerinen ryhmittely (Post-drainage development of vegetation in southern Finnish peatlands studied by numerical analysis). Suo 43:1–10

Houghton JT, Jenkins GJ, Ephraums JJ (1990) (eds) Climate change. IPCC Sci Assess, Cambridge Univ Press, Cambridge

Huikari O (1952) Suotyypin määritys maa-ja metsätaloudellista käyttöarvoa silmällä pitäen (On the determination of mire types, specially considering their drainage value for agriculture and forestry). Silva Fenn 75:1–22

Huikari O (1953) Tutkimuksia ojituksen ja tuhkalannoituksen vaikutuksesta eräiden soiden pieneliöstöön (Studies on the effect of drainage and ash fertilization upon the microbes of some swamps). Commun Inst For Fenn 42(2):1–18

Huikari O (1956) Primäärisen soistumisen osuudesta Suomen soiden synnyssä (Untersuchungen über den Anteil der primären Versumpfung an der Entstehung der Finnischen Moore). Commun Inst For Fenn 46(6):1–79

Huikari O (1958) Metsäojituksen koneellistamisesta (Über die Mechanisierung der Waldentwässerung). Commun Inst For Fenn 50(5):1–93

Huikari O (1959a) Kenttämittaustuloksia turpeiden vedenläpäisevyydestä (Feldmessungsergebnisse über die Wasserdurchlässigkeit von Torfen). Commun Inst For Fenn 51(1):1–26

Huikari O (1959b) Metsäojitettujen turvemaiden vesitaloudesta (Über den Wasserhaushalt waldentwässerter Torfböden). Commun Inst For Fenn 51(2):1–45

Huikari O (1959c) On the effect of anaerobic media upon the roots of birch, pine and spruce seedlings. Commun Inst For Fenn 50(9):1–28

Huikari O (1961) Koetuloksia metsäojitettujen soiden ravinnetalouden keinollisesta parantamisesta. Metsätal Aikak 1961(5):212–216

Huikari O (1972) Marjojen ja sienien tuotanto metsäojitusalueella (Berry and mushroom production in forest drainage areas). Lapin tutkimusas vuosikirja 13:33–37

Huikari O (1973) Koetuloksia metsäojitettujen soiden lannoituksesta (Results of fertilization experiment on peatlands drained for forestry). Metsäntutkimusl suontutkimusosaston tied 1/1973: 1–154

Huikari O (1974) Site quality estimation of forest land. Proc Int Symp For Drainage, Jyväskylä-Oulu, Finland, pp 15–24

Huikari O (1977a) Micro-nutrient deficiencies cause growth disturbances in trees. Silva Fenn 11(3):251–255

Huikari O (1977b) Use of different N, P and K fertilizers in forests growing on drained peat. In: Int Symp For Fertiliz. Ministere de l'agriculture, Paris 1973, FAO-IUFRO, Paris, pp 391–404

Huikari O, Paarlahti K, Paavilainen E, Ravela H (1966) Sarkaleveyden ja ojasyvyyden vaikutuksesta suon vesitalouteen ja valuntaan (On the effect of strip-width and ditch-depth on water economy and runoff on peat soil). Commun Inst For Fenn 61(8):1–39

Huttunen A (1983) Hilla- ja karpalosadoista. Metsäntutkimusl tied 90:143–147

Huttunen P, Holopainen AL, Ahtiainen M (1988) Effects of silvicultural measures on production biology of forest brooks. Publ Acad Finl 4:239–248

Hynninen P, Sepponen P (1983) Erään suoalueen ojituksen vaikutus purovesien laatuun Kiiminkijoen vesistöalueella Pohjois-Suomessa (The effect of drainage on the quality of brook waters in the Kiiminkijoki river basin, northern Finland). Silva Fenn 17(1):23–43

Hytönen J, Silfverberg K (1991) Kuivatustehon vaikutus turvemaan lämpöoloihin (Effect of drainage on thermal conditions in peat soils). Folia For 780:1–24

Hyvärinen A (1990) Deposition on forest soils: effect of tree canopy on throughfall. In: Kauppi P, Anttila P, Kenttämies K (eds) Acidification in Finland. Springer, Berlin Heidelberg New York, pp 199–213

Hyvärinen V, Vehviläinen B (1980) The effects of climatic fluctuations and man on discharge in Finnish river basins. Int Assoc Hydrol Sci, IAHS-AISH Publ 130:97–103

Illner K (1983) Die Moore in der DDR und ihre Standorteigenschaften. In: Moorstandortkunde und Moormelioration. Humboldt-Univ, Berlin, Ber 8:3–12

Ilomets M (1982) The productivity of *Sphagnum* communities and the rate of peat accumulation in Estonian bogs. In: Masing V (ed) Peatland ecosystems. Valgus, Tallinn, pp 102–116

Ingram HAP (1967) Problems of hydrology and plant distribution in mires. J Ecol 55:711–724

Ingram HAP (1983) Hydrology. In: Gore AJP (ed) Mires: swamp, bog, fen and moor. Ecosystems of the world, vol 4A. Elsevier, Amsterdam, pp 67–158

Ingram HAP (1992) Introduction to the ecohydrology of mires in the context of cultural perturbation. In: Bragg OM, Hulme PD, Ingram HAP, Robertson RA (eds) Peatland ecosystems and man: an impact assessment. Univ Dundee, Dep Biol Sci, pp 67–93

Ipatiev V (1990) Vodnyj regim i potreblenije elementov pitanija v meliolirovannyh sosnovyh fitotsenozah. Nauka i technika. Minsk, 127 pp

Ipatiev V, Paavilainen E (1975) Lannoituksen vaikutuksen kestoaika vanhassa tupasvilLarämeen männikössä (Duration of the effect of fertilization in an old pine stand on a cottongrass pine swamp). Folia For 241:1–13

Ipatiev VA, Smoljak LP, Blintsov IK (1984) Vedenie lesnovo hozjaistva na osusennich zemljach. Lesnaja promyshlennost. Moscow, 144 pp

Irwin RW (1968) Soil water characteristics of some Ontario peats. Proc 3rd Int Peat Congr, Nat Res Council Canada pp 219–223

Isomäki A, Niemistö P (1990) Ajourien vaikutus puuston kasvuun Etelä-Suomen nuorissa kuusikoissa (Effect of strip roads on the growth and yield of young spruce stands in southern Finland). Folia For 756:1–36

Isotalo A (1951) Studies on the ecology and physiology of cellulose decomposing bacteria in raised bogs. Acta Agric Fenn 74:1–106

Issakainen J (1978) Metsänparannusalueet riistan tuotannossa. Metsäntutkimusl Pyhäkosken tutkimusas tied 17:3–19

Ivanov KE (1981) Water movement in mirelands. Academic Press, London, 276 pp

Jack WH (1965) Experiments on tree growing on peat in northern Ireland. Forestry 38(1):20–40

Järvinen O (1986) Laskeuman laatu Suomessa 1971–1982. Vesihallituksen monistesarja 408: 1–142

Jeglum JK (1974) Relative influence of moisture-aeration and nutrients on vegetation and black spruce growth in northern Ontario. Can J For Res 4:114–126

Jeglum JK (1985) The status of peatland site classification for forestry in Ontario. Suo 36:33–44

Jeglum JK (1987) Alternate strip clearcutting in upland black spruce. II. Factors affecting regeneration in first-cut strips. For Chron 63(6):439–445

Jeglum JK (1990) Peatland forestry in Canada: an overview. In: Hånell B (ed) Biomass production and element fluxes in forested peatland ecosystems. Swed Univ Agric Sci, Umeå, Dep For Site Res, pp 19–28

Jeglum JK (1991a) Concluding comments. In: Jeglum JK, Overend RP (eds) Peat and peatlands – diversification and innovation. Can Soc Peat Peatlands 1:223–224

Jeglum JK (1991b) Definition of trophic classes in wooded peatlands by means of vegetation types and plant indicators. Ann Bot Fenn 28:175–192

Jeglum JK (1991c) The Wally Creek area forest drainage project in Ontario's Clay Belt: Progress report. In: Jeglum JK, Overend RP (eds) Peat and peatlands – diversification and innovation. Can Soc Peat Peatlands 1:47–53

Jeglum JK, Kennington DJ (1993) Strip clearcutting in black spruce: a guide for the practicing forester. For Can, Ontario Region, Great Lakes For Cent, 102 pp

Jeglum JK, Wehrhahn CF, Swan JMA (1971) Comparisons of environmental ordinations with principal component vegetational ordinations for sets of data having different degrees of complexity. Can J For Res 1:99–112

Jeglum JK, Haavisto VF, Boissonneau AN (1974) Toward a wetland classification for Ontario. Can For Serv Inf Rep O-X-215:1–54

Jeglum JK, Haavisto VF, Groot A (1982) Peatland forestry in Ontario: an overview. In: Sheppard JD, Musial J, Tibbetts TE (eds) A symposium on peat and peatlands, Shippegan, New Brunswick. Int Peat Soc, Can Nat Comm, pp 127–167

Jeglum JK, Rothwell RL, Berry GJ, Smith GKM (1991) New volumetric sampler increases speed and accuracy of peat surveys. Frontline, For Res Appl, For Can, Tech Note 9:1–4

Jerven O, Wisth OM (1967) Skogproduksjon på myr. Nor Skogselskap, Oslo, 104 pp

Joensuu S (1990) Protection of watercourses from peatlands utilized for forestry: Sedimentation ponds. In: Hånell B (ed) Biomass production and element fluxes in forested peatland ecosystems. Swed Univ Agric Sci, Umeå, Dep For Site Res, pp 157–165

Joensuu S (1992) Peatland forest ditch maintenance and runoff water quality: the effect of sedimentation ponds. Proc 9th Int Peat Congr, Int Peat Soc, Swed Nat Comm, Uppsala, 2:423–433

Johansson I (ed) (1984) Nordic glossary of hydrology. Almqvist & Wiksell, Stockholm, 224 pp

Johnson CW (1985) Bogs of the northeast. Univ Press New England, Hanover, 269 pp

Jones RK, Pierpoint G, Wickware GM, Jeglum JK, Arnup RW, Bowles JM (1983) Field guide to forest ecosystem classification for the Clay Belt, Site Region 3e. Ont Minist Nat Res, Maple, Ont, 123 pp

Joosten JHJ (1992) Bog regeneration in the Netherlands: a review. In: Bragg OM, Hulme PD, Ingram HAP, Robertson RA (eds) Peatland ecosystems and man: an impact assessment. Univ Dundee, Dep Biol Sci, pp 367–373

Juusela T, Kaunisto S, Mustonen S (1969) Turpeesta tapahtuvaan haihduntaan vaikuttavista tekijöistä (On factors affecting evapotranspiration from peat). Commun Inst For Fenn 67(1):1–45

Kaakinen E, Salminen P, Vasander H (1990) Skotlannin peittosuot ja niiden suojelu (Scottish blanket mires and their conservation). Suo 41:1–9

Kaila S, Päivänen J (1981) Metsämaanmuokkauksen suoritemäärät ja konekalusto vuosina 1976–1979 (Forest soil cultivation areas and machinery in 1976–1979). Silva Fenn 15(3):332–346

Kaitera P (1939) Lumen kevätsulamisesta ja sen vaikutuksesta vesiväylien purkautumissuhteisiin Suomessa. PhD Thesis, Tech Univ, Helsinki, Finl, 255 pp

Kalmari A (1982) Combustion technology. In Laine J (ed) Peatlands and their utilization in Finland. Finn Peatland Soc, Int Peat Soc, Finn Nat Comm, Helsinki, pp 89–93

Kangas PC (1990) Long-term development of forested wetlands. In: Lugo AE, Brinson M, Brown S (eds) Forested wetlands. Ecosystems of the world, vol 15. Elsevier, Amsterdam, pp 25–51

Kanninen M (ed) (1992) Muuttuva ilmakehä. Ilmasto, luonto ja ihminen. VAPK-kustannus, Helsinki, 163 pp

Kanninen M, Anttila P (eds) (1992) The Finnish Research Programme on Climate Change, Progress Report. Publ Acad Finl 3:1–308

Karesniemi K (1972) Dependence of humification degree on certain properties of peat. Proc 4th Int Peat Congr,Otaniemi, Finland, 2:273–282

Karsisto K (1967) Eri ajankohtina annetun NPK-lannoituksen aiheuttamista reaktioista rämeen männyntaimistoissa (The effects of NPK fertilization at different dates on pine seedling stands). Suo 18:49–55

Karsisto K (1968) Eri fosforilannoitelajien soveltuvuus suometsien lannoitukseen (Using various phosphatic fertilizers in peatland forests). Suo 19:104–111

Karsisto K (1974a) Ojituksen ja metsänlannoituksen vaikutus vesien saastumiseen. Metsäntutkimusl Pyhäkosken tutkimusas tied 13:1–19,33

Karsisto K (1974b) On the duration of fertilization influence in peatland forests with special reference to the results obtained from experiments with different phosphorus fertilizers. Proc Int Symp For Drainage, Jyväskylä-Oulu, Finland, pp 309–327

Karsisto K (1976a) Fosforilannoitelajit suometsien lannoituksessa. Lic Thesis, Univ Helsinki, Helsinki, Finland 252 pp

Karsisto K (1976b) Puuston elpyneisyyden vaikutus lannoitustulokseen. Pyhäkosken tutkimusas tied 15:29–36

Karsisto K (1977) Kotimaisten fosforirikasteiden käyttökelpoisuus suometsien lannoituksessa. Suo 28:43–46

Karsisto K, Issakainen J (1974) Riistan tuottaminen metsänparannusalueilla. Metsäntutkimusl Pyhäkosken tutkimusas tied 11:1–30

Karsisto K, Ravela H (1971) Washing away of phosphorus and potassium from areas drained for forestry and topdressed at different times of the year. Acta Agric Fenn 123:54–69

Karsisto M (1979a) Effect of forest improvement measures on activity of organic matter decomposing micro-organisms in forested peatlands. Part I. Effect of drainage and NPK fertilization in the spruce and pine swamps at Kivalo and Vilppula. Suo 30:49–58

Karsisto M (1979b) Maanparannustoimenpiteiden vaikutuksista organista ainetta hajottavien mikrobien aktiivisuuteen suometsissä. Osa II. Tuhkalannoituksen vaikutus (Effect of forest improvement measures on activity of organic matter decomposing micro-organisms in forested peatlands. Part II. Effect of ash fertilization). Suo 30:81–91

Kaunisto S (1971) Lannoituksen, muokkauksen ja vesipinnan etäisyyden vaikutus kylvötaimien ensi kehitykseen turvealustalla. Kasvihuoneessa suoritettu tutkimus (Effect of fertilization, soil preparation and distance of water level on the initial development of Scots pine and Norway spruce seedlings on peat. A study performed in the greenhouse). Commun Inst For Fenn 75(2):1–70

Kaunisto S (1974) On direct seeding of open peatlands. Proc Int Symp For Drainage, Jyväskylä-Oulu, Finland, pp 139–143

Kaunisto S (1977) Ojituksen tehokkuuden ja lannoituksen vaikutus männyn viljelytaimistojen kehitykseen karuilla avosoilla (Effect of drainage intensity and fertilization on the development of pine plantations in oligotrophic treeless *Sphagnum* bogs). Folia For 317:1–31

Kaunisto S (1979) Alustavia tuloksia palaturpeen kuivatuskentän ja suonpohjan metsityksestä (Preliminary results on afforestation of sod drying fields and peat cut-over areas). Folia For 404:1–14

Kaunisto S (1982) Development of pine plantations on drained bogs as affected by some peat properties, fertilization, soil preparation and liming. Commun Inst For Fenn 109:1–56

Kaunisto S (1984a) Peat properties in estimating need for nitrogen fertilization on pine plantations on bogs. Proc 7th Int Peat Congr, Irish Nat Peat Comm, Dublin, Ireland, 3:327–341

Kaunisto S (1984b) Suometsien uudistaminen turvekangasvaiheessa. Metsäntutkimusl tied 137: 7–21

Kaunisto S (1985a) Lannoituksen, ilman lämpösumman ja eräiden kasvualustan ominaisuuksien vaikutus mäntytaimikoiden kasvuun turvemailla (Effect of fertilization, temperature sum and some peat properties on the height growth of young pine sapling stands on peatlands). Folia For 616:1–27

Kaunisto S (1985b) Suonpohjien metsätaloudellinen käyttö. Metsäntutkimusl tied 184:4–8

Kaunisto S (1986) Peatlands before and after peat harvesting. In: Socio-economic impacts of the utilization of peatlands in industry and forestry. Int Peat Soc, Finn Nat Comm, Helsinki, pp 241–246

Kaunisto S (1987a) Effect of refertilization on the development and foliar nutrient contents of young Scots pine stands on drained mires of different nutrient regimes. Commun Inst For Fenn 140:1–58

Kaunisto S (1987b) Lannoituksen ja muokkauksen vaikutus männyn ja rauduskoivun istutustaimien kasvuun suonpohjilla (Effect of fertilization and soil preparation on the development of Scots pine and silver birch plantations on peat cut-over areas). Folia For 681:1–23

Kaunisto S (1989) Jatkolannoituksen vaikutus puuston kasvuun vanhalla ojitusalueella (Effect of refertilization on tree growth in an old drainage area). Folia For 724:1–15

Kaunisto S (1991) Maa-analyysin käyttö kasvupaikan ravinnetilan arvioimiseksi eräillä Alkkian metsitetyillä suopelloilla (Soil analysis as a means of determining the nutrient regime on some afforested peatland fields at Alkkia). Folia For 778:1–32

Kaunisto S (1992) Effect of potassium fertilization on the growth and nutrition of Scots pine. Suo 43:45–62

Kaunisto S, Norlamo M (1976) On nitrogen mobilization in peat. I. Effect of liming and rotavation in different incubation temperatures. Commun Inst For Fenn 88(2):1–27

Kaunisto S, Paavilainen E (1977) Response of Scots pine plants to nitrogen refertilization on oligotrophic peat. Commun Inst For Fenn 92(1):1–54

Kaunisto S, Paavilainen E (1988) Nutrient stores in old drainage areas and growth of stands. Commun Inst For Fenn 145:1–39

Kaunisto S, Päivänen J (1985) Metsänuudistaminen ja metsittäminen ojitetuilla turvemailla. Kirjallisuuteen perustuva tarkastelu (Forest regeneration and afforestation on drained peatlands. A literature review). Folia For 625:1–75

Kaunisto S, Tukeva J (1984) Kalilannoituksen tarve avosoille perustetuissa riukuasteen männiköissä (Need for potassium fertilization in polestage pine stands established on bogs). Folia For 585:1–40

Kaunisto S, Tukeva J (1986) Kasvatustiheyden vaikutus männyn istutustaimikoiden kehitykseen turvemailla (Effect of tree spacing on the development of pine plantations on peat). Folia For 646:1–36

Kaunisto S, Viinamäki T (1991) Lannoituksen ja leppäsekoituksen vaikutus mäntytaimikon kehitykseen ja suonpohjaturpeen ominaisuuksiin Aitonevalla (Effect of fertilization and alder (*Alnus incana*) mixture on the development of young Scots pine (*Pinus sylvestris*) trees and the peat properties in a peat cut-over area at Aitoneva, southern Finland). Suo 42:1–12

Kaunisto S, Ferm A, Kokkonen J (1991) Suometsät ja niiden tutkimus Virossa ja Latviassa (Peatland forestry research in Estonia and Latvia). Suo 42:61–70

Kaunisto S, Moilanen M, Issakainen J (1993) Apatiitti ja flogopiitti fosfori- ja kaliumlannoitteina suometsissä (Apatite and phlogopite as phosphorus and potassium fertilizers in peatland pine forests). Folia For 810:1–30

Keltikangas M (1971) Sarkaleveyden vaikutus ojitusinvestoinnin taloudelliseen tulokseen (Effects of drain spacing on the economic results of forest drainage investments). Acta For Fenn 123: 1–70

Keltikangas M (1990) Ojitettujen soiden merkitys Suomen puuhuollolle nyt ja vastaisuudessa. Sem Rep 13.3.1990, Univ Joensuu, Joensuu, Finland

Keltikangas M, Seppälä K (1973) Metsänlannoituksen edullisuuden vaihtelu (Variations in the profitability of forest fertilization). Silva Fenn 7(3):192–235

Keltikangas M, Seppälä K (1977) Ojitusalueiden hieskoivikoiden kasvatus taloudellisena vaihtoehtona (The economics of growing birch stands on drained peatlands). Silva Fenn 11(1): 49–68

Keltikangas M, Laine J, Puttonen P, Seppälä K (1986) Vuosina 1930–1978 metsäojitetut suot: ojitusalueiden inventoinnin tuloksia (Peatlands drained for forestry during 1930–1978: results from field surveys of drained areas). Acta For Fenn 193:1–94

Kenttämies K (1977) Utlakningen av fosfor och kalium från dikade och gödslade torvmarker. 13th Nord Symp Vattenforsk, Diffuse vannforurensninger, tilförsel och transport. Röros, Maj 1977, Nordfors, Miljövårdssekr Pub 2:271–280

Kenttämies K (1980) The effects on water quality of forest drainage and fertilization in peatlands. Int Assoc Hydrol Sci IAHS-AISH Publ 130:277–284

Kenttämies K (1981) The effects on water quality of forest drainage and fertilization in peatlands. Publ Water Res Inst, Nat Board Waters Finl 43:24–31

Kenttämies K, Laine J (1984) The effects on water quality of forest drainage and phosphate fertilization in a peatland area in central Finland. Proc 7th Int Peat Congr, Irish Nat Peat Comm, Dublin, 3:342–354

Ketcheson DE, Jeglum JK (1972) Estimates of black spruce and peatland areas in Ontario. Can Fort Serv Inf Rep O-X-172:1–29

King JA, Smith KA, Pyatt DG (1986) Water and oxygen regimes under conifer plantations and native vegetation on upland peaty gley soil and deep peat soils. J Soil Sci 37:485–497

Kivinen E (1933) Suokasvien ja niiden kasvualustan kasvinravintoainesuhteista (Untersuchungen über den Gehalt an Pflanzennährstoffen in Moorpflanzen und an ihren Standorten). Acta Agric Fenn 27:1–140

Kivinen E (1948) Suotiede. Werner Söderström, Porvoo-Helsinki, 219 pp

Kivinen E (1981) Utilization of peatlands in some countries. Bull Int Peat Soc 12:21–30

Kivinen E, Pakarinen P (1980) Peatland areas and the proportion of virgin peatlands in different countries. Proc 6th Int Peat Congr, Duluth, pp 52–54

Kivinen E, Pakarinen P (1981) Geographical distribution of peat resources and major peatland complex types in the world. Ann Acad Sci Fenn Ser A III Geol Geogr 132:1–28

Klemetti V, Keys D (1983) Relationship between dry density, moisture content, and decomposition of some New Brunswick peats. In: Jarrett PM (ed) Testing of peats and organic soils. ASTM Spec Tech Publ 820:72–82

Klemetti V, Scholz A, Selin P, Nyrönen T (1988) Preliminary results of a mole drainage experiment in Piipsanneva peat harvesting site. Publ Acad Finl 4:277–284

Knighton MD, Stiegler JH (1980) Phosphorus release following clearcutting of a black spruce fen and a black spruce bog. Proc 6th Int Peat Congr, Duluth, pp 577–583

Knize AA, Rubtsov VG (1974) The pecularities of drained forest areas evaluation. Proc Int Symp For Drainage, Jyväskylä-Oulu, Finland, pp 291–295

Koivisto IJ (1985) Kokemuksia Wally Creek – metsäojitusprojektista Pohjois-Ontariossa, Kanadassa (The Wally Creek area forest drainage experiment in northeastern Ontario, Canada). Suo 36:52–57

Kolabinski VS (1991) Effects of cutting method and seedbed treatment on black spruce regeneration in Manitoba. For Can, North For Cent NOR-X-316:1–21

Kolari KK (ed) (1983) Growth disturbances of forest trees. Commun Inst For Fenn 116:1–208

Kolari KK (1988) Metsäpuiden kasvuhäiriöt. Metsäntutkimusl tied 310:1–35

Konstantinov VK (1979) Ekspluatatsija lesoosushitelnyh cistem. Lesnaja promyshlennost. Moscow, 151 pp

Konstantinov VK, Stoliarov DP (1988) Effectiveness of hydrotechnical forest reclamation in the USSR. Proc 8th Int Peat Congr, Leningrad, 3:28–34

Konstantinov VK, Krasilnikov NA, Elpatievsky MM, Smolyanitsky LY, Kalinin MW (1983) Problems of operation and improvement of forest hydromelioration systems. Proc Int Symp For Drainage, Tallin, Estonian SSR, pp 95–102

Koponen S (1980) Effects of drainage upon the spider fauna (*Aranae*) of the ground layer of mires. In: Dindal DL (ed) Soil biology as related to land use practices. Proc 7th Int Soil Zool Coll, Washington, 61:19–22

Koponen S (1985) Soiden hämähäkkilajiston muutoksista (On changes in the spider fauna of bogs). Mem Soc Fauna Flora Fenn 61:19–22

Korhola A (1992) Mire induction, ecosystem dynamics and lateral extension on raised bogs in the southern coastal area of Finland. Fennia 170(2):25–94

Korhonen K-M (1987) Damage caused by the root vole (*Microtus aeconomus*) to Scots pine in man-made habitats in northern Finland. Commun Inst For Fenn 144:1–61

Korpijaakko M (1981) Uusi kairatyyppi tilavuustarkkojen turvenäytteiden ottamiseen (A piston sampler for undisturbed peat samples). Suo 32:7–8

Korpijaakko M, Radforth NW (1972) Studies on the hydraulic conductivity of peat. Proc 4th Int Peat Congr, Otaniemi, Finland, 3:323–334

Korpijaakko M, Häikiö J, Leino J (1981) Vesipitoisuuden ja maatuneisuuden vaikutus turpeen kuivatilavuuspainoon (Effect of water content and degree of humification on dry density of peat). Suo 32:39–43

Kortesharju J (1979) Kokemuksia hillan puoliviljelystä. Metsäntutkimusl Rovaniemen tutkimusas tied 21:19–22

Koski V (1973) On selfpollination, genetic load and subsequent inbreeding in some conifers. Commun Inst For Fenn 78(10):1–42

Kozlovskaya LS (1974) The effect of drainage on the change in the biological activity of forest peat soil. Proc Int Symp For Drainage, Jyväskylä-Oulu, Finland, pp 57–62

Kozlovskaya LS, Medvedeva VM, Pjavchenko NI (1978) Dinamika organicheskogo vechevstva v prochesse torfoobrazovanija. Nauka, Leningrad, 177 pp

Kozlowski TT (1982) Water supply and tree growth. Part II Flooding. For Abstr 43(3):145–161

Kozlowski TT (1986) Soil aeration and growth of forest trees (Review article). Scand J For Res 1:113–123

Kramer PJ (1949) Plant and soil water relationships. McGraw-Hill, New York, 347 pp

Kubin E (1992) The effect of clear cutting, waste wood collecting and site preparation on the nitrate nitrogen leaching to groundwater. In: Krecek J, Haigh MJ (eds) Proc Symp Environmental regeneration in headwaters, Environm Probl Consult, Prague, pp 80–85

Kulmala A, Leinonen L, Säynätkari T (1990) Tausta-asemien ilmanlaatu Suomessa 1980–1986 (Air quality at background stations in Finland 1980–1986). Ilmatieteen laitos, Ilmansuojelun julk 7:1–201

Kuntze H (1984) Hoch- und Niedermoor – zwei verschiedene Moortypen. In: Kuntze H (ed) Bewirtschaftung und Düngung von Moorböden. Ber Bodentechnol Inst Niedersächsischen Landesamtes Bodenforsch, pp 7–13

Kuntze H, Roeschmann G, Schwerdtfeger G (1988) Bodenkunde. Ulmer, Stuttgart, 568 pp

Kurimo H (1983) Surface fluctuation in three virgin pine mires in eastern Finland. Silva Fenn 17(1):45–64

Kurkela T (1973) *Godronia multisporan* aiheuttama tauti raudus- ja hieskoivun taimissa eräissä metsänlannoituskokeissa turvemaalla (A disease caused by *Godronia multispora* groves on young *Betula pendula* Ehr and *B pubescens* Ehr on fertilized peatlands). Suo 24:8–15

Kurkela T (1983) Fungal diseases associated with nutritional growth disturbances of Scots pine. Commun Inst For Fenn 116:73–77

Kurki M (1963) Suomen peltojen viljavuudesta (Über die Fruchtbarkeit des finnischen Ackerbodens auf Grund der in den Jahren 1955–1960 durchgeführten Bodenfruchtbarkeitsuntersuchungen) Viljavuuspalvelu Oy, Helsinki, 107 pp

Kuusela K, Salminen S (1983) Metsävarat Etelä-Suomen kuuden pohjoisimman piirimetsälautakunnan alueella 1979–1982 sekä koko Etelä-Suomessa 1977–1982. (Forest resources in the six northernmost Forestry Board Districts of South Finland, 1979–1982, and in the whole South Finland, 1977–1982). Folia For 568:1–79

Kuzila MS, Rundquist DC, Green JA (1991) Methods for estimating wetland loss: the Rainbasin region of Nebraska, 1927–1981. J Soil Water Conserv 46:441–446

Lääperi A, Löyttyniemi K (1988) Hirvituhot vuosina 1973–1982 perustetuissa männyn viljelytaimikoissa Uudenmaan-Hämeen metsälautakunnan alueella (Moose (*Alces alces*) damage in pine plantations established during 1973–1982 in the Uusimaa-Häme Forestry Board District). Folia For 719:1–13

Lähde E (1966) Vertical distribution of biological activity in peat of some virgin and drained swamp types. Acta For Fenn 81 (6):1–15

Lähde E (1969) Biological activity in some natural and drained peat soils with special reference to oxidation-reduction conditions. Acta For Fenn 94:1–69

Lähde E (1992) Regeneration potential of all-sized spruce-dominated stands. In: Hagner M (ed) Silvicultural alternatives. Swed Univ Agric Sci, Umeå, Dep Silvicult, Rep 35:111–116

Lähde E, Laiho O, Norokorpi Y, Saksa T (1992) Alternative silvicultural treatments as applied to advanced stands – research plan. In: Hagner M (ed) Silvicultural alternatives. Swed Univ Agric Sci, Umeå, Dep Silvicult, Rep 35:66–73

Lähdesmäki P, Pietiläinen P (1990) Seasonal variation in nitrate reductase activity and concentration of NO_3^-, NO_2^- and NH_4^+ in buds and needles of Scots pine. Aquilo Ser Bot 26:7–11

Laiho R (1991) Metsäojituksen vaikutus nevamaisten rämeiden turpeen ravinnevaroihin. Lic Thesis, Univ Helsinki, Dep Peatland For, Helsinki, Finland, 90 pp

Laiho R, Laine J (1990) Post-drainage nutrient stores in peat. In: Hånell B (ed) Biomass production and element fluxes in forested peatland ecosystems. Swed Univ Agric Sci, Umeå, Dep For Site Res, pp 81–91

Laiho R, Laine J (1992a) Effect of forest drainage on soil frost conditions. In: Bragg OM, Hulme PD, Ingram HAP, Robertson RA (eds) Peatland ecosystems and man: an impact assessment. Univ Dundee, Dep Biol Sci, pp 149–152

Laiho R, Laine J (1992b) Potassium stores in peatlands drained for forestry. Proc 9th Int Peat Congr, Int Peat Soc, Swed Nat Comm, Uppsala, 1:158–169

Laiho R, Laine J (1994) Nitrogen and phosphorus stores in peatlands drained for forestry in Finland. Scand J For Res 9:251–260

Laine J (1979) Kontortamännyn alkukehitys ojitetulla karulla avosuolla (Initial development of *Pinus contorta* on a nutrient poor open bog in Finland). Suo 30:27–32

Laine J (1984) Estimation of evapotranspiration from peatlands by means of daily water table hydrographs. Univ Helsinki, Publ Dep Peatland For, Helsinki, Finland, 5:1–100

Laine J (1986) Kuivatustekniikan, kuivatussyvyyden ja puuston kasvun välisiä vuorosuhteita 25 vuotta vanhoilla rämeojitusalueilla. Tutkimussopimushankkeen "Metsäojitettujen soiden ekologia" loppuraportti. Univ Helsinki, Dep Peatland For, Helsinki, Finland, 49 pp

Laine J (1989) Metsäojitettujen soiden luokittelu (Classification of peatlands drained for forestry). Suo 40:37–51

Laine J, Laiho R (1992) Effect of forest drainage on the carbon balance and nutrient stores of peatland ecosystems. In: Kanninen M, Anttila P (eds) The Finnish research programme on climate change. Progress report. Publ Acad Finl 3:205–210

Laine J, Mannerkoski H (1975) Tensiometrin käyttö turvemaiden kosteusolojen kuvauksessa (On the use of tensiometers in describing moisture conditions of peat soils). Suo 26:17–24

Laine J, Mannerkoski H (1980) Lannoituksen vaikutus männyntaimikon kasvuun ja hirvituhoihin karuilla ojitetuilla nevoilla (Effect of fertilization on tree growth and elk damage in young Scots pine stands planted on drained, nutrient-poor open bogs). Acta For Fenn 166:1–45

Laine J, Päivänen J (1982) Water content and bulk density of peat. Proc Int Symp IPS Comm IV and II "Peat, its properties and perspectives of utilization", Minsk, Byelorussian SSR, pp 422–430

Laine J, Päivänen J (1992) Carbon balance of peatlands and global climatic change: summary. In: Kanninen M, Anttila P (eds) The Finnish research programme on climate change. Progress report. Publ Academy Finl 3:189–193

Laine J, Starr M (1979) An analysis of the post-drainage stand increment in relation to the peatland site type classification in Finland. In: Classification of peat and peatlands. Int Peat Soc, Helsinki, pp 147–159

Laine J, Vanha-Majamaa I (1992) Vegetation ecology along a trophic gradient on drained pine mires in southern Finland. Ann Bot Fenn 29:213–233

Laine J, Vasander H (1990) Suotyypit. Kirjayhtymä, Helsinki, 80 pp

Laine J, Päivänen J, Schneider H, Vasander H (1986) Site types at Lakkasuo mire complex. Field guide. Univ Helsinki, Publ Dep Peatland For, Helsinki, Finland, 8:1–35

Laine J, Vasander H, Puhalainen A (1992a) A method to estimate the effect of forest drainage on the carbon store of a mire. Suo 43:227–230

Laine J, Vasander H, Puhalainen A (1992b) Effect of forest drainage on the carbon balance of mire ecosystems. Proc 9th Int Peat Congr, Int Peat Soc, Swed Nat Comm, Uppsala, 1:170–181

Lappalainen E (1982) Peat resources. In: Laine J (ed) Peatlands and their utilization in Finland. Finn Peatland Soc, Int Peat Soc, Finn Nat Comm, Helsinki, pp 12–13

Lappalainen E (1991) Kiina panostaa soiden ja turpeen käyttöön. Peat Indust 4:28–29

Lappalainen E (1992) Turve – fossiilista ainetta vai biomassaa? Suo ja turve. Peat Indust 1:14–17

Lappalainen E, Uhlgren S (1991) Turvetutkimusta 50 vuotta. Geologian tutkimuskeskus, Turvetutkimukset 1941–1991, Kuopio, 139 pp

Latja A, Kurimo H (1988) Temperature changes in the soil and close to the ground on wetlands drained for forestry. Publ Acad Finl 4:46–51

Lauhanen R (1992) PATU M 100 -kaivuri metsäojituksessa (PATU M 100 excavator in forest drainage). Metsäntutkimusl tied 401:1–23

Leikola M (1975) Verhopuuston vaikutus metsikön lämpöoloihin Pohjois-Suomessa (The influence of the nurse crop on stand temperature conditions in northern Finland). Commun Inst For Fenn 85(7):1–33

Leikola M, Pylkkö P (1969) Verhopuuston tiheyden vaikutus metsikön minimilämpötiloihin hallaöinä (On the influence of stand density on the minimum temperatures during frost nights). Silva Fenn 3(1):20–32

Lévesque M, Morita H, Schnitzer M, Mathur SP (1980) The physical, chemical, and morphological features of some Quebec and Ontario peats. Agric Can, Res Branch, Chem Biol Res Inst Contrib 1155:1–70

Lieffers VJ, Macdonald SE (1990) Growth and foliar nutrient status of black spruce and tamarack in relation to depth of water table in some Alberta peatlands. Can J For Res 20:805–809

Lindholm T, Vasander H (1988) Effect of readily and slowly soluble PK and NPK fertilizers on the growth of Scots pine on a drained raised bog in southern Finland. Proc 8th Int Peat Congr, Leningrad, 3:144–152

Lindsay RA (1992) Peatland conservation: the Cinderella syndrome. In: Bragg OM, Hulme PD, Ingram HAP, Robertson RA (eds) Peatland ecosystems and man: an impact assessment. Univ Dundee, Dep Biol Sci, pp 331–344

Lipas E, Levula T (1980) Urealannoitus eri vuodenaikoina (Urea fertilization at different times of the year). Folia For 421:1–14

Lohmus E (1981) Anthropogenous forest site types on drained peatlands. In: Laasimer L (ed) Anthropogenous changes in the plant cover of Estonia. Tartu, pp 77–90

Lorenz W-D, Sauerbrey R, Eschner D, Lehrkamp H, Zeitz J (1992) Zustand der landwirtschaftlich genutzten Niedermoore in der ehemaligen DDR. Wasser Boden 2/1992:58–61

Lugo AE (1990) Introduction. In: Lugo AE, Brinson M, Brown S (eds) Forested wetlands. Ecosystems of the world, vol 15. Elsevier, Amsterdam, pp 1–14

Lugo AE, Brown S, Brinson MM (1990) Concepts in wetland ecology. In: Lugo AE, Brinson M, Brown S (eds) Forested wetlands. Ecosystems of the world, vol 15. Elsevier, Amsterdam, pp 53–85

Lukin J (1988) Ojitustekniikan ja puuston vaikutus pohjavedenpinnan tasoon 25 vuotta vanhoilla rämeojitusalueilla. MSc Thesis, Univ Helsinki, Dep Peatland For, Helsinki, Finland, 67 pp

Lukkala OJ (1929) Tutkimuksia soiden metsätaloudellisesta ojituskelpoisuudesta erityisesti kuivatuksen tehokkuutta silmälläpitäen (Untersuchungen über die waldwirtschaftliche Entwässerungsfähigkeit der Moore). Commun Inst For Fenn 15(1):1–278

Lukkala OJ (1931) Metsäojituksen oppikirja. Keskusmetsäseura Tapio, Helsinki, 258 pp

Lukkala OJ (1936) Rämemäntyjen siementen laadusta (Über die Veschaffenheit der Moorkiefernsamen). Commun Inst For Fenn 22(3):1–45

Lukkala OJ (1939) Soiden metsäojituskelpoisuus. Keskusmetsäseura Tapio, Helsinki, 48 pp

Lukkala OJ (1946) Korpimetsien luontainen uudistaminen (Die natürliche Verjüngung der Bruchwälder). Commun Inst For Fenn 34(3):1–150

Lukkala OJ (1948) Metsäojien kunnossapito (Die Instandhaltung der Waldgräben). Commun Inst For Fenn 36(1):1–56

Lukkala OJ (1949) Soiden turvekerroksen painuminen ojituksen vaikutuksesta (Über die Setzung des Moortorfes als Folge der Entwässerung). Commun Inst For Fenn 37(1):1–67

Lukkala OJ (1951) Kokemuksia Jaakkoinsuon koeojitusalueelta (Experiences from Jaakkoinsuo experimental drainage area). Commun Inst For Fenn 39(6):1–53

Lukkala OJ (1952) Rämemäntyjen perinnöllisyydestä (On the heredity of räme-pines). Commun Inst For Fenn 40(1):1–22

Lukkala OJ, Kotilainen MJ (1945) Soiden ojituskelpoisuus. Keskusmetsäseura Tapio, Helsinki, 56 pp

Lumme I (1988) Early effects of peat ash on growth and mineral nutrition of the silver birch (*Betula pendula*) on a mined peatland. Silva Fenn 22(2):99–112

Lundin L (1984) Torvmarksdikning. Hydrologiska konsekvenser för Docksmyren (Peatland drainage. Effects on the hydrolgy of the mire Docksmyren). Univ Uppsala, Dep Phys Geogr, Hydrol Div, Rep Ser A 3:1–75

Lundin L (1990) Vattenmiljön vid våtmarksdikning (Hydrological effects of forest wetland drainage). Sver Skogsvårdsförb Tidskr 1:37–47

Lundmark T (1987) Föryngring på dikade marker. III. Kungl Skogs o Lantbr-akad tidskr 126: 283–296

Lüttig G (1983) Peat – lignite – coal in the present discussion on energy resources strategy. In: Lüttig GW (ed) Recent technologies in the use of peat. Schweizerbart'sche Verlagsbuchhandlung, Stuttgart, pp 1–9

Lüttig G (1989) History of the utilization of peatlands in Europe with special reference to Germany. Suo 40:169–175

Lüttig G (1991) Peatbogs and peatlands in Germany – the state of the art. Bull Int Peat Soc 22: 38–42

Lutz HJ, Chandler RF (1965) Forest soils. Wiley, New York, 514 pp

Lyly O, Saksa T (1992) The effect of stand density on moose damage in young *Pinus sylvestris* stands. Scand J For Res 7:393–403

Mackenzie JM, Thomson JH, Wallis KE (1976) Control of heather by 2,4-D. For Comm Leaflet 64:1–19

Makarevskii MF (1992) Stocks and balance of organic carbon in forest and marsh biogeocenoses in Karelia. Sov J Ecol 22(3):133–139

Maki TE (1959) Improving site quality by wet-land drainage. In: Burns PY (ed) Southern forest soils. Louisiana State Univ, Baton Rouge, pp 106–113

Mäkitalo A (1985) Tree growth in relation to site characteristics on selected peatland sites in central Alberta. MSc Thesis, Univ Alberta, Dep For Sci, 80 pp

Malcolm DC (1975) The influence of heather on silvicultural practice – an appraisal. Scott For 29(1):14–24

Malcolm DC (1979) The future development of even-aged plantations: silvicultural implications. In: Ford ED, Malcolm DC, Atterson J (eds) The ecology of even-aged forest plantations. Inst Terr Ecol, Cambridge, pp 481–504

Malcolm DC (1991) Afforestation in Britain – a commentary. Scott For 45(4):259–274

Malcolm DC, Cuttle SP (1983) The application of fertilizers to drained peat. 1. Nutrient losses in drainage. Forestry 56(2):155–174

Mälkönen E (1974) Annual primary production and nutrient cycle in some Scots pine stands. Commun Inst For Fenn 84(5):1–87

Mälkönen E (1977) Annual primary production and nutrient cycle in a birch stand. Commun Inst For Fenn 91(5):1–35

Malm D, Möller G (1974) Skillnader i volymtillväxtökning efter gödsling med urea resp ammonium nitrat (Differences in volume growth increase after fertilization with urea and ammonium nitrate fertilizer). Skogsförbättr Årsbok, pp 46–47

Malmer N (1985) Remarks to the classification of mires and mire vegetation – Scandinavian arguments. Aquilo Ser Bot 21:9–17

Malmström C (1928) Våra torvmarker ur skogsdikningssynpunkt (Our peat areas from the point of forest-draining. Medd Stat Skogsförsöksanstalt 24(9):251–372

Malmström C (1952) Svenska gödslingsförsök för belysande av de näringsekologiska villkoren för skogsväxt på torvmark. Commun Inst For Fenn 40(17):1–27

Malmström C (1958) Våra torvmarkers utnyttjande för skogsproduktion. Statens Skogsforskningsinstitut, Uppsatser 59:1–6

Maltby E (1986) Waterlogged wealth. Earthscan, London, 200 pp

Maltby E (1988a) Global wetlands – history, current status and future. In: Hook DD, McKee WH, Smith HK et al. (eds) The ecology and management of wetlands. Croom Helm, Timber Press, Portland, 1:3–14

Maltby E (1988b) Properties, geomorphology and classification of peatlands. Publ Acad Finl 5: 5–13

Malterer TJ, Verry ES, Erjavec J (1992) Peat classification in relation to several methods used to determine fiber content and degree of decomposition. Proc 9th Int Peat Congr, Int Peat Soc, Swed Nat Comm, Uppsala, 1:310–318

Mannerkoski H (1971) Lannoituksen vaikutus kylvösten ensi kehitykseen turvealustalla (Effect of fertilization on the development of Scots pine and Norway spruce plantations established by sowing on peat). Silva Fenn 5(2):105–128

Mannerkoski H (1973) Ekologisia havaintoja eräältä ojitetulta suolta (Ecological investigations in a drained peatland area). Silva Fenn 7(2):129–151

Mannerkoski H (1975) Vanhan ojitusalueen uudistaminen mätästysmenetelmällä (Hummock-building method in reforestation of an old drainage area). Suo 26:65–68

Mannerkoski H (1976) Puuston ja pintakasvillisuuden kehitys ojituksen jälkeen saraisella suolla (Changes in the tree cover and ground vegetation of a sedge bog following drainage). Suo 27:97–102

Mannerkoski H (1979) Comparison of methods in classification of vegetation on a drained peatland area. In: Classification of peat and peatlands. Int Peat Soc, Helsinki, pp 109–120

Mannerkoski H (1985) Effect of water table fluctuation on the ecology of peat soil. Univ Helsinki, Publ Dep Peatland For, Helsinki, Finland, 7:1–190

Mannerkoski H (1988) Effect of drainage on the temperature of surface peat. Publ Acad Finl 4: 60–65

Mannerkoski H (1991) Relation between tree roots and soil aeration on drained peatlands. In: Jeglum JK, Overend RP (eds) Peat and peatlands – diversification and innovation. Can Soc Peat Peatlands 1:109–114

Mannerkoski H, Möttönen V (1990) Maan vesitalous ja ilmatila metsäaurausalueilla (Soil water conditions and air-filled porosity on ploughed reforestation areas). Silva Fenn 24(3):279–301

Marachi ND, Dayton DJ, Dare CT (1983) Geotechnical properties of peat in San Joaquin Delta. In: Jarrett PM (ed) Testing of peats and organic soils. ASTM Spec Tech Publ 820:207–217

Markkula T (1986) Comparison of the communities of the oribatids (*Acari: Cryptostigmata*) of virgin and forest-ameliorated pine bogs. Ann Zool Fenn 23:33–38

Markov VD, Khoroshev PI (1986) The peat resources of the USSR and prospects for their utilization. Int Peat J 1:41–47

Marshall TJ (1959a) Relations between water and soil. Commonw Bur Soils Tech Commun 50:1–89

Marshall TJ (1959b) The diffusion of gases through porous media. J Soil Sci 10:79–82

Martikainen PJ, Nykänen H, Silvola J (1992) Emissions of methane and nitrous oxide from peat ecosystems. Publ Acad Finl 3:199–204

Martikainen PJ, Nykänen H, Alm J, Silvola J (1994) Changes in vegetation and fluxes of carbon dioxide, methane and nitrous oxide due to forest drainage of mire sites of different trophy. Plant and Soil (in print)

Masing V (1975) Mire typology of the the Estonian SSR. In: Laasimer L (ed) Some aspects of botanical research in the Estonian SSR. Acad Sci Estonian SSR, Tartu, pp 123–136

Mathur SP, Dinel H, Levesque M, Brown A, Buttler A (1991) The role of methane gas in peatland hydrology: a new concept. In: Jeglum JK, Overend RP (eds) Peat and peatlands – diversification and innovation. Can Soc Peat Peatlands 1:153–157

Matthews E (1990) Global distribution of forested wetlands (Addendum to map on inside covers). In: Lugo AE, Brinson M, Brown S (eds) Forested wetlands. Ecosystems of the world, vol 15. Elsevier, Amsterdam

Matthews E, Fung I (1987) Methane emission from natural wetlands: global distribution, area, and environmental characteristics of sources. Global Biogeochem Cycles 1(1):61–86

Mattila E, Penttilä T (1987) Lapin ja Koillis-Suomen metsälautakuntien suometsät vuosina 1952–1984 (Peatland forests of Lappi and Koillis-Suomi forestry board districts, North Finland, 1952–1984). Folia For 703:1–49

McDonald A (1973) Some views on the effect of peat drainage. Scott For 27(4):315–327

Melin E (1917) Studier över de norrländska myrmarkernas vegetation med särskild hänsyn til deras skogsvegetation efter torrläggning. Norrländskt Handbibliotek 7:1–426

Meshechok B (1960) Om grøfteavstand og grøftedybde ved myrgrøfting (On depth and distance by ditching swamps). Nor Skogbruk 10:1–7

Meshechok B (1968) Om startgjødsling ved skogkultur på myr (Initial fertilization when afforesting open swamps). Medd Nor Skogforsøksv 25:1–140

Meshechok B (1969) Tørrlegging av myr ved ulik grøfteavstand og grøftedybbe (Drainage of swamps at different ditch distances and ditch depths). Medd Nor Skogforsøksv 27:227–294

Meshechok B (1971) Kalkning ved skogkultur på nedbørsmyr (Application of $CaCO_3$ when afforesting raised bogs). Medd Nor Skogforsøksv 114:241–259

Metsänterveysopas – Metsätuhot ja niiden torjunta (1988). Samerka Oy, Vaasa, Finland, 168 pp

Metsä 2000 -ohjelman tarkistustoimikunnan mietintö (1992) Komiteanmietintö 5:1–112

Meyer JL, Tate CM (1983) The effects of watershed disturbance on dissolved organic carbon dynamics of a stream. Ecology 64:33–44

Micka J, Fiola L, Chalupa J (1985) Humic acids in the inflows to the Flaje reservoir after deforestation of its watershed. Acta Hydrochim Hydrobiol 13(1):47–52

Mikola P (1973) Koivu suopuuna. Suo 24:1–3

Mikola P (1982) Application of vegetation science to forestry in Finland. In: Jahn G (ed) Handbook of vegetation science, vol 12. Junk, The Hague, pp 199–224

Mikola P, Mikola I (1958) Suon metsittäminen polttoturpeen noston jälkeen (Reforestation of bogs after peat harvesting). Suo 9:44–47

Miyazawa T (1992) Mires and their utilization and conservation in Japan. Suo 43:77–92

Moilanen M (1984) Tuloksia suursararämeen männikön jatkolannoituksesta Pohjois-Pohjanmaalla ja Kainuussa (Results on refertilization of large sedge swamp pine stands in the North Ostrobothnia and Kainuu area). Suo 35:102–105

Moilanen M (1985) Lannoituksen ja harvennuksen vaikutus hieskoivun kasvuun ohutturpeisilla, ojitetuilla rämeillä (Effect of fertilization and thinning on the growth of birch (*Betula pubescens*) on the drained mires with thin peat layer). Folia For 629:1–29

Moilanen M (1993) Lannoituksen vaikutus männyn ravinnetilaan ja kasvuun Pohjois-Pohjanmaan ja Kainuun soilla (Effect of fertilization on the nutrient status and growth of Scots pine on drained peatlands in northern Ostrobothnia and Kainuu). Folia For 820:1–37

Moilanen M, Issakainen J (1985) Lannoitusvaikutuksen riippuvuus levitysajankohdasta nuorissa rämemänniköissä. Metsäntutkimusl tied 204:1–32

Moilanen M, Penttilä T (1988) Pääravinnelannoitus. Metsäntutkimusl tied 308:85–101

Moilanen M, Ferm A, Issakainen J (1987) Kasvihuonekokeita erilaisten jäteaineiden vaikutuksesta hieskoivun alkukehitykseen turvealustalla. Metsäntutkimusl tied 281:1–46

Möller G (1978) Femåriga effekter av gödsling på tallrismossar. Inst Skogsförbättring Årsbok 1978, pp 57–78

Moore JJ, Dowding P, Healy B (1975) Glenamoy, Ireland. Ecol Bull (Stockholm) 20:321–343

Moore PD (1984) The classification of mires: an introduction. In: Moore PD (ed) European mires. Academic Press, London, pp 1–10

Moore PD, Bellamy DJ (1974) Peatlands. Elek, London, 221 pp

Moore PD, Merryfield DL, Price MDR (1984) The vegetation and development of blanket mires. In: Moore PD (ed) European mires. Academic Press, London, pp 203–235

Mugasha AG, Pluth DJ, Higginbotham KO, Takyi SK (1991) Foliar responses of black spruce to thinning and fertilization on a drained shallow peat. Can J For Res 21:152–163

Mugasha AG, Pluth DJ, Hillman GR (1993) Foliar responses of tamarack and black spruce to drainage and fertilization of a minerotrophic peatland. Can J For Res 23:166–180

Multamäki SE (1936) Über den Grundwasserstand in versumpften Waldböden vor und nach der Entwässerung. 5th Hydrol Konf der baltischen Staaten, Finland 4A:1–16

Multamäki SE (1942) Kuusen taimien paleltuminen ja sen vaikutus ojitettujen soiden metsittymiseen (Das Erfrieren der Fichtenpflanzen in seiner Wirkung auf die Bewaldung der entwässerten Moore). Acta For Fenn 51(1):1–353

Multamäki SE (1962) Die Wirkung von Waldentwässerung auf die Ablaufverhältnisse von Torfböden. Commun Inst For Fenn 55(23):1–16

Mustonen SE (1965) Meteorologisten ja aluetekijöiden vaikutuksesta valuntaan (Effects of meteorologic and basin characteristics on runoff). Soil Hydrotech Inv 12:1–109

Mustonen SE, Laikari H (1961) Ojituksen vaikutuksesta valuntaan Huhtisuon havaintoalueella. Maataloushallituksen insinööriosasto, Maa- ja Vesitekn Tied Toim, Tiedotus 2:1–53

Mustonen SE, Seuna P (1971) Metsäojituksen vaikutuksesta suon hydrologiaan (Influence of forest draining on the hydrology of peatlands). Nat Board of Waters, Finland, Publ Water Res Inst 2:1–63

Mustonen SE, Seuna P (1972) Influence of forest draining on the hydrology of an open bog in Finland. Int Symp Hydrol Marsh-Ridden Areas, Minsk, Byelorussian SSR, 12 pp

Myllys M (1992) Physical properties of agricultural peat soils and their effects on drainage. Proc 9th Int Peat Congr, Int Peat Soc, Swed Nat Comm, Uppsala, 2:352–360

Myllys M (1994) Effect of mineral soil addition on hydraulic conductivity of peat. Proc Int Symp Conserv Manage Fens, Int Peat Soc, Polish Nat Comm, Warsaw, Poland (in press)

Naucke W (1990) Chemie von Moor und Torf. In: Göttlich K (ed) Moor- und Torfkunde. Schweizerbart'sche Verlags, Stuttgart, pp 237–261

Neily P (1991) Forest drainage in Nova Scotia. In: Jeglum JK, Overend RP (eds) Peat and peatlands – diversification and innovation. Can Soc Peat Peatlands 1:45–46

Neishtadt MI (1968) Sapropel deposits in the USSR. Trans 2th Int Peat Congr, Leningrad, USSR, 1:223–229

Nelson DG, Quine CP (1990) Site preparation for restocking. For Comm Res Inf Note 166:1–8

Nesterenko IM (1990) Changes in the hydrophysical properties of peat soil following their drainage. In: Whigham DF et al. (eds) Wetland ecology and management: case studies. Kluwer, Dordrecht, pp 49–54

Neustein SA (1977) A history of plough development in British forestry. Scott For 31(1):2–12

Nichols DS, Brown JM (1980) Evaporation from *Sphagnum* moss surface. J Hydrol 48:289–302

Niedersächssisches Moorschutzprogram (1981) Der Niedersächsische Minister für Ernährung, Landwirtschaft und Forsten, Hannover. 33 pp

Niemelä P, Danell K (1988) Comparison of moose browsing on Scots pine (*Pinus sylvestris*) and lodgepole pine (*P. contorta*). J Appl Ecol 25:761–775

Nieminen M, Ahti E (1993) Talvilannoituksen vaikutus ravinteiden huuhtoutumiseen karulta suolta (Leaching of nutrients from an ombrotrophic peatland area after fertilizer application on snow). Folia For 814:1–22

Niemistö P (1991) Hieskoivikoiden kasvatustiheys ja harvennusmallit Pohjois-Suomen turvemailla (Growing density and thinning models for *Betula pubescens* stands on peatlands in northern Finland). Folia For 782:1–36

Nilsson T (1992) Effects of drainage intensity and PK fertilization on the growth and needle nutrient contents of Scots pine on a cutover peat area in west central Sweden. Proc 9th Int Peat Congr, Int Peat Soc, Swed Nat Comm, Uppsala, 2:315–325

Nilsson T, Olsson M, Lundin L (1987) Markförbättring och intensiv skogsproduktion på en torvmark (Soil improvements and intense forest production at a peatland). Swed Univ Agric Sci Rep For Ecol For Soils 58:1–51

Niskanen M (1980) Forest drainage. In: Forest improvement – techniques and machinery. For Improvement Foundation, Arvi A Karisto, Hämeenlinna, pp 17–25

Njøs A (1978) Physical properties of peats and their importance in cultivated peatlands. In: Landscaping of cut-over peatlands and soil conservation on cultivated peatlands. Int Peat Soc, Norw Nat Comm, pp 43–71

Nömmik H, Möller G (1981) Nitrogen recovery in soil and needle biomass after fertilization of a Scots pine stand and growth responses obtained. Stud For Suecica 159:1–37

Numminen E (1983) Black spruce (*Picea mariana*) and white spruce (*Picea glauca*) proveniences suitable for northern Finland. Proc Int Symp For Drainage, Tallin, Estonian SSR, pp 77–87

Nuorteva M (1982) Metsätuholaiset. Kirjayhtymä, Helsinki, 91 pp

O'Carroll N (1984) Peatland afforestation in the Republic of Ireland. Proc 7th Int Peat Congr, Irish Nat Peat Comm, Dublin, 3:450–461

Obbarius CL (1845) Lärobok i skogsvetenskapen. Nyköpings Läns Hushållningsällskaps Handlingar

Odin H, Magnusson B, Bäckström P-O (1984) Effect of low shelterwood on minimum temperature near the ground. In: Perttu K (ed) Ecology and management of forest biomass production systems. Swed Univ Agric Sci, Uppsala, Dep Ecol Environ Res, Rep 15:77–99

Oechel WC (1989) Nutrient and water flux in a small Arctic watershed: an overview. Holarct Ecol 12:229–237

Ogilvie JF (1990) Forestry and wetland – conservation of the border mires. Scott For 44(1):10–18

Oikarinen M, Pyykkönen J (1981) Harvennuksen ja lannoituksen vaikutus turvekankaan hieskoivun kehitykseen Pohjois-Pohjanmaalla (The effect of fertilization and thinning on the growth of pubescent birch (*Betula pubescens*) on drained *Myrtillus* spruce swamp in Ostrobothnia). Folia For 486:1–15

Okruszko H (1960) Muck soils of valley peat bogs and their chemical and physical properties. Rocz Nauk Roln 74:5–89 (transl from Polish)

Okruszko H (1972) Differentiation of peat soils from the viewpoint of their physico- hydrological properties. Proc 4th Int Peat Congr, Otaniemi, Finland, 3:311–322

Okruszko H (1976) Peat soils and their utilization. In: Peatlands and their utilization in Poland. Int Peat Soc, Pol Nat Comm, Warszawa, pp 76–104

Okruszko H (1979) Classification of peats. In: Supplement to Proc of the Int Symp on Classification of peat and peatlands, Int Peat Soc, Helsinki, pp 41–47

Olkinuora M (1990) Kunnostusojituksen vaikutus pohjavesipinnan tasoon ja puuston kasvuun. MSc Thesis, Univ Helsinki, Dep Peatland For, Helsinki, Finland, 54 pp

Olson JS, Watts JA, Allison LJ (1983) Carbon in live vegetation of major world ecosystems. DOE/NBB-0037, Nat Tech Inf Serv, Springfield, VA, 152 pp

Overbeck F (1975) Botanische-geologische Moorkunde. Wachholtz, Neumünster, 719 pp

Paarlahti K (1964) Havaintoja pohjaveden korkeuden vaikutuksesta selluloosan hajaantumiseen rämeen ja korven turpeessa. Lic Thesis, Univ Helsinki, Dep Peatland For, Helsinki, Finland

Paarlahti K (1967) Lannoitusajankohdan vaikutus rämemännikön kasvureaktioihin (On the influence of the time of fertilization on the growth reactions in a pine stand on peat soil). Commun Inst For Fenn 63(4):1–20

Paarlahti K (1980) Tuhkan tuotanto ja ominaisuudet. Metsäntutkimusl Muhoksen tutkimusas tied 20:13–15

Paarlahti K, Karsisto K (1968) Koetuloksia kaliummetafosfaatin, raakafosfaatin, hienofosfaatin ja superfosfaatin käyttökelpoisuudesta suometsien lannoituksessa (On the usability of potassium metaphosphate, raw phosphate and superphosphate in fertilizing peatland forests). Folia For 551:1–17

Paarlahti K, Paavilainen E (1985) Turvemaan varttuneiden kuusikoiden ja koivikoiden lannoitus. Ennakkotuloksia (The fertilization of natural spruce and birch stands on peat soils. Preliminary results). Metsäntutkimusl tied 206:4–18

Paarlahti K, Takamaa H (1984) Ojituksen vaikutus puuston tuotokseen. Metsäntutkimusl tied 156:22–28

Paarlahti K, Reinikainen A, Veijalainen H (1971) Nutritional diagnosis of Scots pine stands by needle and peat analysis. Commun Inst For Fenn 74(5):1–58

Paavilainen E (1966a) Maan vesitalouden järjestelyn vaikutuksesta rämemännikön juuristosuhteisiin. (On the effect of drainage on root systems of Scots pine on peat soils). Commun Inst For Fenn 61(1):1–110

Paavilainen E (1966b) On the relationships between the root systems of white birch and Norway spruce and the ground water table. Commun Inst For Fenn 62(1):1–15

Paavilainen E (1967) Männyn juuriston suhteesta turpeen ilmatilaan. (Relationships between the root system of Scots pine and air content of peat). Commun Inst For Fenn 63(5):1–21

Paavilainen E (1968) Juuristotutkimuksia Kivisuon metsänlannoituskentällä (Root studies at the Kivisuo forest fertilization area). Commun Inst For Fenn 66(1):1–33

Paavilainen E (1969a) Juuristojen ja kasvualustan hivenainepitoisuuksien välisistä suhteista suometsissä (On the correlation between the contents of trace elements in roots and growth substratum in certain peatland types). Suo 20:25–29

Paavilainen E (1969b) Tutkimuksia levitysajankohdan vaikutuksesta nopealiukoisten lannoitteiden aiheuttamiin kasvureaktioihin suometsässä (Influence of the time of application fast-dissolving fertilizers on the response of trees growing on peat). Folia For 75:1–24

Paavilainen E (1970) Astiakokeita pintalannoituksen vaikutuksesta koivun, kuusen ja männyn kylvön onnistumiseen muokkaamattomalla kasvualustalla (On the effect of top dress fertilization on succesful seeding of birch, spruce and pine. Vessel experiments in soil with an untreated surface). Commun Inst For Fenn 72(1):1–37

Paavilainen E (1971) The effect of various nitrogen fertilizers on the leader growth of pine on peatlands. Acta Agric Fenn 123:197–205

Paavilainen E (1972) Reaction of Scots pine on various nitrogen fertilizers on drained peatlands. Commun Inst For Fenn 77(3):1–46

Paavilainen E (1973) Studies on the fertilizer nitrogen by Scots pine using 15_N labelled urea. Influence of peat thickness and application time. Commun Inst For Fenn 79(2):1–47

Paavilainen E (1974) The use of nitrogen in fertilizing peatland forests. Proc Int Symp For Drainage, Jyväskylä-Oulu, Finland, pp 337–345

Paavilainen E (1975a) Koetuloksia lannoituksen vaikutuksesta korpikuusikossa (On the response to fertilizer application of Norway spruce growing on peat). Folia For 239

Paavilainen E (1975b) Urea suometsien lannoitteena. Metsä ja Puu 12:26–27

Paavilainen E (1976) Taimistojen lannoitus niukkaravinteisilla soilla. Metsäntutkimusl Parkanon tutkimusas tied 3:1–8

Paavilainen E (1977a) Helppoliukoisten lannoitteiden vaikutuksen riippuvuus levitysajankohdasta turvemaalla (Effect of application time on growth response to easily dissolving fertilizers on peatlands). Folia For 300:1–16

Paavilainen E (1977b) Jatkolannoitus vähäravinteisilla rämeillä. Ennakkotuloksia (Refertilization on oligotrophic pine swamps. Preliminary results). Folia For 327:1–32

Paavilainen E (1978) PK-lannoitus Lapin ojitetuilla rämeillä. Ennakkotuloksia (PK fertilization on drained pine swamps in Lapland. Preliminary results). Folia For 343:1–17

Paavilainen E (1979a) Jatkolannoitus runsastyppisillä rämeillä. Ennakkotuloksia (Refertilization on nitrogen-rich pine swamps. Preliminary results). Folia For 414:1–23

Paavilainen E (1979b) Metsänlannoitusopas. Kirjayhtymä, Helsinki, 112 pp

Paavilainen E (1980) Effect of fertilization on plant biomass and nutrient cycle on a drained dwarf shrub pine swamp. Commun Inst For Fenn 98(5):1–71

Paavilainen E (1984a) Litter fall in birch and Scots pine stands after refertilization on nitrogen-rich peat soil. Proc 7th Int Peat Congr, Irish Nat Peat Comm, Dublin, 3:462–475

Paavilainen E (1984b) Typpi ja hivenravinteet ojitettujen rämeiden jatkolannoituksessa (Nitrogen and micronutrients in the refertilization of drained pine swamps). Folia For 589:1–28

Paavilainen E (1986) Turpeen paksuus ja ojituksen ajankohta typpilannoituksen tarpeeseen vaikuttavina tekijöinä karuilla rämeillä. Metsäntutkimusl tied 235:4–14

Paavilainen E (1987) Effect of fertilization on the litter fall of *Pinus sylvestris* and *Betula pubescens* on drained peatlands. Scand J For Res 2:59–75

Paavilainen E (1990a) Effect of refertilization of pine and birch stands on a drained fertile mire. Silva Fenn 24(1):83–92

Paavilainen E (1990b) The SNS Project (Nr 23 näringsämnenas fördelning och dynamik i skogsekosystem på torvmark i olika klimatförhållanden). An introduction. In: Hånell B (ed) Biomass production and element fluxes in forested peatland ecosystem. Swed Univ Agric Sci, Umeå, Dep For Site Res, pp 49–56

Paavilainen E (1991) Puuntuotannon mahdollisuudet Etelä-Suomen soilla. Metsäntutkimusl tied 394:3–7

Paavilainen E, Norlamo M (1975) Effect of various nitrogen fertilizers on the initial development of birch, spruce and pine. Commun Inst For Fenn 86(2):1–43

Paavilainen E, Pietiläinen P (1983) Foliar responses caused by different nitrogen rates at refertilization of fertile pine swamps. Commun Inst For Fenn 116:91–104

Paavilainen E, Simpanen J (1976) Tutkimuksia typpilannoituksen tarpeesta Pohjois-Suomen ojitetuilla rämeillä (Studies concerning the nitrogen fertilization requirements on drained pine swamps in North Finland). Commun Inst For Fenn 86(4):1–70

Paavilainen E, Tiihonen P (1988) Suomen suometsät vuosina 1951–1984 (Peatland forests in Finland in 1951–84). Folia For 714:1–29

Paavilainen E, Virtanen J (1977) Metsänlannoituksen vaikutuksen riippuvuus levitysmenetelmistä (Effect of spreading method on forest fertilization results). Folia For 297:1–15

Paavilainen E, Virtanen J (1979) Metsänlannoituksen vaikutuksen riippuvuus levitysmenetelmistä turvemaalla (Effect of spreading method on forest fertilization results on peatlands). Folia For 382:1–9

Päivänen J (1964) Menetelmä pohjavesikertoimen ja pintakasvillisuuden haihdunnan määrittämiseksi (A method to determine the ground water coefficient and the ground vegetation transpiration). Suo 15:88–91

Päivänen J (1966) Sateen jakaantuminen erilaisissa metsiköissä (The distribution of rainfall in different types of forest stands). Silva Fenn 119(3):1–37

Päivänen J (1968) Pohjavesipinta ja turpeen vesipitoisuus rahkamättäisellä lyhytkortisella nevalla (Ground-water level and water content of peat in an open low-sedge swamp with *Sphagnum fuscum* hummocks). Suo 19:17–24

Päivänen J (1969) The bulk density of peat and its determination. Silva Fenn 3:1–19

Päivänen J (1970) Hajalannoituksen vaikutus lyhytkortisen nevan pintakasvillisuuden kenttäkerrokseen (On the influence of broadcast fertilization on the field layer of the vegetation of open low-sedge bog). Suo 21:18–24

Päivänen J (1972) The effect of fertilization on throughfall and ground water table in peatlands drained for forestry. Proc 4th Int Peat Congr, Finland, 3:469–477

Päivänen J (1973a) Harvennuksen vaikutus lumi- ja routasuhteisiin nuoressa turvemaan männikössä (The effect of thinning on the snow cover and soil frost conditions in a young Scots pine stand on drained peat). Silva Fenn 7(2):114–128

Päivänen J (1973b) Hydraulic conductivity and water retention in peat soils. Acta For Fenn 129:1–70

Päivänen J (1974a) Hydrological effects of clear cutting in peatland forests. Proc Int Symp For Drainage, Jyväskylä-Oulu, Finland, pp 219–228

Päivänen J (1974b) Nutrient removal from Scots pine canopy on drained peatland by rain. Acta For Fenn 139:1–19

Päivänen J (1974c) Sarkaleveyden ja naveroinnin vaikutus pohjavesipinnan syvyyteen ja männyntaimiston kehitykseen lyhytkortisella nevalla (The effect of ditch spacing and furrowing on the depth of the ground water table and on the development of a Scots pine plantation on small-sedge bog). Silva Fenn 8(4):215–224

Päivänen J (1975) Männyn istutuksen ajankohta ojitetuilla avosoilla. Pääteinventoinnin tuloksia (Planting date of Scots pine on drained open peatlands. Results from final inventories). Suo 26:95–100

Päivänen J (1976) Effect of different types of counter ditches on the hydrology of an open bog. Proc 5th Int Peat Congr, Poznan, Poland, 1:93–106

Päivänen J (1980a) The effect of silvicultural treatments on the ground water table in Norway spruce and Scots pine stands on peat. Proc 6th Int Peat Congr, Minn, USA, pp 433–438

Päivänen J (1980b) The effect of silvicultural treatments on the hydrology of old forest drainage areas on peat. Int Assoc Hydrol Sci, IAHS-AISH Publ 130:137–140

Päivänen J (1980c) Treatment of young stands. In: Forest improvement – techniques and machinery. For Improve Found, Arvi A Karisto, Hämeenlinna, pp 68–73

Päivänen J (1982a) Hakkuun ja lannoituksen vaikutus vanhan metsäojitusalueen vesitalouteen (The effect of cutting and fertilization on the hydrology of an old forest drainage area). Folia For 516:1–19

Päivänen J (1982b) Physical properties of peat samples in relation to shrinkage upon drying. Silva Fenn 16(3):247–265

Päivänen J (1983) Initial development of *Picea mariana* on drained, nutrient poor open bog in Finland. Proc Int Symp For Drainage, Tallin, USSR, pp 50–64

Päivänen J (1984a) Increasing the land base and yield through drainage. Univ Alberta, For Indust Lect Ser 13:1–22

Päivänen J (1984b) The effect of runoff regulation on tree growth on a forest drainage area. Proc 7th Int Peat Congr, Irish Nat Peat Comm, Dublin, 3:476–488

Päivänen J (1985) Soiden metsätaloudellisista käyttömahdollisuuksista Kanadassa (Potential of peatlands for forest drainage in Canada). Suo 36:45–52

Päivänen J (1988) Development of mire originated Scots pine on mineral soil. Proc 8th Int Peat Congr, Leningrad, 3:137–143

Päivänen J (1990) Suometsät ja niiden hoito. Kirjayhtymä, Helsinki, 231 pp

Päivänen J, Ahti E (1988) Ditch cleaning and additional ditching in peatland forestry – effect on ground water level. Publ Acad Finl 4:184–189

Päivänen J, Paavilainen E (1990) Managing and protecting forested wetlands. IUFRO World Congr, Montreal, Canada, Div 1, 1:432–443

Päivänen J, Seppälä K (1968) Hajalannoituksen vaikutus lyhytkortisen nevan pintakasvillisuuteen (Effect of broadcast fertilizer on the ground vegetation of a low sedge swamp). Suo 19:51–56

Päivänen J, Vasander H (1994) Carbon balance in mire ecosystems. World Resource Rev 6: 102–111

Päivänen J, Wells ED (1977) Guidelines for the development of peatland drainage systems for forestry in Newfoundland. Newfoundland For Res Cent, St.John's, Newfoundland, Inf Rep N-X-156:1–44

Pakarinen P (1973) Moss production and decomposition, a world model. Devon Island Proj, Rep March 1973, Edmonton, 8 pp

Pakarinen P (1976) Bogs as peat-producing ecosystems. Bull Int Peat Soc 7:51–54

Pakarinen P (1979) Ecological indicators and species groups of *Bryophytes* in Boreal peatlands. In: Classification of peat and peatlands. Int Peat Soc, Helsinki, pp 121–134

Pakarinen P (1984) Definitions of peats and organic sediments. Bull Int Peat Soc 15:40–46

Pakarinen P, Ruuhijärvi R (1978) Ordination of northern Finnish peatland vegetation with factor analysis and reciprocal averaging. Ann Bot Fenn 15:147–157

Pakarinen P, Tolonen K (1977) Pääravinteiden sekä sinkin ja lyijyn vertikaalijakautumista rahkaturpeessa. (Vertical distributions of N, P, K, Zn and Pb in *Sphagnum* peat). Suo 28:95–102

Parker WH, Knowles P, Bennett F, Gray A, Krickle T (1983) Habitat-dependent morphological and chemical variation in *Picea mariana* from northwestern Ontario. Can J Bot 61:1573–1579

Pätilä A, Nieminen M (1990) Turpeen emäsravinne- ja rikkitase karuilla ojitetuilla rämeillä laskeuma huomioon ottaen (Base cation nutrients and sulphur status of drained oligotrophic pine mires considering the atmosphere input). Folia For 759:1–16

Payandeh B (1973) Analyses of a forest drainage experiment in northern Ontario. I Growth analysis. Can J For Res 3:387–398

Payandeh B (1986) Predictability of site index from soil factors and lesser vegetation in northern Ontario forest types. Can For Serv, Great Lakes For Cent, Inf Rep O-X-373:1–12

Peat testing manual (1979) Day JH, Rennie, PJ, Stanek, W, Raymond GP (eds) Assoc Comm Geotech Res, Natl Res Counc Can, Tech Mem 125:1–193

Peerlkamp PK, Boekel P (1960) Moisture retention by soils. Reprint Versl Mede Comm Hydrol Onderz TNO 5:122–138

Pelkonen E (1975) Vuoden eri aikoina korkealla olevan pohjaveden vaikutus männyn kasvuun (Effects on Scots pine growth of ground water adjusted to the ground surface for year). Suo 26:25–32

Penman HL (1956) Estimating evaporation. Trans Am Geophys Union 37(1):43–50

Penttilä T (1984) Jatkolannoitus Lapin viljavilla rämeillä (Refertilization on mesotrophic pine swamps in northern Finland). Suo 35:106–110

Pessi Y (1956) Studies on the effect of the admixture of mineral soil upon the thermal conditions of cultivated peat land. State Agric Res Publ Finl 147:1–89

Pessi Y (1972) Agricultural utilization of peatlands. In: Päivänen J (ed) Finnish peatlands and their utilization. Suoseura ry – Finn Peatland Soc, Helsinki, pp 23–27

Pettersson R (1990) Peat in Sweden – a retrospect and state of the art 1989. Bull Int Peat Soc 21:32–40

Pfadenhauer J, Schneekloth H, Schneider R, Schneider S (1993) Mire distribution. In: Heathwaite AL (ed) Mires: process, exploitation and conservation. Wiley, Chichester, pp 77–121

Pietilä M, Kuusipuro P, Pietiläinen P, Lähdesmäki P (1989) Specificity and seasonal variation of arginase, glutamate synthase and nitrate reductase activities in Scots pine needles. Plant Sci 64:153–160

Pietiläinen P, Kuivamäki A (1982) Ravinteiden puutteen vaikutus männyn neulasten valonheijastusominaisuuksiin. Metsäntutkimusl tied 70:14–21

Pietiläinen P, Lähdesmäki P (1986) Free amino acid and protein levels, and -glutamyltransferase activity in *Pinus sylvestris* apical buds and shoot during the growing season. Scand J For Res 1:387–395

Pietiläinen P, Lähdesmäki P (1988) Effect of various concentrations of potassium nitrate and ammonium sulphate on nitrate reductase activity in the roots and needles of Scots pine seedlings in N Finland. Ann Bot Fenn 25:201–206

Pihelgas E (1963) Seemnete päritolu mojust männikultuuride kasvule (Über den Einfluss der Samenherkunft auf den Wuchs der Kiefernkulturen). Eesti Pollumajanduse Akadeemia. Metsamajandusalased Tööd 33:31–37

Plamondon AP, Belleau P (1991) Estimation of the hydraulic conductivity of peat from bulk density and von Post decomposition. In: Jeglum JK, Overend RP (eds) Peat and peatlands – diversification and innovation. Can Soc Peat Peatlands 1:146–152

Pollett FC, Bridgewater PB (1973) Phytosociology of peatlands in central Newfoundland. Can J For Res 3:433–442

Poulovassilis A, Childs EC (1971) The hysteresis of pore water: Th non-independence of domains. Soil Sci 112:301–312

Proposal for an international classification of peat (1979) In: Suppl Proc Int Symp Classification Peat Peatlands. Int Peat Soc, Helsinki, 48 pp

Puustjärvi V (1962a) Suometsien fosforiravitsemuksesta ja neulasten P/N-suhteesta neulasanalyysin valossa (On the phosphorus nutrition of wet peatland forests and on the P/N ratio in their needles). Suo 13:21–24

Puustjärvi V (1962b) Suometsien kaliumravitsemuksesta ja neulasten N/K-suhteesta neulasanalyysin valossa (On the potassium nutrition of wet peatland forests and on the N/K ratio of the needles in the light of needle analysis). Suo 13:36–40

Puustjärvi V (1962c) Turpeen typen mobilisoitumisesta ja sen käyttökelpoisuudesta suometsissä neulasanalyysin valossa (On the mobilization of nitrogen in peat and its usefulness in the light of needle analysis). Suo 13:2–11

Puustjärvi V (1968) Standards for peat used in peat culture. Peat & Plant News 1:19–26

Puustjärvi V (1969) Water-air relationships of peat in peat culture. Peat & Plant News 2:43–55

Puustjärvi V (1970) Degree of decomposition. Peat & Plant News 3:48–52

Pyatt DG (1990a) Forest drainage. For Comm Res Div, Res Inf Note 196:1–4

Pyatt DG (1990b) Long term prospects for forests on peatland. Scott For 44(1):19–25

Pyatt DG, Craven MM (1979) Soil changes under even-aged plantations. In: Ford ED, Malcolm, DC, Atterson J (eds) The ecology of even-aged forest plantations. Inst Terrest Ecol, Cambridge, pp 369–386

Quine CP, White IMS (1993) Revised windiness scores for the windthrow hazard classification: the revised scoring method. For Comm Res Inf Note 230:1–6

Radforth NW (1969) Muskeg as an engineering problem. In: MacFarlane IC (ed) Muskeg engineering handbook. Univ Press, Toronto, pp 3–30

Raitio H (1985) Hallavauriot männyllä. Metsäntutkimusl tied 184:25–34

Raitio H, Huttunen A (1976) Turpeen maatumisasteen määritysmenetelmistä (Methods of determining the humification degree of peat). Suo 27:19–23

Raitio H, Rantala E-M (1977) Männyn kasvuhäiriön makro- ja mikroskooppisia oireita (Macroscopic and microscopic symptoms of a growth disturbance in Scots pine). Commun Inst For Fenn 91(1):1–30

Rajala P, Linden H (1982) Rämeiden ojituksen ja lannoituksen vaikutuksista riistaeläinten esiintymisrunsauteen. Suom Riista 29:93–97

Ramberg L (1981) Increase in stream pH after a forest drainage. Ambio 10(1):34–35

Rantonen H, Päivänen J (1989) Kasvatusmetsien metsänhoidollinen tila ojitusalueilla puunkorjuun jälkeen (Silvicultural condition of tree stands after thinning on drained peatlands). Silva Fenn 23(1):33–50

Reader RJ, Stewart JM (1972) The relationship between net primary production and accumulation for a peatland in southeastern Manitoba. Ecology 53:1024–1037

Reinikainen A (1967) The appearance of nutrient deficiency in plants growing in the experimental area for forest fertilization at Kivisuo. Proc Vth Colloq For Fertiliz, Jyväskylä, Finland, Int Potash Inst, Bern, Switzerland, pp 345–361

Reinikainen A (1968) Ravinteiden puuteoireita puulajeilla. In: Jamalainen EA (ed) Kasvien puutostaudit. Kirjayhtymä, Helsinki, pp 101–109

Reinikainen A (1976) Suoekosysteemi tutkimuskohteena. (How to study a mire ecosystem). Suo 27:9–18

Reinikainen A (1981) Metsänparannustoimenpiteiden vaikutuksesta suoekosysteemin kasvibiomassaan ja perustuotantoon (Effect of drainage on plant biomass and primary production in mire ecosystem). Suo 32:110–113

Reinikainen A (1983) Problems concerning the classification of drained peatlands in Finland. Proc Int Symp For Drainage, Tallin, USSR, pp 203–215

Reinikainen A (1988a) Metsäojitettujen soiden kasvupaikkaluokituksen suunnanhakua (The need of improving the site classification of mires drained for forestry). Suo 39:61–71

Reinikainen A (1988b) Lannoitustarpeen määritysmenetelmät. Metsäntutkimusl tied 308:68–84

Reinikainen A (1990) The peatland ecosystem in change. In: Hånell B (ed) Biomass production and element fluxes in forested peatland ecosystems. Swed Univ Agric Sci, Umeå, Dep For Site Res, pp 167–175

Reinikainen A, Vasander H, Lindholm T (1984) Plant biomass and primary production of southern boreal mire-ecosystems in Finland. Proc 7th Int Peat Congr, Irish Nat Peat Comm, Dublin, 4: 1–20

Rekolainen S (1989) Phosphorus and nitrogen load from forest and agricultural areas in Finland. Aqua Fenn 19:95–107

Repo S, Löyttyniemi K (1985) Lähiympäristön vaikutus männyn viljelytaimikon hirvivahinkoalttiuteen (The effect of immediate environment on moose (*Alces alces*) damage in young Scots pine plantations). Folia For 626:1–14

Reynolds WD, Brown DA, Mathur SP, Overend RP (1992) Effect of *in situ* gas accumulation on the hydraulic conductivity of peat. Soil Sci 153:397–408

Richard F (1963) Wasserhaushalt und Entwässerung von Weideböden. Mitt Schweiz Anst Forstl Versuchsw 39:245–269

Riley LF (1973) Silviculture treatment of seeded stands. In: Cayford JH (ed) Direct seeding symposium. Dep Environ Can For Serv 1339:139–155

Robinson WO (1930) Some chemical phases of submerged soil conditions. Soil Sci 30:197–217

Roderfeld H (1992) Erste Ergebnisse zur Bewertung von Bunkerde für die Hochmoorregeneration. Telma 22:217–233

Roderfeld H (1993) Raised bog regeneration after peat harvesting in north-west Germany. Suo 44:43–51

Romanov VV (1968) Hydrophysics of bogs. Isr Progr Sci Transl Ltd, IPST 1967:299

Roosaluste E (1982) Growth forms of the Scots pine growing in bogs. In: Masing V (ed) Peatland ecosystems. Acad Sci Estonian SSR 9:121–127

Rosén K (1984) Effect of clear-felling on runoff in two small watersheds in central Sweden. For Ecol Manage 9:267–281

Rosen M (1987) The cost of peatland forest drainage in the Claybelt of Ontario. On Line, Ont Minist Nat Res 3(1):4–5

Rosen M (1989) Forest drainage manual. Ont Minist Nat Res Sci Technol Ser 3:1–25

Roulet NT (1991) Stormflow production in a headwater basin swamp. Nord Hydrol 22:161–174

Rowe JS (1972) Forest regions of Canada. Can For Serv Publ 1300:172

Rubec CDA, Lynch-Stewart P, Wickware GM, Kessel-Taylor I (1988) Wetland utilization in Canada. In: Wetlands of Canada. National Wetlands Working Group, Env Can, Ecol Land Classification Ser 24:381–412

Rudin D, Eriksson G, Rasmuson M (1977) Inbreeding in a tree stand of *Pinus sylvestris* L. in northern Sweden. A study by the aid of the isozyme technique. Royal Coll For, Garpenberg, Sweden, Dep For Genetics, Res Notes 25:1–45

Rummukainen A (1984) Peatland properties and their evaluation for wood harvesting. Univ Helsinki, Dep Logging and Utilization of Forest Products, Helsinki, Finland, Res Notes 45:1–119

Ruseckas J (1991) Theoretical background for forest hydromelioration. Leningrad Forestry Academy, Leningrad. 32 pp (In Russian)

Ruuhijärvi R (1982a) Mire complex types in Finland. In: Laine J (ed) Peatlands and their utilization in Finland. Finn Peatland Soc, Finn Natl Comm Int Peat Soc, Helsinki, pp 24–28

Ruuhijärvi R (1982b) Mire preservation in Finland In: Laine J (ed) Peatlands and their utilization in Finland. Finn Peatland Soc, Finn Natl Comm Int Peat Soc, Helsinki, pp 109–113

Ruuhijärvi R (1983) The Finnish mire types and their regional distribution. In: Gore AJP (ed) Mires: swamp,bog, fen and moor. Ecosystems of the world, vol 4B. Elsevier, Amsterdam, pp 47–67

Ruuhijärvi R (1989) Soidensuojelun kehitys Suomessa (The development of mire preservation in Finland). Suo 40:93–98

Rybnicek K (1985) A Central-European approach to the classification of mire vegetation. Aquilo Ser Bot 21:19–31

Rycroft DW, Williams DJA, Ingram HAP (1975a) The transmission of water through peat I. Review. J Ecol 63:535–556

Rycroft DW, Williams DJA, Ingram HAP (1975b) The transmission of water through peat II. Field experiments. J Ecol 63:557–568

Saarilahti M (1982) Tutkimuksia radioaaltomenetelmien soveltuvuudesta turvemaiden kulkukelpoisuuden arvioimiseen (Studies on the possibilities of using radar techniques in detecting the trafficability of peatlands). Acta For Fenn 176:1–105

Saarinen M (1988) Suonpohjien metsitystutkimukset. Metsäntutkimusl tied 300:10–14

Saarinen M (1989) Metsien uudistaminen vanhoilla ojitusalueilla (Forest regeneration in old forest drainage areas). Suo 40:31–36

Saarinen T, Tolonen K, Vasander H (1992) Use of 14C labelling to measure below-ground biomass of mire plants. Suo 43:245–247

Sabo ED (1983) Choice of draining rates under the new requirements of forest hydroamelioration. Proc Int Symp For Drainage, Tallin, USSR, pp 42–49

Sabo ED (1988) Hydrotechnical forest reclamation fund in the USSR. Proc 8th Peat Congr, Leningrad, 3:58–63

Sadler R (1992) Fight for peat bogs. European, Feb 13–19, UK

Sakovets VV, Germanova N (1992) Changes in the carbon balance of forested mires in Karelia due to drainage. Suo 43:249–252

Sallantaus T (1986) Soiden metsä- ja turvetalouden vesistövaikutukset – kirjallisuuskatsaus. Maa- ja metsätalousministeriö, Luonnonvarajulkaisuja 11:1–203

Sallantaus T (1988) Water quality of peatlands and man's influence on it. Publ Acad Finl 5:80–98

Sallantaus T (1992) Leaching in the material balance of peatlands – preliminary results. Suo 43: 253–258

Sallantaus T, Pätilä A (1983) Runoff and water quality in peatland drainage areas. Proc Int Symp For Drainage, Tallinn, USSR, pp 183–202

Salminen O (1984) Ojitetun korpikuusikon kertalannoituksen kannattavuus (The profitability of fertilization on drained spruce swamps). Suo 35:127–131

Salminen P (1991) Irlannin suot ja niiden suojelu (Irish mires and their conservation). Suo 42: 25–32

Salo H (1988) Kaivukoneiden ajankäyttö ja tuottavuus metsäojituksessa (Time consumption and productivity of excavators in forest drainage). Suo 39:51–59

Salo K (1979) Mushrooms and mushroom yields on transitional peatlands in Central Finland. Ann Bot Fenn 16:181–192

Salo K (1980) Luonnontilaisen rämeen ja rämemuuttumien sienilajistosta (Mushroom species on virgin, drained and fertilized pine bogs in Parkano, central Finland). Suo 31:67–71

Salo K (1981) Metsänparannustoimenpiteiden vaikutus rämeiden sienisatoon (The effect of forest amelioration on mushroom yield on pine bogs). Suo 32:1–6

Salo K (1982) Metsänhoitotoimenpiteet ja marja-ja sienisadot. Pohjois-Karjalan Luonto 1982. Pohjois-Karjalan luonnonsuojelupiiri, Joensuu, Finland, pp 14–18

Salo K (1988) Marjat ja sienet. Metsäntutkimusl tied 308:187–198

Saramäki J (1977) Ojitettujen turvemaiden hieskoivikoiden kehitys Kainuussa ja Pohjanmaalla (Development of white birch (*Betula pubescens* Ehrh) stands on drained peatlands in northern central Finland). Commun Inst For Fenn 91(2):1–51

Sarasto J (1952) Metsäojituksen aiheuttamista aluskasvillisuuden muutoksista eräissä suotyypeissä (Über Veränderungen in der Untervegetation einiger Moortypen als Folge der Waldentwässerung). Commun Inst For Fenn 40(13):1–32

Sarasto J (1957) Metsän kasvattamiseksi ojitettujen soiden aluskasvillisuuden rakenteesta ja kehityksestä Suomen eteläpuoliskossa (Über Struktur und Entwicklung der Bodenvegetation auf für Walderziehung entwässerten Mooren in der südlichen Hälfte Finnlands). Acta For Fenn 65(7):1–108

Sarasto J (1960) Turpeen maatumisasteen määrittämisestä. v Postin maatumisasteen ja Pjavtshenkon maatumisprosentin vertailu (Zur Bestimmung der Zersetzung des Torfes). Acta For Fenn 71(2):1–16

Sarasto J (1961a) Ojitettujen soiden luokittelusta (How the drained peatlands are classified). Suo 12:75–77

Sarasto J (1961b) Über die Klassifizierung der für Walderziehung entwässerten Moore. Acta For Fenn 74(5):1–47

Sarasto J, Seppälä K (1964) Männyn kylvöistä ojitettujen soiden sammal- ja jäkäläkasvustoihin (On sowing of pine in moss and lichen vegetation on drained swamps). Suo 15:54–58

Sarjala T, Kaunisto S (1993) Needle polyamine concentrations and potassium nutrition in Scots pine. Tree Physiol 13:87–96

Särkkä M (1970) Metsänlannoituksen vaikutus vesistöissä (On the influence of forest fertilization on watercourses). Suo 21:67–74

Sarvas R (1967) The annual period of development of forest trees. Proc Finn Acad of Science and Letters 1967, Helsinki, Finland, pp 211–231

Saura M, Frisk T, Sallantaus T, Bilaletdin Ä, Huttula T (1993) Metsänlannoitus vesistöjen rehevöittäjänä. In: Saukkonen S, Kenttämies K (eds) Metsätalouden vesistöhaitat ja niiden torjunta. Metve-projektin väliraportit. Vesi- ja ympäristöhallituksen monistesarja 455:105–120

Schalin I (1967) On the effect of nitrogen fertilization on the bacteria and microfungi in humus layer. Silva Fenn 1(121):1–11

Schlaudt EA (1955) Drainage in forestry management in the South. USDA, Yearbook of Agriculture, Washington, DC, pp 564–568

Schlesinger WH (1977) Carbon balance in terrestrial detritus. Annu Rev Ecol Syst 8:51–81

Schmidt W, Scholz A (1993) Das Niedermoor Friedländer Grosse Wiese, landschaftsökologische Zielstellung und angelaufene Massnahmen zur Erhaltung und Renaturierung. Naturschutz Landschaftspflege Brandenburg, Sonderheft Niedermoore, pp 41–46

Schneekloth H, Schneider S (1976) Proposal for classification of the decomposition of peat. Trans Working Group for Classification of Peat. Int Peat Soc, Helsinki, pp 11–14

Schneider S (1976) Verteilung der Moore auf der Erde. In: Göttlich K (ed) Moor- und Torfkunde. Stuttgart, pp 27–30

Scholz A (1986) Anzustrebende Grundwasserstände auf tiefgründigem Niedermoor und Möglichkeiten ihrer Realisierung mittels der rohrlosen Dränung. Diss B – Berlin, Akad Landwirtschaftswiss DDR, 128 pp

Schothorst CJ (1982) Drainage and behaviour of peat soils. In: de Bakker, van den Berg (eds) Proc Symp on Peat lands below sea level. ILRI, Wageningen, 30:130–168

Segeberg H (1956) Über den Zusammenhang zwischen den "Humositätsgraden" nach v Post, den "Vertorfungsgraden" nach Keppeler und den spezifischen Gewichten von Hochmoortorfen. Z Düngung Bodenkd 73:74–86

Sepp M (1990) Metsakuivenduse projekteerimisjuhend. EPA maaparanduse kateeder, Tartu, 38 pp

Seppälä K (1968) Välituloksia ojitettujen soiden viljelykokeista (Results on seeding and planting experiments in peatland). Suo 19:30–38

Seppälä K (1969) Kuusen ja männyn kasvun kehitys ojitetuilla turvemailla (Post-drainage growth rate of Norway spruce and Scots pine on peat). Acta For Fenn 93:1–88

Seppälä K (1971) Metsityslannoituksessa käytetyn lannoitemäärän ja levitysajan merkitys istutustaimien alkukehitykselle ojitetuilla avosoilla (On the quantity of fertilizer and application methods used in afforestation of open bogs). Silva Fenn 5:61–69

Seppälä K (1972) Ditch spacing as a regulator of post-drainage stand development in spruce and in pine swamps. Acta For Fenn 125:1–21

Seppälä K (1976) Relationship between stem diameter of trees at time of draining and their post-drainage increment. Suo 27:55–60

Seppälä K, Keltikangas M (1978) Alikasvostaimistot Pohjanmaan ojitusalueiden hieskoivikoissa (Occurrence of understorey seedlings in drained *Betula pubescens* stands in Ostrobothnia). Suo 29:11–16

Seppälä K, Westman CJ (1976) Results of some fertilization experiments in North-Eastern Finland. Proc 5th Int Peat Congr, Poznan, Poland, 3:199–210

Sepponen P (1979) Pohjois-Suomen suot marjasoina. Metsäntutkimusl Rovaniemen tutkimusas tied 21:4–9

Seuna P (1980) Long-term influence of forestry drainage on the hydrology of an open bog in Finland. Int Assoc Hydrol Sci, IAHS-AISH Publ 130:141–149

Seuna P (1981) Long-term influence of forestry drainage on the hydrology of an open bog in Finland. Publ Water Res Inst Natl Board Waters Finl 43:3–14

Seuna P (1982) Influence of forestry draining on runoff and sediment discharge in the Ylijoki basin, North Finland. Aqua Fenn 12:3–16

Seuna P (1986) Ihmistoiminnan vaikutus hydrologiseen kiertoon. In: Mustonen S (ed) Sovellettu hydrologia. Vesiyhdistys ry, Helsinki, pp 387–420

Seuna P (1988) Effects of clear-cutting and forestry drainage on runoff in the Nurmes-study. Publ Acad Finl 4:122–134

Shier CW (1985) Tropical peat resources – an overview. In: Tropical peat resources – prospects and potential. Int Peat Soc, Helsinki, pp 29–46

Silfverberg K (1984) Kuivatustehon ja lannoituksen vaikutus rämemännikön kehitykseen (Effects of drainage intensity and fertilization in a pine bog stand). Suo 35:86–90

Silfverberg K (1991) Träaska, PK gödsel och markförbättringsmedel på dränerade tallmyrar (Wood ash, PK-fertilizer and two soil amelioration additives on drained pine mires). Suo 42: 33–44

Silfverberg K, Huikari O (1985) Tuhkalannoitus metsäojitetuilla turvemailla (Wood-ash fertilization on drained peatlands). Folia For 633:1–25

Silvola J (1986) Carbon dioxide dynamics in mires reclaimed for forestry in eastern Finland. Ann Bot Fenn 23:59–67

Silvola J (1988) Ojituksen ja lannoituksen vaikutus turpeen hiilen vapautumiseen ja ravinteiden mineralisoitumiseen (Effect of drainage and fertilization on carbon output and nutrient mineralization of peat). Suo 39:27–37

Silvola J, Alm J (1992) Dynamics of greenhouse gases in virgin and managed peatlands. Publ Acad Finl 3:193–198

Silvola J, Alm J, Ahlholm U (1992) The effect of plant roots on CO_2 release from peat soil. Suo 43:259–262

Simola H, Lodenius M (1982) Recent increase in mercury sedimentation in a forest lake attributable to peatland drainage. Bull Environ Contam Toxicol 29:298–305

Simonsson P (1987) Skogs- och myrdikningens miljökonsekvenser (Environmental effects of draining wetland and forest). Naturvårdsverket, Sverige, Rapp 3270:1–196

Sjörs H (1948) Myrvegetation i Bergslagen (Mire vegetation in Bergslagen). Acta Phytogeogr Suec 21:1–299

Sjörs H (1981) The zonation of northern peatlands and their importance for the carbon balance of the atmosphere. Int J Ecol Environ Sci 7:11–14

Sjörs H (1983) Mires of Sweden. In: Gore AJP (ed) Mires: swamp, bog, fen and moor. Ecosystems of the world, vol 4B. Elsevier, Amsterdam, pp 69–94

Skeates DA, Haavisto VF (1987) Black spruce cone and seed production: yields from Ontario collections. Ont Minist Nat Res For Res Rep 119:1–16

Skoropanov SG (1968) Reclamation and cultivation of peat-bog soils. Isr Progr Sci Transl, Jerusalem, 234 pp (transl from Russian)

Smolander A (1990) Frankia in forest soils. PhD Thesis, Univ Helsinki, pp 1–33

Soil taxonomy. Soil Conservation Services. (1975) USDA, Agric Handb 436:1–736

Sokolov BN, Yampolski AL (1986) Socio-economic consequences of industrial development of peat deposits in the USSR. In: Socio-economic impacts of the utilization of peatlands in industry and forestry. Finn Nat Comm, Int Peat Soc, Helsinki, pp 69–71

Solantie R (1986) Suoyhdistymätyypeistä suhteessa ilmastollisiin ja hydrologisiin vyöhykkeisiin (The zonation of peatland complex types in relation to climatic and hydrological zones). Suo 37: 73–85

Sorteberg A (1978) Subsidence in peat soil after drainage. In: Landscaping of cut-over peatlands and soil conservation on cultivated peatlands. Int Peat Soc, Norw Nat Comm, pp 31–42

Stanek W (1968a) A forest drainage experiment in northern Ontario. Pulp Paper Mag Can 69: 58–62

Stanek W (1968b) Development of black spruce layers in Quebec and Ontario. For Chron 44(2): 1–4

Stanek W (1970) Amelioration of water-logged terrain in Quebec. I Description of the area, hydrology and drainage. Can For Serv, For Res Lab Quebec, Inf Rep Q-X-17:1–18

Stanek W (1975) The role of layerings in black spruce forests on peatlands in the Clay Belt of northern Ontario. In: Black spruce Symp. Can For Serv, Symp Proc O-P-4:242–249

Stanek W (1979) Amelioration of water-logged terrain in Quebec. I Description of the area, hydrology and drainage. Can For Serv, For Res Lab Quebec, Inf Rep Q-X-17:1–34

Stanek W, Silc F (1976) An instrument and technique for collecting peat core samples in peatland surveys. Can J Soil Sci 56(1):59–61

Stanek W, Worley IA (1983) A terminology of virgin peat and peatlands. In: Fuchsman CH, Spigarelli SA (ed) Proc Int Symp Peat Utilization, Oct 10–13, Bemidji State Univ, Bemidji, Minn, pp 75–102

Starr MR (1984) Non-metric multidimensional scaling of a peatland vegetation data set. Suo 35: 57–61

Starr MR (1986) Forest site classification and evaluation in Finland. In: Groenewoud H (ed) Proc IUFRO Workshop For site classification meth, Fredericton, New Brunswick, Canada, 1985, pp 155–170

Starr MR, Laine J (1988) Interactions among peat properties, nutrient contents and water chemistry at Lakkasuo mire, central Finland. Proc 8th Int Peat Congr, Leningrad, 2:140–148

Starr MR, Päivänen J (1981) The influence of peatland forest drainage on runoff peak flows. Suo 32:79–84

Starr M, Westman CJ (1978) Easily extractable nutrients in the surface peat layer of virgin sedge pine swamps. Silva Fenn 12(2):65–78

State standard of the USSR (1976) Peat. Method of the determination of the degree of decomposition. Trans Working Group for Classification of Peat. Int Peat Soc, Helsinki, pp 59–67

Stewart GG (1958) Preliminary results of experiments in drain deepening in two Border forests. Reproduced from Rep For Res for the year ending March 1958, pp 131–137

Stoeckeler JH (1961) Soil and water management for increased forest and range production. Soil Sci Soc Am Proc 25:446–451

Sturges DL (1968) Hydrologic properties of peat from a Wyoming mountain bog. Soil Sci 106: 262–264

Sundin T (1982) Stabilitetsjämförelse mellan *Pinus silvestris* och *P contorta* på organogen jord. In: Rotdeformationer hos skogsplantor. Nord Symp, Sveriges Lantbruksuniversitet, Inst skogsproduktion, Rapp 8:137–143

Sundström E (1991) Afforestation of open peatlands. In: Jeglum JK, Overend RP (eds) Peat and peatlands – diversification and innovation. Can Soc Peat Peatlands 1:93–99

Sundström E (1992a) Effects of removal of the paperpot container on growth and development of black spruce seedlings on a drained peatland clearcut in northeastern Ontario. For Can, Great Lakes For Cent, Inf Rep O-X-418:1–20

Sundström E (1992b) Five-year growth response in drained and fertilized black spruce peatlands. I. Permanent growth plot analysis. For Can, Great Lakes For Cent, Inf Rep O-X-417

Sundström E (1992c) Growth response of black spruce five years after drainage and fertilization in the Clay Belt of northern Ontario, Canada. Proc 9th Int Peat Congr, Int Peat Soc, Swed Nat Comm, Uppsala, 2:434–444

Sundström E, Jeglum JK (1992) Five-year growth response in drained and fertilized black spruce peatlands. II. Stem analysis. For Can, Great Lakes For Cent, Inf Rep O-X-420

Suoninen A (1982) Peat production. In: Laine J (ed) Peatlands and their utilization in Finland. Finn Peatland Soc, Finn Nat Comm, Int Peat Soc, Helsinki, pp 67–83

Swank WT, Swift Jr LW, Douglass JE (1988) Streamflow changes associated with forest cutting, species conversions, and natural disturbances. In: Swank WT, Crossley DA Jr (eds) Forest hydrology and ecology at Coweeta. Ecological studies 66. Springer, Berlin Heidelberg New York, pp 297–312

Swanson LE, Rothwell RL (1989) Substrate freeze-thaw in a drained Alberta fen. Can J For Res 19:1024–1029

Tamm CO (1964) Determination of nutrient requirement of forest stands. Int Rev For Res 1: 115–170

Tamm CO (1965) Some experiences from forest fertilization trials in Sweden. Silva Fenn 117(3): 1–24

Tanttu A (1915) Tutkimuksia ojitettujen soiden metsittymisestä (Studien über die Aufforstungsfähigkeit der entwässerten Moore). Acta For Fenn 5(2):1–211 + 32

Taylor JA (1983) The peatlands of Great Britain and Ireland. In: Gore AJP (ed) Mires: swamp, bog, fen and moor. Ecosystems of the world, vol 4B. Elsevier, Amsterdam, pp 1–46

Taylor SA (1949) Oxygen diffusion in porous media as a measure of soil aeration. Soil Sci Soc Am Proc 14:55–61

Teivainen T (1979) Metsäpuiden taimien myyrätuhot metsänuudistusaloilla ja metsitetyillä pelloilla Suomessa vuosina 1973–76 (Vole damage to forest tree seedlings in reforested areas and fields in Finland in the years 1973–76). Folia For 387:1–23

Terkamp H (1992) Public discussion on protection and utilisation of peatland in Germany. Proc 9th Int Peat Congr, Int Peat Soc, Swed Nat Comm, Uppsala, 1:188–204

Terry TA, Hughes JH (1978) Drainage of excess water – why and how? In: Balmer WE (ed) Proc Soil moisture site productivity Symp. Myrtle Beach, SC, Nov 1–3, 1977, USDA – For Serv, pp 148–166

Thompson DA (1979) Forest drainage schemes. For Comm Leaflet 72:1–14

Thompson K, Hamilton AC (1983) Peatlands and swamps of the African continent. In: Gore AJP (ed) Mires: swamp, bog, fen and moor. Ecosystems of the world, vol 4B. Elsevier, Amsterdam, pp 331–373

Thomson JH, Neustein SA (1973) An experiment in intensive cultivation of an upland heath. Scott For 27:211–221

Thurmann-Moe P (1941a) Om bedommelse av myr og vannsyk skogsmark til planteproduksjo (Über die Beurteilung verschiedener Moor- und sumpfiger Waldböden zur Pflanzenproduktion). Meld Nor LandbruksHøgsk 21:1–89

Thurmann-Moe P (1941b) Om skogens innflytelse på jordens vannforråd med spesielle undersokelser over dens drenerende evne (Über den Einfluss des Waldes auf die Wasservorräte des Bodens mit besonderen Untersuchungen über die entwässernde Fähigkeit desselben). Meld Nor LandbruksHøgsk 21:217–300

Tibbetts TE (1968) Peat resources of the world – a review. Proc 3rd Int Peat Congr, Nat Res Council Canada, pp 8–9

Timmer VR (1985) Responses of a hybrid poplar clone to soil acidification and liming. Can J Soil Sci 65:727–735

Timmer VR, Morrow LD (1984) Predicting fertilizer growth response and nutrient status of jack pine by foliar diagnosis. In Stone (ed) Proc 6th For Soil Conf, Univ Tennessee, Knoxville, pp 335–351

Timmer VR, Ray PN (1988) Evaluating soil nutrient regime for black spruce in the Ontario Claybelt by fertilization. For Chron 64:40–46

Timmer VR, Stone EL (1978) Comparative foliar analysis of young balsam fir fertilized with nitrogen, phosphorus, potassium and lime. Soil Sci Soc Am J 42:125–130

Tirkkonen OOJ (1952) Suomen metsäojitus 1900-luvun alkupuoliskolla (Forest swamp drainage in Finland in the first half of the 20th century). Silva Fenn 72:1–103

Todd DK (1959) Ground water hydrology. Wiley, New York, 336 pp

Toleman RDL (1973) A peat classification for forest use in Great Britain. In: Classification of peat and peatlands. Int Peat Soc, Helsinki, pp 10:1–14

Tolonen K (1977) Turvekertymistä ja turpeen tilavuuspainoista kolmessa etelä-suomalaisessa keidassuossa (On dry matter accumulation and bulk density values in three South Finnish raised bogs). Suo 28:1–8

Tolonen K, Hosiaisluoma V (1978) Chemical properties of surface-water in Finnish ombrotrophic mire complexes with special reference to algal growth. Ann Bot Fenn 15:55–72

Tolonen K, Ijäs L (1982) Comparison of two peat samplers used in estimating of dry peat yield in field inventories. Suo 33:33–42

Tolonen K, Saarenmaa L (1979) The relationship of bulk density to three different measures of the degree of peat humification. In: Classification of peat and peatlands. Int Peat Soc, Helsinki, pp 227–238

Tolonen K, Vasander H (1992) Rate of net accumulation of carbon in peat layers, with special emphasis on the past few hundred years. Publ Acad Finl 3:219–225

Tolonen K, Keys D, Klemetti V (1982) Predicting energy content of *in situ* peats by means of their moisture content and bulk density. Suo 33:17–24

Tolonen K, Davis RB, Widoff L (1988) Peat accumulation rates in selected Maine peat deposits. Maine Geol Survey Dep Conserv Bull 33:1–99

Toth J, Gillard D (1988) Experimental design and evaluation of a peatland drainage system for forestry by optimization of synthetic hydrographs. Can J For Res 18(3):353–373

Trettin CC (1986) Application of water management techniques to hardwood silviculture. In: Mroz GD, Reed DD (eds) Proc Conf Northern Harwood Resource: Management and Potential. Mich Tech Univ, School of Forestry and Wood Products, pp 183–192

Trettin CC, Johnson JR, Misiak RD (1991) Hydrologic effects of a prescription drainage system on a forested wetland in northern Michigan. In: Jeglum JK, Overend RP (eds) Peat and peatlands – diversification and innovation. Can Soc Peat Peatlands 1:175–183

Trottier F (1991a) Draining wooded peatlands: expected growth gains. In: Jeglum JK, Overend RP (eds) Peat and peatlands – diversification and innovation. Can Soc Peat Peatlands 1:78–82

Trottier F (1991b) Prescriptions and programs for forest drainage in Quebec. In: Jeglum JK, Overend RP (eds) Peat and peatlands – diversification and innovation. Can Soc Peat Peatlands 1: 207–212

Tuokko K (1992) Metsänparantajat kansakunnan asialla 1908–1988. Saarijärvi, Finland, 509 pp

Tuominen L (1981) Selluloosan hajoaminen eräillä luonnontilaisilla räme- ja nevatyypeillä (Decomposition of cellulose in the peat of some pine bogs and fens). Suo 32:130–133

Tuovinen J-P, Kangas L, Nordlund G (1990) Model calculations of sulphur and nitrogen deposition in Finland. In : Kauppi P, Anttila P, Kenttämies K (eds) Acidification in Finland. Springer, Berlin Heidelberg New York, pp 167–197

Turvetuotanto ennätystasolle (1992) Suo ja turve. Assoc Finn Peat Indust 3:15–16

Uhden O (1956) Neue Erkenntnisse über das Sacken und Atmen der Hochmoore. Wasserwirtschaft 56:261–265

Uhden O (1972) Gebirgshochmoore und Wasserwirtschaft am Beispiel des Brockenfeldmoores im Oberharz. Wasser und Boden, Hamburg, 175 pp

Urvas L, Sillanpää M, Erviö R (1979) The chemical properties of major peat types in Finland. In: Classification of Peat and Peatlands, Int Peat Soc, Helsinki, pp 184–189

Valk U (1967) Rabamänni pärilikest omadustest (On hereditary properties of pines growing on raised bogs). Metsanduslikud uurimused 5:107–130

Valk U (1980) The effect of fertilization on tree growth on drained oligotrophic peatlands in the Estonian SSR. Proc Int Peat Congr, Duluth, Minnesota, pp 488–492

Valk U (1988) Eesti sood (Estonian peatlands). Tallinn, Valgus, 343 pp

Valmari A (1982) Agricultural utilization of peatlands. In: Laine J (ed) Peatlands and their utilization in Finland. Finn Peatland Soc, Finn Nat Comm Int Peat Soc, Helsinki, pp 42–49

Van Beers WFJ (1958) The auger-hole method. A field measurement of the hydraulic conductivity of soil below the water table. Int Inst Land Reclam Improv Bull 1:1–32

Van de Griendt HF (1992) Peat developments in the Netherlands. Bull Int Peat Soc 23:32–36

Van Eck H, Govers A, Lemaire A, Schaminee J (1984) Irish bogs. A case for planning. Catholic Univ Nijmegen, Netherlands, 333 pp

Vasander H (1981) Luonnontilaisen keidasrämeen sekä lannoitetun ojikon ja muuttuman ravinnevarat (Nutrients in an ombrothropic bog ecosystem in the virgin state and after forest-improvements). Suo 32:137–141

Vasander H (1982) Plant biomass and production in virgin, drained and fertilized sites in a raised bog in southern Finland. Ann Bot Fenn 19:103–125

Vasander H (1984) Effect of forest amelioration on diversity in an ombrotrophic bog. Ann Bot Fenn 21:7–15

Vasander H (1990) Plant biomass, its production and diversity on virgin and drained southern boreal mires. Univ Helsinki, Publ Dep Bot, Helsinki, Finland, 18:1–16

Vasander H, Lindholm T (1991) Use of needle analysis to evaluate the nutrient status of peatland forests. In: Jeglum JK, Overend RP (eds) Peat and peatlands – diversification and innovation. Can Soc Peat Peatlands 1:123–128

Vasander H, Starr M (eds) (1992) Proc Int Worksh Carbon cycling in boreal peatlands and climate change. Suo 43:117–292

Vasander H, Leivo A, Tanninen T (1992) Rehabilitation of a drained peatland area in the Seitseminen National Park in southern Finland. In: Bragg OM, Hulme PD, Ingram HAP, Robertson RA (eds) Peatland ecosystems and man: an impact assessment. Univ Dundee, Dep Biol Sci, pp 381–387

Vasander H, Kuusipalo J, Lindholm T (1993) Vegetation changes after drainage and fertilization in pine mires. Suo 44:1–9

Vasari Y (1988) The role of peatlands and flooded meadows in the economic history of Kuusamo. Oulanka Rep 8:96–102

Veihmeyer FJ, Hendrickson AH (1949) Methods of measuring field capacity and permanent wilting percentage of soils. Soil Sci 68:75–94

Veijalainen H (1974a) Berries, mushrooms and wild life on drained peatlands. Proc Int Symp For Drainage, Jyväskylä-Oulu, Finland, pp 395–404

Veijalainen H (1974b) Metsänparannettu ja luonnontilainen suo eläinten talvisena elinympäristönä (Winter habitat preferences of wild animals on improved and virgin peatland). Metsäntutkimusl suontutkimusosaston tied 3:1–10

Veijalainen H (1975) Kasvuhäiriöistä ja niiden syistä metsäojitusalueilla (Dieback and fertilization on drained peatland). Suo 26:87–92

Veijalainen H (1977) Use of needle analysis for diagnosing micronutrient deficiensis of Scots pine on drained peatlands. Commun Inst For Fenn 92(4):1–32

Veijalainen H (1979) Suomarjat ja metsänparannus. Metsäntutkimusl Rovaniemen tutkimusas tied 21:28–33

Veijalainen H (1980) Eräiden hivenlannoitteiden käyttökelpoisuus suometsien lannoituksessa. Neulasanalyysiin perustuva tarkastelu (Usability of some micronutrient fertilizers in peatland forests). Folia For 443:1–15

Veijalainen H (1981) Hivenlannoituksen vaikutus istutusmännikön kehitykseen turvemaalla (Long-term responses of Scots pine to micronutrient fertilization on acid peat soil). Folia For 477:1–15

Veijalainen H (1984) Lannoitustarpeen määritys metsäojitusalueilla (Diagnosis of nutrient deficiencies on drained peatlands). Suo 35:94–97

Veijalainen H (1988) Hivenlannoitus. Metsäntutkimusl tied 308:102–105

Veijalainen H, Reinikainen A, Kolari KK (1984) Metsäpuiden ravinneperäinen kasvuhäiriö Suomessa (Nutritional growth disturbances of forest trees in Finland). Folia For 601:1–41

Verry ES (1972) Effect of an aspen clearcutting on water yield and quality in northern Minnesota. In: Csallony SC, McLaughlin TG, Striffer WD (eds) Watersheds in transition. Proc Symp Fort Collins, Colorado, Colorado State Univ and Am Water Resour, ASSOC (AWRA) Ser 14: 276–284

Verry ES (1976) Estimating water yield differences between hardwood and pine forests: an application of net precipitation data. USDA For Serv Res Rep NC 128:1–12

Verry ES (1980) Water table and streamlow changes after stripcutting and clearcutting on an undrained black spruce bog. Proc 6th Int Peat Congr, Duluth, Minn, USA, pp 493–498

Verry ES (1986) Forest harvesting and water: the Lake States experience. Water Resour Bull 22(6): 1039–1047

Verry ES (1988) The hydrology of wetlands and man's influence on it. Publ Acad Finl 5:41–61

Verry ES (1991) Concrete frost in peatlands and mineral soils: Northern Minnesota. In: Grubich DN, Malterer TJ (eds) Proc Int Peat Symp Peat and Peatlands: the resource and its utilization, Duluth, Minn, USA, pp 121–141

Verry ES, Boelter DH (1972) The influence of bogs on the distribution of streamflow from small bog-upland watersheds. Int Symp Hydr Marsh-ridden areas, Minsk, Byelorussian SSR, 11 pp

Verry ES, Elling AE (1978) Two years necessary for successful natural seeding in nonbushy black spruce bogs. USDA – For Serv Res Note NC–229:1–3

Verta M (1990) Mercury in Finnish forest lakes and reservoirs: Anthropogenic contribution to the load and accumulation in fish. Publ Water Res Inst, Natl Board Waters Environ, Finl 6:1–33

Verta M, Rekolainen S (1985) Ilmaperäisen elohopean, metsäojituksen ja tekojärvien rakentamisen vaikutus kalojen elohopeapitoisuuteen. Vesihallituksen monistesarja 320:1–42

Vincent AB (1965) Black spruce: a review of its silvics, ecology and silviculture. Can Dep For, Ottawa, Ont, Publ 1100:79 pp

Virgo K (1975) Practical aspects of regeneration on Ontario's Clay Belt peatlands. In: Black spruce Symp. Can For Serv, Symp Proc O-P-4:265–276

Virta J (1966) Measurement of evapotranspiration and computation of water budget in treeless peatlands in the natural state. Comm Phys Math Soc Sci Fenn 32(11):1–70

Vompersky SE (1987) Problems of draining excessively moist forests in the temperate zone of the USSR. In: The temperate forest ecosystem, ITE Symp 20:80–85

Vompersky SE (1991) Current status of forest drainage in the USSR and problems in research. In: Jeglum JK, Overend RP (eds) Peat and peatlands – diversification and innovation. Can Soc Peat Peatlands 1:13–18

Vompersky SE, Eizerman NI (1990) Substantiation of forest drainage standards on the basis of a thermodynamic approach. Sov For Sci 5:17–24

Vompersky SE, Ivanov AI (1993) Classification of mires in the forestry in the former USSR. Suo 44:57–62

Vompersky SE, Smagina MV (1984) The impact of hydroreclamation of forests on peat accumulation. Proc 7th Int Peat Congr, Irish Nat Peat Comm, Dublin, 4:86–95

Vompersky SE, Nazarov NA, Sirin AA (1991) Influence of forest drainage on river runoff regime: main concepts and assesment possibilities. In: Jeglum JK, Overend RP (eds) Peat and peatlands – diversification and innovation. Can Soc Peat Peatlands 1:161–167

Vompersky SE, Smagina MV, Ivanov AI, Glukhova TV (1992) The effect of forest drainage on the balance of organic matter in forest mires. In: Bragg OM, Hulme PD, Ingram HAP, Robertson RA (eds) Peatland ecosystems and man: an impact assessment. Univ Dundee, Dep Biol Sci, pp 17–22

Von Post L (1922) Sveriges geologiska undersöknings torvinventering och några av dess hittills vunna resultat. Sv MosskulturförTidskr 1:1–27

Vuokila Y, Väliaho H (1980) Viljeltyjen havumetsiköiden kasvatusmallit (Growth and yield models for conifer cultures in Finland). Commun Inst For Fenn 99(2):1–271

Vuollekoski M (1983) Hydrostaattisella voimansiirrolla varustetun kaivurin soveltuvuus metsäojien perkaukseen (Evaluation of a specially developed excavator for forest ditch cleaning). Folia For 578:1–13

Wang ZM, Macdonald SE (1993a) Peatland and upland black spruce populations in Alberta, Canada: isozyme variation and seed germination ecology. Silvae Genet 41:117–122

Wang ZM, Macdonald SE (1993b) Peatland and upland black spruce population in Alberta: morphology and ecophysiology. Can J For Res 23:33–40

Warner BG, Clymo RS, Tolonen K (1993) Implications of peat accumulation at Point Escuminac, New Brunswick. Quater Res 39:245–248

Watkins C (1991) Nature conservation and the new lowland forests. Nature Conserv Counc, Peterborough, UK, 74 pp

Weber A, Sundman V (1986) Nitrogen fixation in coniferous bark litter. Plant Soil 90:419–425

Wells ED (1991) Effects of refertilization of an 18-year-old Japanese larch (*Larix leptolepis*) peatland plantation in western Newfoundland, Canada. In: Jeglum JK, Overend RP (eds) Peat and peatlands – diversification and innovation. Can Soc Peat Peatlands 1:129–138

Westman CJ (1981) Fertility of surface peat in relation to the site type and potential stand growth. Acta For Fenn 172:1–77

Westman CJ (1987) Site classification in estimation of fertilization effects on drained mires. Acta For Fenn 198:1–55

Wheeler ICT, Hooker JE, Crowe A, Berrie AMM (1986) The improvement and utilization in forestry of nitrogen fixation by actinophizal plants with special reference to *Alnus* in Scotland. Plant Soil 90:393–406

Wiklander G (1977) Skogsgödslingens inverkan på mark och vatten. Sver Skogsvårdsförb Tidskr 75(2–3):179–187

Wiklander G, Nömmik H (1987) Net mineralization of nitrogen in a fen peat soil, Central Sweden. Acta Agric Scand 37:189–198

Wojahn E, Schmidt W (1987) Ergebnisse und Probleme der landwirtschaftlichen Moornutzung in der DDR. In: Int Sym Bodenentwicklung auf Niedermoor und Konsequenzen für die landwirtschaftliche Nutzung, Acad Landwirtschaftswiss DDR, 1:4–47

Wood JE, Jeglum JK (1984) Black spruce regeneration trials near Nipigon, Ontario: planting versus seeding, lowlands versus uplands, clearcut versus strip cut. Can For Serv, Great Lakes For Res Cent, Inf Rep O-X-361:1–19

Yearbook of forest statistics 1990–91 (1992) Folia For 790:1–281

Yli-Vakkuri P (1958) Tutkimuksia ojitettujen turvemaiden kulotuksesta (Untersuchungen über Absengen als Waldbauliche Massnahme auf entwässerten Torfböden). Acta For Fenn 67(4):1–33

Zalitis P (1989) Produktivität der Kiefernreinbestände und Birken-Kiefern-Mischbestände und ihr Wasserhaushalt. In: Bush KK, Zalitis PP, Bisenieks YP, Krastinsh MA, Bush HK (eds) Berioza b sosnjakach. Riga "Zincte" 29–39 (in Russian with German Summary)

Zalitis P (1990a) Markförbättring med kalium i Lettland. Sver Skogsvårdsförb Tidskr 2/1990: 38–41

Zalitis P (1990b) Yield capacity on drained forest sites in Latvia. In: Annu Rep summarized, Res Product Assoc "Silava" Latvian For Res Inst, Salaspils, pp 4–5

Zehetmayr JWL (1954) Experiments in tree planting on peat. For Comm Bull 22:1–110

Zhao K, Päivänen J (1986) Peatlands and their utilization in northeast China. Suo 37:57–65

Zobel M (1992) Estonian mires and their utilization. Suo 43(1):25–33

Zoltai SC, Pollett FC (1983) Wetlands in Canada: their classification, distribution, and use. In: Gore AJP (ed) Mires: swamp, bog, fen and moor. Regional studies. Ecosystems of the world, vol 4B. Elsevier, Amsterdam, pp 245–268

Zoltai SC, Pollett FC, Jeglum JK, Adams GD (1973) Developing a wetland classification for Canada. Proc 4th North Am For Soils Conf, pp 497–511

Subject Index

Ecological Studies

Volumes published since 1989

Volume 74
Inorganic Contaminants in the Vadose Zone (1989)
B. Bar-Yosef, N. J. Barrow, and J. Goldshmid (Eds.)

Volume 75
The Grazing Land Ecosystems of the African Sahel (1989)
H. N. Le Houérou

Volume 76
Vascular Plants as Epiphytes: Evolution and Ecophysiology (1989)
U. Lüttge (Ed.)

Volume 77
Air Pollution and Forest Decline: A Study of Spruce (*Picea abies*) on Acid Soils (1989)
E.-D. Schulze, O. L. Lange, and R. Oren (Eds.)

Volume 78
Agroecology: Researching the Ecological Basis for Sustainable Agriculture (1990)
S. R. Gliessman (Ed.)

Volume 79
Remote Sensing of Biosphere Functioning (1990)
R. J. Hobbs and H. A. Mooney (Eds.)

Volume 80
Plant Biology of the Basin and Range (1990)
B. Osmond, G. M. Hidy, and L. Pitelka (Eds.)

Volume 81
Nitrogen in Terrestrial Ecosystems: Questions of Productivity, Vegetational Changes, and Ecosystem Stability (1991)
C. O. Tamm

Volume 82
Quantitative Methods in Landscape Ecology: The Analysis and Interpretation of Landscape Heterogeneity (1990)
M. G. Turner and R. H. Gardner (Eds.)

Volume 83
The Rivers of Florida (1990)
R. J. Livingston (Ed.)

Volume 84
Fire in the Tropical Biota: Ecosystem Processes and Global Challenges (1990)
J. G. Goldammer (Ed.)

Volume 85
The Mosaic-Cycle Concept of Ecosystems (1991)
H. Remmert (Ed.)

Volume 86
Ecological Heterogeneity (1991)
J. Kolasa and S. T. A. Pickett (Eds.)

Volume 87
Horses and Grasses: The Nutritional Ecology of Equids and Their Impact on the Camargue (1992)
P. Duncan

Volume 88
Pinnipeds and El Niño: Responses to Environmental Stress (1992)
F. Trillmich and K. A. Ono (Eds.)

Volume 89
Plantago: A Multidisciplinary Study (1992)
P. J. C. Kuiper and M. Bos (Eds.)

Volume 90
Biogeochemistry of a Subalpine Ecosystem: Loch Vale Watershed (1992)
J. Baron (Ed.)

Volume 91
Atmospheric Deposition and Forest Nutrient Cycling (1992)
D. W. Johnson and S. E. Lindberg (Eds.)

Volume 92
Landscape Boundaries: Consequences for Biotic Diversity and Ecological Flows (1992)
A. J. Hansen and F. di Castri (Eds.)

Ecological Studies

Volumes published since 1989

Volume 93
Fire in South African Mountain Fynbos: Ecosystem, Community, and Species Response at Swartboskloof (1992)
B.W. van Wilgen et al. (Eds.)

Volume 94
The Ecology of Aquatic Hyphomycetes (1992)
F. Bärlocher (Ed.)

Volume 95
Palms in Forest Ecosystems of Amazonia (1992)
F. Kahn and J.-J. DeGranville

Volume 96
Ecology and Decline of Red Spruce in the Eastern United States (1992)
C. Eagar and M.B. Adams (Eds.)

Volume 97
The Response of Western Forests to Air Pollution (1992)
R.K. Olson, D. Binkley, and M. Böhm (Eds.)

Volume 98
Plankton Regulation Dynamics (1993)
N. Walz (Ed.)

Volume 99
Biodiversity and Ecosystem Function (1993)
E.-D. Schulze and H.A. Mooney (Eds.)

Volume 100
Ecophysiology of Photosynthesis (1994)
E.-D. Schulze and M.M. Caldwell (Eds.)

Volume 101
Effects of Land Use Change on Atmospheric CO_2 Concentrations: South and South East Asia as a Case Study (1993)
V.H. Dale (Ed.)

Volume 102
Coral Reef Ecology (1993)
Y.I. Sorokin

Volume 103
Rocky Shores: Exploitation in Chile and South Africa (1993)
W.R. Siegfried (Ed.)

Volume 104
Long-Term Experiments With Acid Rain in Norwegian Forest Ecosystems (1993)
G. Abrahamsen et al. (Eds.)

Volume 105
Microbial Ecology of Lake Plußsee (1993)
J. Overbeck and R.J. Chrost (Eds.)

Volume 106
Minimum Animal Populations (1994)
H. Remmert (Ed.)

Volume 107
The Role of Fire in Mediterranean-Type Ecosystems (1994)
J.M. Moreno, W.C. Oechel

Volume 108
Ecology and Biogeography of Mediterranean Ecosystems in Chile, California and Australia (1994)
M.T.K. Arroyo, P.H. Zedler, and M.D. Fox (Eds.)

Volume 109
Mediterranean-Type Ecosystems. The Function of Biodiversity (1995)
G.W. Davis and D.M. Richardson (Eds.)

Volume 110
Tropical Montane Cloud Forests (1995)
L.S. Hamilton, J.O. Juvik, and F.N. Scatena (Eds.)

Volume 111
Peatland Forestry. Ecology and Principles (1995)
E. Paavilainen and J. Päivänen

Volume 112
Tropical Forests: Management and Ecology (1995)
A.E. Lugo and C. Lowe (Eds.)

Volume 113
Arctic and Alpine Biodiversity. Patterns, Causes and Ecosystem Consequences (1995)
F.S. Chapin III and C. Körner (Eds.)

Printing: Mercedesdruck, Berlin
Binding: Buchbinderei Lüderitz & Bauer, Berlin